AF597964

Neuromethods

Series Editor
Wolfgang Walz
University of Saskatchewan
Saskatoon, SK, Canada

For further volumes:
http://www.springer.com/series/7657

Visualization Techniques

From Immunohistochemistry to Magnetic Resonance Imaging

Edited by

Emilio Badoer

Professor of Pharmacology, School of Medical Sciences, RMIT University, Bundoora, VIC, Australia

Editor
Emilio Badoer
School of Medical Sciences
RMIT University
Bundoora, VIC, Australia

ISSN 0893-2336 e-ISSN 1940-6045
ISBN 978-1-61779-896-2 e-ISBN 978-1-61779-897-9
DOI 10.1007/978-1-61779-897-9
Springer New York Dordrecht Heidelberg London

Library of Congress Control Number: 2012939062

© Springer Science+Business Media, LLC 2012
All rights reserved. This work may not be translated or copied in whole or in part without the written permission of the publisher (Humana Press, c/o Springer Science+Business Media, LLC, 233 Spring Street, New York, NY 10013, USA), except for brief excerpts in connection with reviews or scholarly analysis. Use in connection with any form of information storage and retrieval, electronic adaptation, computer software, or by similar or dissimilar methodology now known or hereafter developed is forbidden.
The use in this publication of trade names, trademarks, service marks, and similar terms, even if they are not identified as such, is not to be taken as an expression of opinion as to whether or not they are subject to proprietary rights.
While the advice and information in this book are believed to be true and accurate at the date of going to press, neither the authors nor the editors nor the publisher can accept any legal responsibility for any errors or omissions that may be made. The publisher makes no warranty, express or implied, with respect to the material contained herein.

Printed on acid-free paper

Humana Press is part of Springer Science+Business Media (www.springer.com)

Preface to the Series

Under the guidance of its founders Alan Boulton and Glen Baker, the Neuromethods series by Humana Press has been very successful since the first volume appeared in 1985. In about 17 years, 37 volumes have been published. In 2006, Springer Science + Business Media made a renewed commitment to this series. The new program will focus on methods that are either unique to the nervous system and excitable cells or which need special consideration to be applied to the neurosciences. The program will strike a balance between recent and exciting developments like those concerning new animal models of disease, imaging, in vivo methods, and more established techniques. These include immunocytochemistry and electrophysiological technologies. New trainees in neurosciences still need a sound footing in these older methods in order to apply a critical approach to their results. The careful application of methods is probably the most important step in the process of scientific inquiry. In the past, new methodologies led the way in developing new disciplines in the biological and medical sciences. For example, Physiology emerged out of Anatomy in the nineteenth century by harnessing new methods based on the newly discovered phenomenon of electricity. Nowadays, the relationships between disciplines and methods are more complex. Methods are now widely shared between disciplines and research areas. New developments in electronic publishing also make it possible for scientists to download chapters or protocols selectively within a very short time of encountering them. This new approach has been taken into account in the design of individual volumes and chapters in this series.

Wolfgang Walz

Preface

Visualization of chemicals in tissues has seen an incredible advance in the last few years. The array of visualization techniques has expanded to include immunohistochemistry for multiple neurochemicals, detecting expression levels of neurochemicals, following cellular processes and ionic movement, and identifying polysynaptic pathways subserving physiological responses to identifying functional changes in vivo. The present volume provides practical advice as well as an excellent overview of some of the advances in visualization that have been made in recent years. In Chap. 1, well-established procedures for multiple-labelling immunofluorescence in peripheral neurons are described, including the necessary and important critical controls in all stages of the process. In Chap. 2, a procedure for combining non-radioactive in situ hybridization histochemistry with multi-label fluorescence immunohistochemistry on rat brain tissue is presented to facilitate the visualization of multiple mRNA and protein targets located within subcellular compartments. Chapter 3 provides details of a novel method that combines radioactive in situ hybridization histochemistry with immunofluorescence. This method will be particularly useful for investigators looking to identify cell populations producing mRNAs expressed in low abundance. In Chap. 4, confocal laser scanning microscopy and multiple immunofluorescence analysis is described to study constitutive $GABA_B$ receptor internalization and intracellular trafficking. In Chap. 5, a new powerful application of fluorescence microscopy called Total Internal Reflection Fluorescence Microscopy is described. This method allows selective imaging of fluorescent molecules that are either in or close to the plasma membrane of a cell, such as the trafficking and exocytosis of the glucose transporter, GLUT4, in response to insulin stimulation. In Chap. 6, a visualization protocol for automated temporal analysis of mitochondrial position in living cells is described, and how it can be used for computer-assisted quantification of mitochondrial morphology and membrane potential is discussed.

In Chap. 7, the advantages, shortcomings, and possible developments of two-photon microscopy applied to imaging neurons and cells in living tissue are presented. Exciting recent developments in high-speed imaging of 3D objects are also given particular attention. In Chap. 8, techniques that have been optimized to measure $[Ca^{2+}]$ dynamics in neocortical dendrites in response to physiological patterns of APs are described, and the importance of quantifying the dendritic Ca^{2+} dynamics with a high spatial and temporal resolution using two-photon imaging is detailed. In Chap. 9, the technique of extracellular recording combined with juxtacellular labelling is described, and its application to the characterization of cardiorespiratory brainstem neurons is discussed, although this technique could be applied to any neuron. In Chap. 10, the development of transgenic mice expressing fluorescent proteins under the control of specific neuropeptide promoters describes how such animals enable direct visualization of specific cell population within the central nervous system. Techniques aimed at detecting activated neurons, using the protein cFos as a marker, can be combined with tissue from these animals to provide a powerful method of detecting peptides expressed at low levels in activated neuronal populations. In Chap. 11, an overview of guidelines for the use of pseudorabies virus for transneuronal tracing is provided. This fabulous technique exploits the abilities of neurotropic viruses to invade neurons and generate infectious progeny that cross synapses to infect other neurons within a circuit. The method has become increasingly popular with the development of recombinant strains of virus that are reduced in virulence and express unique reporters allowing different pathways to different organs to be explored in the same animal. In Chap. 12, the use of digital infrared thermography is discussed to detect changes in skin vasoconstriction, body temperature, brown adipose tissue thermogenesis, and nociception in the conscious and unrestrained rat. All these physiological phenomena are measured remotely through emitted infrared radiation. Finally, in Chap. 13, the method known as dynamic susceptibility-contrast magnetic resonance imaging is discussed in regard to assessment of brain perfusion and tissue hemodynamics. The typical steps involved in this technique are described. The problems and the approaches that have been developed to counter them are discussed.

The authors and I believe you will find this volume invaluable in gaining an understanding of the practical skills, strengths, and pitfalls that these wonderful and exciting visualization techniques provide.

Bundoora, VIC, Australia

Emilio Badoer
(*Professor of Pharmacology*)

Contents

Contributors

EMILIO BADOER • *School of Medical Sciences, RMIT University, Bundoora, VIC, Australia*
DOMINIC BASTIEN • *Laboratory of Endocrinology and Genomics, CHUL Research Centre, Québec City, QC, Canada; Department of Molecular Medicine, Université Laval, Québec City, QC, Canada*
PAOLO BIANCHINI • *Italian Institute of Technology (IIT), Genoa, Italy*
AUDREY BONNAN • *Institut National de la Santé et de la Recherche Médicale (INSERM) Unité 862, Circuit and dendritic mechanisms underlying cortical plasticity, NeuroCentre Magendie, Bordeaux, France*
KARINE BOUYER • *Centre de Psychiatrie et Neurosciences, UMR894 INSERM, Université Paris Descartes, Sorbanne Paris Cité, Paris, France*
BELINDA R. BOWMAN • *The Australian School of Advanced Medicine, Macquarie University, Sydney, NSW, Australia*
JAMES G. BURCHFIELD • *The Garvan Institute of Medical Research, Sydney, NSW, Australia*
FERNANDO CALAMANTE • *Fernando Calamante, Brain Research Institute, Florey Neuroscience Institutes, Melbourne Brain Centre, Heidelberg, Victoria, Australia*
J. PATRICK CARD • *Department of Neuroscience, University of Pittsburgh, Pittsburgh, PA, USA*
PASCAL CARRIVE • *School of Medical Sciences, University of New South Wales, Sydney, NSW, Australia*
THOMAS DENEUX • *The Weizmann Institute of Sciences, Rehovot, Israel*
ALBERTO DIASPRO • *Italian Institute of Technology (IIT), Genoa, Italy; Department of Physics (DIFI), University of Genoa, Genova, Italy*
LYNN W. ENQUIST • *Department of Molecular Biology, Princeton University, Princeton, NJ, USA*
JACQUES EPELBAUM • *Centre de Psychiatrie et Neurosciences, UMR894 INSERM, Université Paris Descartes, Sorbanne Paris Cité, Paris, France*
MARLEEN FORKINK • *Department of Biochemistry, Nijmegen Centre for Molecular Life Sciences, Radboud University Nijmegen Medical Centre, Nijmegen, The Netherlands*
ANDREAS FRICK • *Circuit and dendritic mechanisms underlying cortical plasticity, NeuroCentre Magendie, Institut National de la Santé et de la Recherche Médicale (INSERM) Unité 862, Bordeaux, France*
IAN L. GIBBINS • *Department of Anatomy and Histology, and Centre for Neuroscience, School of Medicine, Flinders University, Adelaide, SA, Australia*
ANN K. GOODCHILD • *The Australian School of Advanced Medicine, Macquarie University, Sydney, NSW, Australia*

Benjamin Grewe • *Department of Neurophysiology, Brain Research Institute, University of Zurich, Zurich, Switzerland*
Rim Hassouna • *Centre de Psychiatrie et Neurosciences, UMR894 INSERM, Université Paris Descartes, Sorbanne Paris Cité, Paris, France*
William E. Hughes • *The Garvan Institute of Medical Research, Sydney, NSW, Australia; Department of Medicine, St. Vincent's Hospital, Sydney, NSW, Australia*
Attila Kaszás • *Institute of Experimental Medicine, Hungarian Academy of Sciences, Budapest, Hungary; Pázmány Péter Catholic University, Budapest, Hungary*
Gergely Katona • *Institute of Experimental Medicine, Hungarian Academy of Sciences, Budapest, Hungary*
Werner J.H. Koopman • *Department of Biochemistry, Nijmegen Centre for Molecular Life Sciences and Centre for Systems Biology and Bioenergetics, Radboud University Nijmegen Medical Centre, Nijmegen, The Netherlands*
Natasha N. Kumar • *The Australian School of Advanced Medicine, Macquarie University, Sydney, NSW, Australia*
Steve Lacroix • *Laboratory of Endocrinology and Genomics, CHUL Research Centre, Québec City, QC, Canada; Department of Molecular Medicine, Université Laval, Québec City, QC, Canada*
Jamie A. Lopez • *The Garvan Institute of Medical Research, Sydney, NSW, Australia; Peter MacCallum Cancer Centre, East Melbourne, VIC, Australia*
Marco Nooteboom • *Departments of Biochemistry and Pediatrics, Nijmegen Centre for Molecular Life Sciences, Nijmegen Centre of Mitochondrial Disorders, Centre for Systems Biology and Bioenergetics, Radboud University Nijmegen Medical Centre, Nijmegen, The Netherlands*
Paola Ramoino • *Department of Territory and Its Resources (DIP.TE.RIS), University of Genoa, Genoa, Italy*
Balazs Rozsa • *Institute of Experimental Medicine, Hungarian Academy of Sciences, Budapest, Hungary; Pázmány Péter Catholic University, Budapest, Hungary*
Ruth L. Stornetta • *Department of Pharmacology, University of Virginia, Charlottesville, VA, USA*
Lucille Tallot • *Centre de Psychiatrie et Neurosciences, UMR894 INSERM, Université Paris Descartes, Sorbanne Paris Cité, Paris, France*
Virginie Tolle • *Centre de Psychiatrie et Neurosciences, UMR894 INSERM, Université Paris Descartes, Sorbanne Paris Cité, Paris, France*
Cesare Usai • *Institute of Biophysics, CNR, Genoa, Italy*
Ivo Vanzetta • *INT, CNRS UMR 7289, Marseille, France; Aix-Marseille Université, Marseille, France*
Daniel M.L. Vianna • *School of Medical Sciences, University of New South Wales, Sydney, NSW, Australia*
Catherine Videau • *Centre de Psychiatrie et Neurosciences, UMR894 INSERM, Université Paris Descartes, Sorbanne Paris Cité, Paris, France*
Odile Viltart • *Laboratoire "Développement et Plasticité du cerveau post-natal", Centre de recherche JPARC, UMR837 INSERM, Lille, France; Université Nord de France (USTL), Lille, France*
Peter H.G.M. Willems • *Department of Biochemistry, Nijmegen Centre for Molecular Life Sciences and Centre for Systems Biology and Bioenergetics, Radboud University Nijmegen Medical Centre, Nijmegen, The Netherlands*

Chapter 1

Multiple Immunohistochemical Labelling of Peripheral Neurons

Ian L. Gibbins

Abstract

This chapter describes well-established procedures for multiple-labelling immunofluorescence as applied to peripheral neurons. Tissues are fixed with a mixture of formaldehyde and picric acid, and then processed through solvents (ethanol and dimethyl sulfoxide or xylene) before preparation as sections (frozen or embedded in polyethylene glycol) or whole mounts. Specimens are exposed to small volumes of antibody mixtures and are then mounted in buffered glycerol prior to examination with fluorescent microscope fitted with appropriate optics for multi-labelling fluorescence. Critical controls are summarised for all stages of the process, including the specificity of the primary antibodies, the secondary antibodies, and the subsequent image acquisition and analysis.

Key words: Immunofluorescence, Multiple labelling, Co-localisation, Peripheral neurons, Fluorescence microscopy, Confocal microscopy

1. Introduction

The development and routine application of indirect immunofluorescence procedures during the 1970s and 1980s has revolutionised the microscopic analysis of neuronal structure and function. With the easy availability of a wide range of fluorophores, improved microscope optics, high-sensitivity detectors and powerful post-acquisition image processing, high-quality multiple-labelling immunofluorescence can be a standard methodology in any neuroscience laboratory. Nevertheless, although the procedures are robust when applied rigorously, there are many variables that will lead to inferior results if not adequately controlled. These variables include tissue fixation and post-fixation processing, choice of antibodies, storage and handling of the specimens, and microscope

Emilio Badoer (ed.), *Visualization Techniques: From Immunohistochemistry to Magnetic Resonance Imaging*, Neuromethods, vol. 70, DOI 10.1007/978-1-61779-897-9_1, © Springer Science+Business Media, LLC 2012

setups. In this chapter, I will present some robust methods that we have been using for more than 20 years now for the immunohistochemical study of peripheral autonomic and sensory neurons in nearly every tissue of the body. Although I will concentrate on labelling mammalian peripheral neurons, we have applied these methods across many species of fish, amphibians, reptiles, birds and mammals.

1.1. Some Basic Theory and Background

Immunohistochemistry is based on the principle that antibodies recognise specific molecular structures (epitopes), typically a short sequence of amino acids from a peptide or protein. High quality immunohistochemical labelling then depends on a combination of the specificity and sensitivity of the interactions between the antibodies and their presumed targets, as well as the methods used to visualise these interactions.

The most common methods of multiple-labelling immunofluorescence use primary antibodies that are raised in different host species (e.g. mouse, rabbit, goat). These primary antibodies are then labelled with secondary antibodies raised in another species (e.g. donkey) against immunoglobins of the primary antibody host species (e.g. a donkey antibody to rabbit IgG). In turn, secondary antibodies are labelled with different fluorophores, each excited and emitting at characteristic wavelengths that can be distinguished with appropriate optical systems.

In principle, multiple labelling can be done with light absorbing markers generated by enzymatic labels such as horseradish peroxidase or alkaline phosphatase. However, their applicability is much more limited, especially for detailed co-localisation analyses, and they will not be discussed further here.

Probably the most critical steps in these methods are (1) the choice of appropriate antibodies; (2) tissue fixation and post-fixation processing and (3) setting up of the fluorescence microscope.

1.2. Choice of Appropriate Primary Antibodies

Investigators are now faced with a vast array of primary antibodies that are potentially suitable for immunofluorescence. Unfortunately, even after matching protocols and tissues with those suggested by the suppliers, only a relatively small proportion work as advertised. Most commonly, no labelling at all is seen (1). Alternatively, any apparent labelling that does occur is clearly artefactual (2, 3). When selective labelling occurs, the signal-to-noise ratio maybe too poor for effective use, either because the labelling is weak compared with background or because there is a mixture of selective and non-selective labelling. Finally, even if an apparently high level of selectivity is found, it is still necessary to show that the labelling observed is sufficiently specific to be of value. Many journals aiming to publish high-quality immunohistochemical studies (most notably the *Journal of Comparative Neurology*) now insist on detailed documentation of the specificity and selectivity of antibodies used in the work (4).

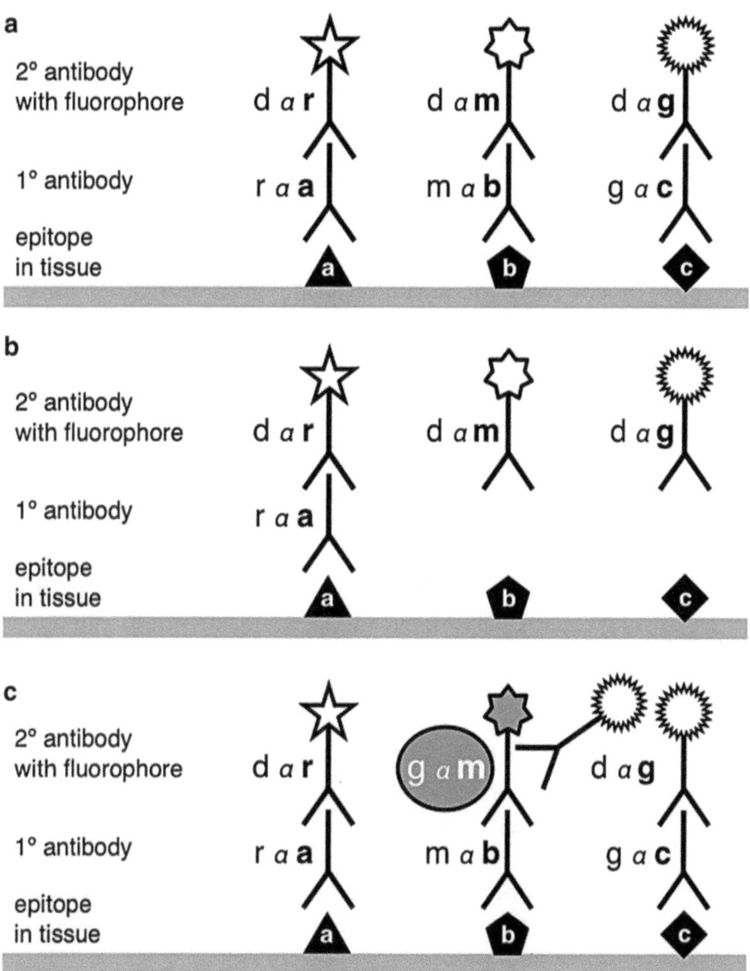

Fig. 1. Basic principles of triple-labelling immunofluorescence. (**a**) Three epitopes (a,b,c) localised with primary antibodies raised in different host species: a rabbit antibody to epitope a (r α a), a mouse antibody to epitope b (m α b) and a goat antibody to c (g α c). All the secondary antibodies are raised in donkeys and each is coupled to a different fluorophore (indicated by different *shaped stars*). (**b**) An example of a cross-reactivity control for secondary antibodies. This test checks that only the donkey anti-rabbit (d α r) secondary antibody binds to the rabbit primary antibody to epitope a (r α a). No labelling should be seen with the other two secondary antibodies. If any labelling is seen, it could be due to cross-reactivity with the inappropriate primary antibody host species, binding to the correct *secondary* antibody already bound to the primary (rare, but it does happen), or the secondary antibody binding directly to tissue components (also rare unless the tissue is from the same species as the immunoglobin targeted by the secondary in question, e.g. a mouse in this example). (**c**) The problem of mixing secondary antibodies raised in different host species. Here the donkey anti-goat immunoglobin secondary (d α g) will bind to the goat anti-mouse immunoglobin secondary antibody (g α m).

When considering multiple-labelling immunofluorescence the choice of host species matters. Ideally, all the primary antibodies need to be raised in different host species, and none of those can be the same the species in which the labelling will be done (Fig. 1). Whilst there are methods for multiple-labelling with conspecific primary antibodies (e.g. elute and re-label; use of Fab fragments as secondary antibodies, (5)), in our hands, none works reliably under routine conditions. Although rarely a problem in the central nervous system, primary antibodies raised in the same species as that being labelled are rarely useful (e.g. mouse monoclonal antibodies used on mouse peripheral tissue). In this case, the secondary antibodies bind to not only the target primary antibody but all the native immunoglobins that permeate the tissue, usually leading to insurmountable background labelling.

A significant potential problem for testing the specificity and sensitivity of antibodies for immunohistochemistry is that the physico-chemical conditions of fixed tissues are not the same as unfixed frozen tissues, and even less like those in an SDS–PAGE gel commonly used for Western Blots or in environment of an immunoassay (e.g. radioimmunoassay or ELISA). Thus, an antibody that is highly selective and sensitive for detecting a particular protein in a Western Blot may be totally ineffective in detecting the same protein in the same source tissue processed for immunohistochemistry (1).

While immunohistochemistry can be done on unfixed tissue, albeit usually after snap freezing at liquid nitrogen temperature, generally tissue is fixed with a cross-linking agent (5). The most common of these is formaldehyde. Cross-linking fixatives optimally bind to amine groups (e.g. on lysine) and create a polymeric meshwork that hopefully preserves the overall spatial localisation of the proteins of interest, whilst also maintaining access of the primary antibodies to the epitopes they are supposed to target. This environment is not only different from the strongly denaturing environment of SDS–PAGE but also is quite different from the natural tissue environment in which the antibodies were generated in the first place.

Testing the specificity of an antibody for multiple-labelling immunohistochemistry can be a tedious exercise. At an absolute minimum, any labelling should be abolished by pre-incubating the antibody with a sample of the peptide sequence or protein against which the antibody was raised (an "absorption test"). A range of concentrations of the "blocking peptide" should be used, and ideally, compared with a different peptide sequence, since high concentrations of peptide (e.g. greater than 0.1 mM) may block the binding of any antibody. In multiple labelling, it is necessary to check all primary antibodies for cross-reactivity with the other sequences being co-labelled (4).

While absorption tests demonstrate that an antibody recognises its intended target sequence, they do not rule out the possibility that it might recognise a similar epitope in an unrelated peptide or protein. For largely unknown reasons, many antibodies bind non-specifically to material in the nucleus or endoplasmic reticulum and Golgi apparatus. The best defence against this outcome is to use at least two different antibodies that have been raised against different epitopes of the same protein of interest. If such a pair of antibodies shows different labelling patterns, then the first assumption must be that one of them (at least!) is binding to a compound other than its intended target.

1.3. Choice of Appropriate Secondary Antibodies

There are two main issues to consider with secondary antibodies for multiple labelling: (1) avoiding cross-reactivity with undesirable target immunoglobins and (2) selecting appropriate fluorophores that match both the tissue characteristics and the microscope setup.

All secondary antibodies need to be tested for cross-reactivity to immunoglobins from non-target species. Buying affinity-purified antibodies (most of them are) does not insure against this problem. However, several manufacturers offer secondary antibodies that have been pre-absorbed against a large range of non-target immunoglobins (e.g. Jackson ImmunoResearch; Rockland). In our experience, these antibodies are very good, but occasional cross-reactions still turn up: just because your donkey anti-rabbit IgG does not cross-react with any of your current antibodies raised in sheep, it does not mean that it will not cross-react with the next one you use!

When doing multiple labelling, it is essential that none of the secondary antibodies is raised in the same species as any of the primaries (Fig. 1). For example, if you have primaries raised in a mouse and in a goat, you cannot use an anti-mouse IgG secondary raised in a goat. Ideally all the secondary antibodies should be raised in the same host species (e.g. donkeys) to minimise the likelihood of them binding to each other.

Fluorophores generally work by absorbing light of a given wavelength and emitting light at a longer wavelength (6). The size of the difference in peak excitation and emission wavelengths is known as Stokes shift. Both the excitation and emission wavelengths can have relatively large effective ranges away from their peaks, and much care needs to be taken to ensure that the fluorophores that are used match the filter combinations available on the microscopes and vice versa. For example, use of common pairs of green and red fluorophores (e.g. fluorescein and rhodamine, or Alexa488 and Cy3) requires narrow band-pass excitation and emission filters to minimise both cross-excitation (i.e. the red fluorophore being excited by the wavelengths intended for the green fluorophore) and bleed-through (e.g. fluorescence from the "red" label being visible through the "green" filter; see Sect. 1.5). For critical co-localisation studies, the use of fluorophores with non-overlapping spectral characteristics (e.g. fluorescein and Cy5) effectively eliminates this risk.

Tissue background fluorescence. The choice of secondary antibody also depends to some degree on tissue background fluorescence. Many tissue components are autofluorescent, often over a large proportion of the visible spectrum. Common offenders include haemoglobin, collagen, elastin, keratin, myelin and lipofuscin. In general, these substances are more fluorescent at shorter excitation wavelengths (e.g. near UV to green). They have much reduced fluorescence at longer excitation wavelengths (e.g. red or near IR), such as those used to excite Cy5. Long-wavelength fluorophores have the additional advantage of reduced fading rates.

Enhanced fluorescence signal. Finally, when signal levels are low, but signal-to-noise ratios are relatively high, additional signal amplification can be achieved via the choice of secondary antibody (5). The easiest and most common approach is to use a biotinylated secondary antibody for the weakest primary label in the combination followed by streptavidin (or a proprietary variant) linked to the fluorophore of choice. Other amplification options include tyramide-enhancement procedures and their variants. Both procedures carry risks of making things worse: many peripheral tissues seem to contain endogenous binding sites for streptavidin; and tyramide amplification also enhances any background binding of the original primary antibody. In each case, signal-to-noise ratios become critical issues that will not be improved by any amplification procedure.

1.4. Tissue Fixation and Post-fixation Processing

Fixatives. As mentioned earlier, most fixatives for immunohistochemistry work primarily by cross-linking proteins (5). The best known cross-linking fixative is formaldehyde. It is a small molecule, allowing rapid tissue penetration, but forms relatively weak cross-links. Glutaraldehyde, a dialdehyde, is an even better cross-linking fixative, which makes it an ideal primary fixative for electron microscopy. However, it has two significant disadvantages for immunofluorescence: it is strongly autofluorescent across a broad range of wavelengths, thereby creating very high background; this characteristic is compounded by the bivalent nature of the fixative, since free aldehyde groups can cross-link antibodies to the tissue, thereby increasing non-specific fluorescence even more.

A very widely used fixative for immunofluorescence (Zamboni's fixative) combines formaldehyde with picric acid (2,4,6-trinitrophenol). Although the chemistry of picric acid fixation is still not known, a combination formaldehyde–picric acid fixative is an excellent general purpose fixative for peripheral tissues that, in our experience, tends to reduce tissue autofluorescence and improve antibody penetration compared with fixation in formaldehyde alone (7).

Cross-linking reagents may not work so well for some types of proteins, especially cell membrane constituents, such as receptors and ion channels. Instead, rapid freezing in isopentane cooled to liquid nitrogen temperatures may be a suitable alternative. Fixation with ice-cold acetone or methanol are other options to explore when aldehyde fixatives fail to preserve immunoreactivity of membrane proteins.

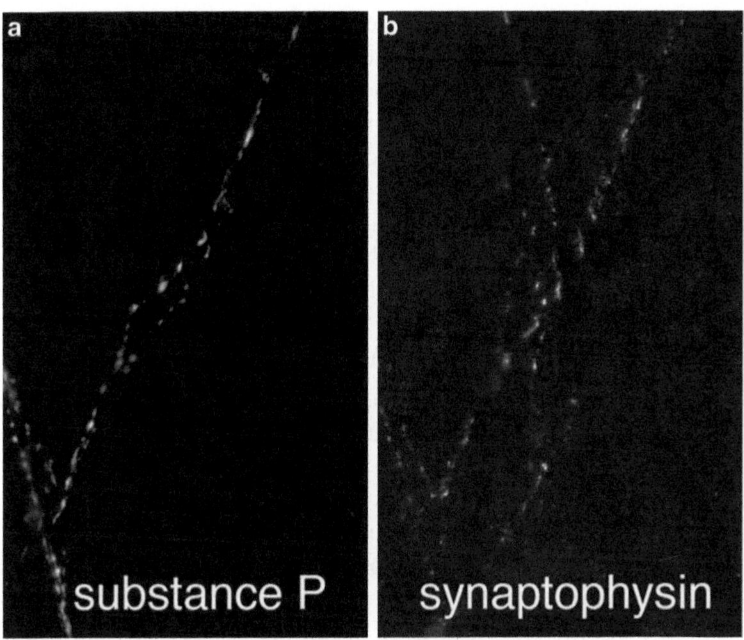

Fig. 2. Whole mount of guinea pig pericardium containing varicose sensory nerve fibres double labelled for the neuropeptide, substance P (**a**) and the synaptic vesicle protein, synaptophysin (**b**). The whole mount preparation clearly shows the overall arrangement of the fibres in a way that would be impossible to determine from sections. In this case, varicosities labelled with each marker probably represent different nerve fibres even though they run in the same nerve fibre bundles. A ×40 objective was used here, wide-field microscopy.

Sections or whole mounts? Traditionally, immunohistochemistry is done on tissue sections, as in conventional histology. However, many peripheral tissues lend themselves to preparation as whole mounts. Whole-mount preparations have the immense advantage of visualising the three-dimensional organisation of peripheral neurons throughout relatively large samples of tissue ((7); Fig. 2). The value of this approach is limited by the distance that antibodies can penetrate the tissue.

Despite the value of whole mounts, some tissues are better examined using sections. Whilst wax embedding as used in conventional histology facilitates serial sectioning, the strongly hydrophobic environment of the wax combined with the relatively high temperatures required even for "low melting point" wax leads to a loss of immunoreactivity compared with other methods. Presumably this is due to variable degrees of protein denaturation. Antigenicity can be restored to various degrees by antigen retrieval methods. However, such methods tend to be adapted to specific situations and are not suitable for routine use.

A well-proven method for sectioning peripheral tissue for multiple-labelling immunofluorescence is the use of a freezing microtome (cryostat). However, freezing tissue blocks and subsequent section sublimation onto glass slides can lead to structural artefacts in tissue, even if appropriate cryoprotectants are used. An alternative for peripheral ganglia (but not peripheral tissues themselves) is polyethylene glycol (PEG) embedding (8). A big advantage of this method is its minimal spatial distortion of the tissue. Once cut, PEG sections are placed into buffered saline which dissolves the PEG, thereby eliminating any support for the tissue by the embedding medium. This means that PEG sections can only be used for tissues that hold themselves together once sectioned (e.g. peripheral ganglia, (8–10)). Most peripheral tissues (e.g. gut, urinary bladder, skin, blood vessels, etc.) will not do this and are better examined with cryostat sections or whole mounts.

Antibody penetration. No matter how tissue is fixed and sectioned, there usually must be a processing step that allows large immunoglobin molecules to penetrate the tissue, and, even more importantly, especially in whole mounts, cross cell membranes. The most widely used solution to this problem is to treat sections with detergent either prior to or together with application of the primary antibodies (5). A superior alternative, that causes less overall tissue damage, is to treat the tissue with solvent (e.g. ethanol, xylene or dimethyl sulfoxide), prior to embedding and sectioning (8). Such pre-treatment is vital for immunohistochemistry of whole mounts (7).

Mounting and storing labelled tissue. Once the tissues have been treated with primary and then secondary antibodies, they need to be mounted in a medium that both clears the tissue (i.e. matches the refractive indices of the tissue and the coverslip glass) and inhibits fading of the fluorophores. Fluorophore fading seems to be due mostly to oxidation. Some commercial mounting agents act to secure the coverslip, match the refractive index and act as an antioxidant. However, whilst reducing fading, they often also reduce peak fluorescence levels. A simple alternative is to use buffered glycerol combined with sealing the edge of the coverslips. Mounted specimens should be stored in the dark to minimise light-induced fading.

1.5. Setting Up the Fluorescence Microscope

Wide-field or confocal? The choice of wide-field or confocal microscopy depends on many factors (apart from the availability of instruments!). Wide-field fluorescence microscopy is usually quicker and allows rapid scanning and documentation of large tissue sections. However, confocal microscopy has real advantages in testing for co-localisation of fluorophores in three-dimensional structures that need to be examined at high magnification (6).

Regardless of the type of microscope used, the optical setup and fluorophores used for immunohistochemistry need to be well matched. Traditionally band-pass filters have been used to select excitation and emission wavelengths suitable for the selective visualisation of each fluorophore. However, in laser scanning confocal microscopy, excitation wavelength is determined by the available laser lines, and emission detection wavelengths can be specified by methods other than simple band-pass filters (6).

Probably the most critical issue is to avoid bleed-through of emission from a single fluorophore into a second detection channel. The usual way to deal with this problem is to select a narrow range of emission wavelengths unique to each fluorophore with some kind of band-pass filtering. While this results in increased selectivity, it is necessarily accompanied by a decreased sensitivity. There are two reasons for this: a narrow band pass reduces the amount of light collected from the sample; and, in order to achieve adequate spectral separation, the band pass is often set off the peak emission band of the fluorophore. In modern microscopes, reduced emission detection is less of a problem due to the increased sensitivity of detectors, better optical systems and increased illumination power.

Increased illumination power presents its own problems. Fading of fluorophores under illumination is a constant problem (6). Increased illumination intensity and illumination time both will increase fading. Many modern fluorophores such as the Alexa Fluor® series (originally developed by Molecular Probes, now Invitrogen) or the CyDye® series (originally developed by Amersham, now GE Healthcare) are generally much more stable and have higher fluorescence yields than earlier labels such as the isothiocyanate salts of fluorescein or tetramethyl rhodamine. Nevertheless, they will fade and it is good practice to use the lowest possible illumination intensity and exposure times consistent with a noise-free image.

Long-wave length fluorophores, typically excited by red light and emitting in the near infrared (e.g. Cy5), offer several advantages for multiple labelling in peripheral tissues. Most usefully, there is relatively little tissue autofluorescence at these wavelengths leading to low background. In critical co-localisation studies, the combination of Cy5 with a green emitting marker (e.g. fluorescein, Alexa488) minimises any possibility of bleed-through or cross-excitation. Finally, these labels are relatively resistant to fading compared with their shorter wavelength congeners.

The minor disadvantage of these fluorophores is that their emission wavelengths are mostly outside the range of normal human vision. However, CCD cameras used on wide-field microscopes and the photomultiplier tubes commonly used on laser scanning confocal microscopes are very sensitive to long wavelengths, allowing Cy5 and similar fluorophores to be easily visualised.

Cameras and acquisition systems. Digital camera technology has greatly facilitated the acquisition and analysis of multiple-labelling immunofluorescence data. There is now a wide range of digital cameras available for wide-field fluorescence microscopy. For low level imaging or for imaging setups that will be used with live cells, a cooled CCD system will provide better signal-to-noise ratios, but such cameras are still expensive compared with non-cooled models.

Apart from this, there are three other important considerations for wide-field fluorescence microscopy cameras.

Resolution: The resolution of the microscope is limited by its optics and not the camera. A camera with a CCD of about 1.5 megapixels provides an optimal match to microscope resolution: a camera with more pixels just provides a larger image file of an intrinsically blurry image generated by the microscope optics.

Grey scale or colour: Colour cameras offer no advantage and considerable disadvantages for multiple-labelling immunofluorescence. First and foremost, they are locked into a conventional RGB (red–green–blue) representation of optical space. They cannot cope with emission in the near infrared, for example. Similarly, they have no way to encode four separate colour channels, as could easily be used in a multiple-labelling experiment with four fluorophores. Indeed, in a microscope setup with selective narrow band-pass emission filters, the light gathered at any particular filter combination is effectively monochromatic. It is much better to collect greyscale images from each filter combination and then false colour them if required post-acquisition.

Bit depth: Bit depth refers to the number of greyscale divisions used to digitally encode the image. For most purposes, 8-bit imaging is sufficient, giving rise to 256 levels. Higher bit depths are available on more expensive cameras, usually with an option of running at 12-bit, providing 4,096 levels. The additional bit depth becomes useful when doing mathematical operations on the image data (e.g. background subtraction or ratiometric comparisons) when it is essential that small intensity values do not round down to zero.

Co-localisation analysis: The analysis of multiple-labelled preparations for co-localisation is rarely straightforward. Assuming appropriate fluorophores and suitably matched optical setups have been used, it is then a matter of deciding what level of "co-localisation" is required and then setting image acquisition parameters to ensure adequate images for analysis are obtained.

Probably the most critical initial decision is the magnification required. If "co-localisation" means determining whether or not two or more markers occur in the same cells (e.g. peripheral neuronal cell bodies, Fig. 3), then only modest magnification is required and the images can be acquired quickly and easily with a suitable wide field microscope. However, if "co-localisation" needs to be tested in individual nerve terminals or in subcellular organelles, then images need to be acquired at maximum spatial and spectral resolution. Such resolution usually is most effectively gained with confocal microscopy.

A common way to identify spatial co-localisation is to overlay two channels of interest each with different false colours, and look

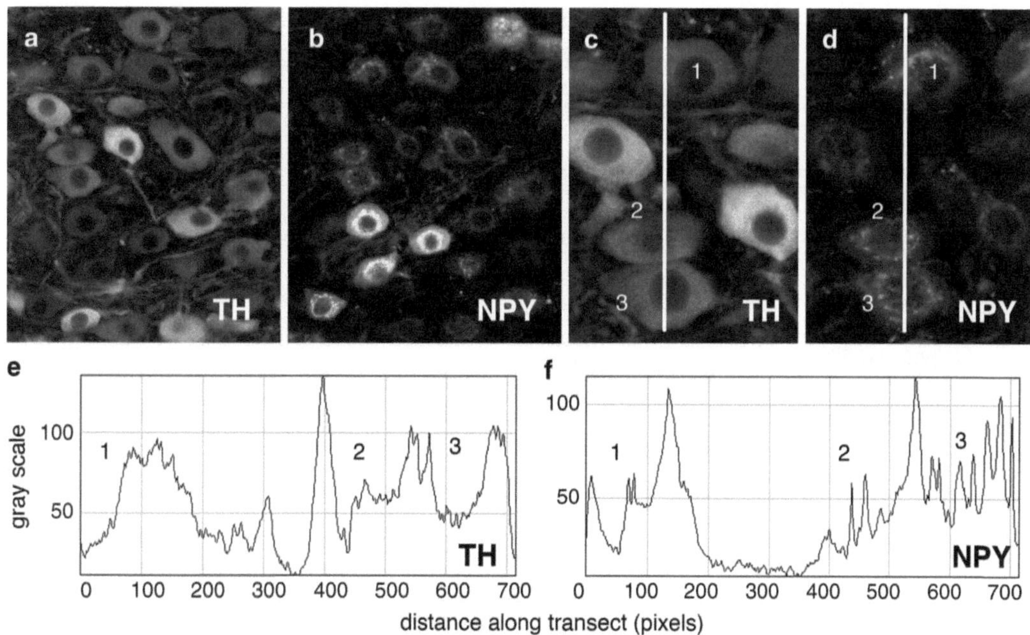

Fig. 3. What do we mean by co-localisation? PEG section of rat coeliac ganglion containing sympathetic neurons double labelled for tyrosine hydroxylase (TH; (**a**, **c**)) and neuropeptide Y (NPY; (**b**, **d**)). Tyrosine hydroxylase is a cytoplasmic enzyme, whereas, within cell bodies, NPY occurs mainly in the Golgi apparatus and endoplasmic reticulum. Although many neurons contain immunoreactivity for both TH and NPY (e.g. cells 1, 2, 3), they are not co-localised in the same subcellular compartments. This can be seen in the enlargements (**c**, **d**) and the greyscale values of cells 1, 2 and 3 shown along the transects (*vertical white lines* in (**c**, **d**); data in (**e**, **f**)). A pixel based co-localisation analysis would be unlikely to indicate significant co-localisation in this example. A ×20 objective was used for these images, wide-field microscopy. Transect data obtained with ImageJ.

for pixels with the summed colour (e.g. combines red and green channels to produce yellow "co-localisation"; or green and magenta to produce white overlaps). In doing so, it is critical to realise that by the very nature of microscope optics, depth resolution (i.e. in *Z*-axis) is significantly less than in the *X*–*Y* plane. This is equally true for confocal and wide-field microscopy (6). Thus, great care needs to be taken in co-localisation analysis to ensure that elements sharing the same (or nearly the same) *X*–*Y* coordinates really have the same *Z*-axis coordinates. In confocal microscopy, for example, this means that co-localisation analysis should not be done on through-focus (*Z*-axis) projections.

Many commercial programs are now available to carry out a wide range of image analyses, including co-localisation analysis. However, all routine image analysis and a wide range of more specialised procedures can be done with the public domain program, *ImageJ*, and its extensive library of plug-ins and macros (see http://rsbweb.nih.gov/ij/).

1.6. Combination Techniques

Multiple-labelling immunofluorescence can be combined with other labelling techniques, including retrograde axonal tracing, intracellular dye fills and electron microscopy. The overall approach to tissue processing is similar but there is insufficient space here to describe the specific procedures required for each combination (for details and examples, see (8–11)).

2. Materials

2.1. Zamboni Fixative (2% Formaldehyde + 0.2% Picric Acid in 0.1 M Phosphate Buffer, pH 7.0)

To make 1 L:

50 mL Formaldehyde (standard stock solution is 35–40% w/v in water, usually containing 15% methanol as preservative)

150 mL Saturated picric acid solution, filtered to remove picric acid crystals

500 mL 0.2 M Phosphate buffer

Part A ($NaH_2PO_4 \cdot 2H_2O$)	195 mL
Part B (Na_2HPO_4)	305 mL

300 mL Distilled water

Store at 4°C.

Hazards:

1. Picric acid is highly explosive when dry. It must always be kept moist, either as a saturated solution or as a working fixative. Any spills must be washed up immediately. Filters used to prepare Zamboni fixative must be washed free of picric acid prior to disposal. Dispose of with copious amounts of water. Since dry picric acid can explode due to friction, never store picric acid solutions in a glass-stoppered or metal-screw-capped container; use glass bottles with plastic screw caps.
2. Both picric acid and formaldehyde are toxic by contact, inhalation or ingestion. Make up the fixative in a fume hood. Wear gloves.

2.2. Items for Holding Fixed Tissue

1. For whole tissue pieces, small vials or containers of glass or plastic with tightly fitting plastic lids.
2. For whole mounts, sheets of thin dental wax, an assortment of stainless steel pins, including fine headless entomology pins (Australian Entomological Supplies; http://www.entosupplies.com.au; Fig. 4), Petri dishes for dissection and storage. Petri dishes with a silicone rubber base (Sylgard®, Dow Corning) are very useful: consider making some dishes with clear Sylgard®, others with black.
3. Containers for fixing tissue should hold a volume of fixative about 10× the tissue volume.

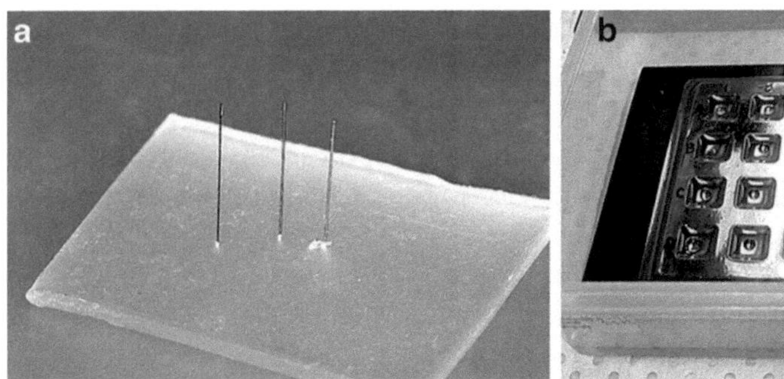

Fig. 4. Examples of items to facilitate efficient tissue preparation for multiple-labelling immunofluorescence, especially for whole mounts. (**a**) Small piece of dental wax with fine headless entomology pins (193 μm diameter, 12.5 mm long). Tissues are stretched and pinned onto the wax and then immersed in fixative. The headless pins allow easy removal of the fixed tissue prior to subsequent processing. (**b**) Serology tray in a plastic clip-top container for incubating whole mounts or free-floating PEG sections with antibody mixtures. Each compartment of the serology tray contains a drop of 5–10 μL of antibody mixture. The individual wells are identified by an *X–Y* grid code. The tray sits on a platform over about 1 cm of water to prevent specimens drying out during incubation. It is essential that the lid of the container is closed firmly: it may be necessary to seal its rim with tape.

2.3. Solvents and Buffers

1. Ethanol, pure, and solutions at 50, 80, 90% in distilled water
2. Dimethyl sulfoxide (DMSO)
3. Xylene
4. Phosphate buffered saline (PBS), pH 7.1

NaCl	8.5 g
Na_2HPO_4 (di-sodium hydrogen orthophosphate, anhydrous)	1.07 g
$NaH_2PO_4 \cdot 2H_2O$ (sodium dihydrogen orthophosphate)	0.39 g
Distilled water	To 1 L

5. PBS–azide (for tissue storage)
   PBS as above containing 0.01% sodium azide (NaN_3).
6. PBS–azide–sucrose (for tissue storage prior to cryostat sectioning)
   PBS–azide as above containing 30% sucrose.

Hazards:

1. Ethanol is a volatile highly flammable organic solvent. Use in a fume hood away from any naked flame.
2. DMSO by itself is not especially toxic, but since it is a "universal solvent" it can easily absorb toxic materials. Consequently, DMSO should be handled in a fume hood as if it were a toxic solution. Avoid contact with the skin. Do not breathe its vapours.

3. Xylene is an organic solvent that is flammable and toxic. It should always be used in a fume hood. Avoid contact with the skin. Do not breathe its vapours.
4. Sodium azide is a metabolic poison. Avoid contact with the skin. Do not inhale or ingest. Avoid contact with acids which can liberate cyanide gas.

2.4. Embedding Materials: Cryostat Sections

1. Liquid nitrogen (for cooling isopentane)
2. Isopentane (for freezing tissue pieces)
3. OCT embedding medium (®Tissue-Tek)
4. Cryomolds (®Tissue-Tek)
5. Phosphorus pentoxide (P_2O_5 for drying sections onto slides)
6. Polyethylenimine (PEI, for coating slides)
7. Strong detergent (e.g. Decon 90®, for cleaning slides)

Hazards:

1. Although nitrogen is inert, liquid nitrogen is extremely cold (–196°C) and can cause severe tissue damage due to freezing on contact. Do not use in a confined space, since cold nitrogen vapours can displace oxygen from the air from ground level up, potentially leading to lethal anoxic environment.
2. Isopentane is highly flammable and toxic. Do not use near a naked flame. Avoid inhaling or ingestion. Avoid contact with skin and eyes. Once cooled with liquid nitrogen, it can cause serious skin damage due to freezing on contact.
3. Phosphorus pentoxide reacts violently and exothermically with water, forming highly corrosive phosphoric acid that can cause serious tissue burns on contact. Avoid contact with skin and eyes. Do not inhale or ingest.

2.5. Embedding Materials: Polyethylene Glycol Sections

1. PEG, 1,000 MW
2. PEG, 1,450 MW
3. Cryomolds (®Tissue-Tek)

2.6. Antibody Diluent: Hypertonic PBS

For 100 mL:

NaCl	1.7 g
Na_2HPO_4	0.107 g
$NaH_2PO_4{\cdot}2H_2O$	0.039 g
NaN_3	0.1 mL of 10% Stock
Distilled water	To 100 mL

2.7. Antibody Incubation

1. Airtight plastic containers with clip on lids (e.g. those used for food storage, Fig. 4). Place about 1 cm of water in the bottom. Make a support to hold the incubation trays above the water.
2. Disposable serology trays for antibody incubations (Fig. 4). If these are unavailable ceramic trays work well enough. Plastic ice cube trays also are a cheap and satisfactory option.

2.8. Buffered Glycerol: For Mounting Sections or Whole Mounts

Solution A	0.5 M Sodium carbonate (Na_2CO_3), 5.3 g/100 mL H_2O, pH 11.5
Solution B	0.5 M Sodium bicarbonate ($NaHCO_3$), 8.4 g/100 mL H_2O, pH 8.3
Solution C	100% Glycerol

1. 0.5 M sodium carbonate buffer, pH 8.6:
   To 50 mL Solution B, add Solution A to adjust pH to 8.6 (about 3 mL)
2. Final mixture:
   Add 2 parts glycerol to 1 part sodium carbonate buffer

2.9. Typical Filter Combinations for Multiple-Labelling Immunofluorescence

Fluorophore	Excitation (nm)	Dichroic (nm)	Emission (nm)
AMCA, Alexa350, DAPI	340–380	400	435–485
FITC, Cy2, Alexa488 GFP	470–490	505	515–545
TRITC, Cy3, Texas Red, Alexa532, Alexa546	515–550	565	575–615
Cy5, Alexa647	590–650	660	665–740

Abbreviations: *AMCA* 7-amino-4-methylcoumarin-3-acetic acid; *DAPI* 4′,6-diamidono-2-phenylindole; *FITC* fluorescein isothiocyanate; *GFP* green fluorescent protein; *TRITC* tetramethyl rhodamine isothiocyanate. Alexa Fluor® dyes are trademarks of Invitrogen; CyDyes® (Cy2, Cy3, Cy5) are trademarks of GE Healthcare. If suitable filters are not available from the microscope manufacturer, custom filters can be produced by companies such as Chroma Technology Corp or Omega Optical.

3. Methods

3.1. Fixation

3.1.1. For Material to Be Sectioned

1. Dissect out tissue samples ensuring that they are kept moist at all times either with PBS or a balanced salt solution suitable for the species under investigation.
2. Small pieces of tissue, less than a few millimetres across, can be dropped directly into fixative (either Zamboni fixative or buffered formaldehyde).
3. Larger pieces of tissue may need to be transferred to a shallow dish (e.g. a plastic or glass Petri dish) filled with PBS or balanced salt solution for finer dissection before fixation. Often it is too difficult to find small peripheral ganglia quickly in situ: transferring a block of tissue to a dish for subsequent dissection is often a much better plan. Dissected tissue pieces can then be dropped into fixative.
4. Once tissue is in fixative, the containers should be sealed to prevent evaporation and spillage, and stored at 4°C for around 24 h (see Note 1).

3.1.2. For Whole Mounts

Every type of tissue must be handled differently to make good whole mounts, but there are some important general steps.

1. Isolate the tissue of interest and transfer it to a suitably sized dish with a silicon polymer base (e.g. Sylgard®, Dow Corning) filled with PBS or balanced salt solution.
2. Assuming you do not wish to look at it, carefully remove as much surrounding adipose tissue as possible. Usually it is best to do this with fine watchmakers' forceps (see Note 2) under a dissecting microscope using a cool light source with flexible fibre-optic arms (see Note 3).
3. Cut pieces of dental wax to a size suitable to hold the whole mounts (Fig. 4). Consider scratching an ID number on one side of the wax.
4. Carefully open out the tissue, if necessary, and stretch it as thin as possible without causing any damage (see Note 4), and pin its periphery onto the wax (see Note 5).
5. Place the wax with pinned tissue into suitably sized containers of fixative and store at 4°C as above. The best way to ensure that the tissue is fully exposed to fixative and will not dry out is to float the wax on the surface of the fixative, tissue side down. Careful positioning of the pins can prevent the samples from tipping over.

3.2. Post-fixation Processing

If tissue has been pinned onto dental wax, carefully remove it before further processing (the solvents will dissolve the wax!). Each piece of tissue should be processed in a separate container.

3.2.1. Processing via DMSO (Suitable for Most Tissues)

1. Wash fixative out of the tissue with repeated changes of 80% ethanol. If Zamboni fixative has been used, keep washing with 80% ethanol until no more yellow colouration appears in the wash (typically four changes of 15 min each, but possibly much more for larger tissues pieces. If so, it is much better to keep changing the solution rather than holding the tissue in any change for a longer time). For formaldehyde-fixed tissue, four washes of 15 min each should be sufficient (see Note 6).
2. Three washes of DMSO about 10 min each.
3. Four washes in PBS about 10 min each.
4. Store in PBS–azide–sucrose if tissue is to be used for cryostat sections, or in PBS–azide if the tissue is going to be prepared for whole mounts (for PEG embedding, see Sect. 3.3.2, and Note 7).

3.2.2. Processing via Xylene (Better for Whole Mounts of Tissues with Higher Levels of Adipose or Connective Tissue, e.g. Blood Vessels)

1. Wash out fixative with 80% ethanol as above
2. 90% Ethanol, 30 min
3. Two changes of 100% ethanol, 30 min each
4. Two changes of xylene, 30 min each
5. Two changes of 100% ethanol, 30 min each
6. Changes through 90, 80 and 50% ethanol, 30 min each
7. Distilled water, 30 min
8. PBS, 30 min
9. Store at 4°C in PBS–azide–sucrose for at least 24 h if tissue is to be used for cryostat sections, or in PBS–azide if the tissue is going be prepared for PEG sections or whole mounts

3.3. Specimen Preparation Prior to Application of Antibodies

3.3.1. Cryostat Sections

1. Collect about 200 mL of liquid nitrogen in a Dewar flask.
2. Add about 30 mL of isopentane to a 100-mL glass beaker, fitted with a small wire holder (you have to make this yourself!) such that the beaker can be lowered into the liquid nitrogen container.
3. Take specimen from its storage in PBS–azide–sucrose and remove excess solution with lint-free paper tissue.
4. Place specimen in a Cryomold and cover it with OCT embedding compound.
5. Float the beaker of isopentane in the liquid nitrogen until the isopentane starts to freeze inside the base of the beaker. Then hold the beaker above the liquid nitrogen while you place the Cryomold with its OCT-embedded specimen on the surface of the isopentane until the OCT solidifies (it turns white after about 20–30 s).
6. Transfer the frozen block to a cryostat for cutting frozen sections immediately or store in a freezer at –20°C until required.

7. Cut frozen sections, typically 10–12 μm thick, and melt them directly onto PEI-coated glass slides.
8. Dry the slides over P_2O_5 in a vacuum desiccator for 30 min.
9. Slides with adhered desiccated sections can be stored in an air-tight, light-tight box at either 4°C or below –20°C for a short time (up to 2 weeks) without loss of labelling quality, but results are much more consistent when freshly cut sections are used.

3.3.2. PEG Sections

1. Transfer specimens from PBS–azide through four changes of 80% ethanol, 10 min each.
2. Two changes of 100% ethanol, 15 min each.
3. Three changes of DMSO, 10 min each.
4. Two changes of 100% ethanol, 15 min each.
5. Place specimen in PEG, 1,000 MW, under vacuum at 55°C for 30 min.
6. Orient specimen in PEG, 1,450 MW, within a Cryomold, and place in a freezer until it solidifies to a wax-like consistency (usually takes several minutes).
7. Remove the PEG block from its mould, trim and attach to a microtome specimen holder with fresh 1,450 MW PEG. Place in freezer to harden. Once set, store at 4°C in a desiccator.
8. Cut sections with a rotary microtome (5–20 μm), collecting them one at a time into shallow well-trays or other small containers containing PBS, which will dissolve the PEG. Use sections immediately or store in PBS–azide at 4°C.

3.3.3. Whole Mounts

1. If required, carefully dissect apart any tissue layers in the specimens (see Note 8), ensuring that all times the tissue is fully immersed in PBS: it must not dry out! (see Note 9).
2. Once the tissues have been separated, and any extraneous material removed, cut the specimens into relatively small pieces, typically no more than 5–10 mm in any direction. This helps ensure the specimens sit flat both during their antibody incubations and when mounted on their slides. It is also difficult to maintain a uniform coverage of antibody solution across larger specimens without using an inordinate amount of antiserum.

3.4. Incubate Specimens with Primary Antibodies

1. Incubation of whole mounts, free-floating PEG sections or cryostat sections on slides should be done in a sealed humidified chamber. It is easy to make a chamber up from an air-tight clip-top plastic food container with a small shelf placed in the bottom (Fig. 4). The container contains a small amount of water (change it regularly to prevent mould growing!). Disposable serology trays with 24 shallow wells are ideal for holding the specimens for their antibody incubations (Fig. 4).

2. For sectioned material, pre-incubate for 30 min with 10% serum from the same species as which your *secondary* antibodies are raised, e.g. donkey (see Note 10). This step is not necessary for whole mounts. The serum is best diluted in hypertonic antibody diluent ((12), see Note 11).
3. The primary antibodies of interest can be mixed together to final working strength in antibody diluent. As a rule, whole mounts require concentrations about four times greater than those required for sections. However, not all antisera follow this rule and each should be tested specifically.

If you use a wide range of antibody combinations, then it is important to have stock working dilutions high enough to allow mixing with two or three other antisera (see Note 12). Antibody stocks can be greatly extended by taking care to use the minimum volume possible for incubating tissue samples. We routinely use a single drop of antibody mixture per sample or section (Fig. 4). Typically this is 5–10 μL droplet, although some large whole mounts or sections may require larger volumes (e.g. 50 μL). If using cryostat sections on slides, this means placing a single droplet of antibody solution on each section.

Good technique is essential here. Do not let tissue samples dry out: process only a small number of samples at a time. Be careful that the specimens are actually *within* the droplet of antibody mixture, not simply floating on top. Ensure there is no cross-contamination of micropipette tips. Check that you have added the proper amounts of all the antibodies in your incubation mixture, and that the correct mixture has been applied to the appropriate tissue samples. Ignore any encouragement to start a conversation while setting up large labelling runs.

Tissue samples can be incubated with primary antisera for 24–72 h. Longer times improve penetration, but beyond 72 h, there is increased risk of the material drying out or becoming mouldy or otherwise contaminated.

3.5. Incubate Specimens with Secondary Antibodies

1. Wash off primary antibody antibodies with several changes of PBS. If using small whole mounts or free-floating PEG sections, you may need to do this step under a dissecting microscope with good lighting so minimise chances of losing your specimens.
2. While the specimens are washing, make up the secondary antibody combinations in antibody diluent. As with the primary antibodies, take care to prevent cross-contamination between stock solutions. If the primary antibody concentrations have been properly tested, the concentrations of each secondary antibody do not need to be changed for different primary antibody combinations. However, it is important that the secondary antibodies have been tested for cross-reactivity with immunoglobins from inappropriate target species (see Note 13).

Tissue samples normally need be incubated with secondary antibodies for 1–2 h. Occasionally, thick whole mounts may require longer incubations (up to 24 h). As with the primary antibody incubations, use the minimum volume of solution required to cover the specimens, while ensuring that the specimens do not dry out.

3. If using a biotinylated secondary antibody (only one can be used per labelling combination), it can be mixed with the other (directly labelled) secondary antibodies. After incubation for 1–2 h, as above, wash the specimens with three changes of PBS, and then incubate with the avidin- or streptavidin-conjugated fluorophore of choice for 1–2 h.

3.6. Mounting Specimens

After the specimens have been incubated with their secondary antibodies, wash them with at least three changes of PBS. Then the specimens must be mounted on slides in buffered glycerol and cover-slipped.

1. *Cryostat sections.* Ensure that as much PBS as possible has been removed from around the sections (use a combination of a Pasteur pipette and small pieces of filter paper to draw up the last drops). Cover the sections with a small amount of buffered glycerol and gently lower a coverslip onto the glycerol such that it spreads out almost to the edge of the coverslip. This takes practice to do properly without either making bubbles or forcing excess glycerol out from under the edges of the coverslip.
2. *Free-floating PEG sections.* There is no easy way to mount PEG sections. One approach is to place a small drop of buffered glycerol on the slide at each location where a section will be placed. Use fine forceps, a stiff haired brush or a pipette (as preferred) to gently transfer each section from its final PBS wash to its own glycerol droplet, and then carefully apply a coverslip. It can be advantageous to pre-wet the section in glycerol before mounting. Alternatively, transfer each section to a slide with a very small amount of accompanying PBS. Put a small drop of glycerol onto a coverslip, invert it and gently place the drop of glycerol onto the section whilst lowering the coverslip. This whole process is best done under a dissecting microscope with good lighting.
3. *Whole mounts.* Place the whole mount onto the slide with just enough PBS to keep it moist. Check the orientation under a dissecting microscope and trim off any rough edges that might prevent the coverslip from sitting flat. To prevent bubbles forming under the coverslip, apply a small drop of buffered glycerol onto the specimen and another to the matching area of the coverslip. Invert the coverslip and gently bring the two glycerol droplets together. Lower the coverslip carefully

(use watchmakers forceps if necessary), ensuring that no bubbles form. If the coverslip is not pulled down flat by surface tension, a gentle tap can help.

4. *Sealing the coverslips.* When using buffered glycerol as a mounting medium, the edges of the coverslips must be sealed to prevent the glycerol leaking out, to stop the coverslips from being dislodged during subsequent handling and to prevent oxygen getting to the specimens. The easiest material to use is clear nail varnish (nail polish; see Note 14). The edge of the coverslip should not extend beyond the edges of the slide and there must not be any glycerol on the exposed surfaces or edges of the glass. Apply just enough varnish to seal between the coverslip and the slide. Let it dry (about 10 min) before examining or storing the slides.

   Preparations are ready for viewing once the coverslips are sealed and dry. However, thicker whole mounts may be better after 24 h wait to allow the glycerol to fully clear the tissue.

5. Do not forget to label the slides unambiguously. Even better, pre-label your slides before the specimens are mounted.

3.7. Imaging

This is not the place to provide instructions on using microscopes, cameras and digital acquisition and processing equipment. However, there are several generic considerations to increase the chances of getting high quality images of your multiple-labelling immunofluorescence.

1. Check that the microscope you intend to use has appropriate optics for your specimens and labels. Are the correct filter combinations available? If you are using a far-red/near infrared fluorophore (e.g. Cy5), does the microscope have a suitable camera system to detect and display this fluorescence which is largely invisible to the naked eye? Are the working distances of the objectives sufficient for the thickness of your specimen, especially if it is a whole mount? Are any of the objectives immersion lenses requiring water, glycerine or oil interfaces?

2. Before placing your slides on the microscope, make sure they are clean. There should be no mounting medium, sealant or any other contaminant anywhere on the slides or coverslips. Any such material runs the risk of contaminating the microscope stage and objectives. It also will decrease resolution and clear focus.

3. Use low magnification objectives to scan your slides to find your specimens and check that they are clean and undamaged. Note that apparent fluorescence intensity increases with increasing numerical aperture and, usually, objective magnification. You may not be able to see much fluorescence at low magnifications.

4. To minimise fading when examining the specimens, use the minimum illumination consistent with good observation. The easiest way to do this is to block the incident light when the specimens are not being actively observed or documented. Most microscopes have a shutter for doing this. If there is none, defocus the stage or shift the objective off-axis so that no light falls on the specimen when not under observation.
5. When recording images for co-localisation analysis, do not change focus between fluorophore channels. The highly corrected objectives of modern microscopes have very little chromatic aberration so there should be no significant focus shift when the filters are changed from one combination to another.
6. If using digital image capture, it is a good general rule to set exposures so that few if any pixels are saturated. Saturation reduces the overall detail in an image. If the resulting image seems too dim or lacking in contrast, it is much better to adjust brightness and contrast after image acquisition.
7. Choose objective magnifications consistent with the size of the structures that will be analysed. For example, it is impossible to do any reliable co-localisation analysis of labels in nerve terminals with objective magnifications less than ×40. However, analysis of co-localisation at the level of neuronal cells bodies could be done with a ×10 or ×20 objective.
8. Quantification of fluorescence labelling levels is difficult to do reliably and is fraught with technical and theoretical issues. If this is to be attempted at all, then good control images need to be acquired during the same sessions as the experimental data images. Depending on the type of analysis, control images include areas of background fluorescence, sections where primary or secondary antibodies have been omitted, and so on (9, 10).
9. Digital images should be saved at the highest resolution compatible with subsequent processing. 8-bit depth is adequate for most purposes, but 12-bit depth is an advantage if there is a wide dynamic range in the image (i.e. both very brightly and dimly labelled structures in the same field). If any degree of post-acquisition quantification is being considered, then it is vital that images are saved in a format that does not change the underlying data. The safest way to do this is to use uncompressed TIFF format. Do not use JPEG formats, which use lossy data compression algorithms that alter the underlying image data at a pixel-to-pixel level (13).

3.8. Co-localisation Analysis

By its very nature co-localisation analysis always will be difficult. The details will vary depending on the biological question, the quality of the labelling, the optical system and the analytical software available to the user. What follows is more of a checklist of things to consider rather than a definitive guide.

1. *What do you mean by "co-localisation"?* Do you mean two or more antigens located in the same cell? The same intracellular compartment (e.g. mitochondria, endosomes)? Or the same location within an organelle (i.e. two antigens bound to each other or to another shared ligand)? If you are interested in either of the last two, then the images need to be gathered at sufficiently high resolution, probably best achieved with confocal microscopy.

   Pixel-by-pixel analysis may not give you the result you expect in any of these circumstances. Most versions of pixel-based analysis place each channel of data into a separate image layer and then compare the intensity levels of the corresponding pixels in each layer. While technically accurate, this type of analysis may not be biologically relevant. Some examples follow.

   (a) *Co-localisation within particular classes of neurons.* If the question is whether or not a particular combination of proteins is expressed by a given class of neurons, such proteins may well be expressed in different subcellular compartments within the neurons. Under such circumstances, it is most unlikely that a pixel-to-pixel comparison between the labelling channels would show "co-localisation" (Fig. 3).

   (b) *Co-localisation at an organelle or molecular level.* In these circumstances, the reliability of pixel-to-pixel co-localisation analysis will depend on the overall resolution of the whole imaging system: antigens that appear "co-localised" at a lower resolution (i.e. their location is encoded within corresponding pixels on the different channel layers) may not end up "co-localised" at a higher resolution, where different locations *within* an organelle may be resolved, for example (10). At the extreme, simple multiple-labelling immunofluorescence never will be able to resolve directly if two antigens are in exactly the same location (i.e. bound to each other or a common ligand). Such analysis requires other techniques such as Fluorescence (or Förster) Resonance Energy Transfer (FRET).

   (c) *Artefactual co-localisation due to compression of depth information.* As mentioned earlier, the depth (*Z*-axis) resolution of light microscopes is considerably less than their planar (*X*–*Y* axis) resolution. This is as true for confocal microscopy as for wide-field microscopy (6). It is entirely possible for two antigens to share their *X*–*Y* planar co-ordinates while having different *Z* co-ordinates. In other words, they sit on top of each other within different planes of the specimen. Any procedure that effectively increases the depth of field (i.e. the apparent depth of the specimen that is in focus at once) will increase the risk of recording artefactual co-localisation in the *Z*-axis. The two most common ways of generating this error are:

(1) In wide-field microscopy, using an objective with too small a numerical aperture, usually due to using a low magnification objective. You should use as high a magnification and numerical aperture as possible. For any given magnification, immersion lenses, especially oil immersion lenses, will have higher numerical apertures. Note that there almost certainly will be a corresponding decrease in working distance.

(2) In confocal microscopy, using through-focus or maximum intensity projection images for co-localisation analysis. One of the great benefits of confocal microscopy is that a series of in-focus optical sections through the *Z*-axis can be combined to generate an image of a relatively thick sample of tissue, within which everything is in focus. However, in doing so, *all depth or Z-axis information is lost!* Consequently, it is completely impossible to determine whether or not pixels that appear labelled in two channel layers are derived from the same point in space: they may share *X*–*Y* planar co-ordinates but they could be separated by several micrometres in the *Z*-axis.

2. *Thresholds: what's positive and what's not?* No matter how good the labelling, there always is a question of determining thresholds for which structures are positive and which are not. Assuming you have checked all your antibodies for specificity and have worked out their optimum concentrations, the next step is to measure the background fluorescence in each detection channel. In theory, there are several components of the background signal: e.g. noise from the camera electronics, stray light from the illumination system, non-specific tissue fluorescence and non-specific binding of the antibodies (primary and secondary) to the tissue. The first two can be measured by taking an image of a microscope field that does not contain any tissue. The total background can be measured by taking an image of a field containing tissue but no specific labelling. In practice, only the second measurement is usually required.

   To determine a reproducible threshold for detection, acquire a series of background images. Using an image-processing program such as ImageJ, select areas of background and determine their mean greyscale value. From a series of such measurements (typically, at least 20), determine the overall mean greyscale value and its standard deviation. In most cases, setting a threshold greyscale value two standard deviations above the mean background level gives a cut-off value that matches with a visual percept of structures being "positive"

(9, 10, 13). A threshold set at three standard deviations above mean background provides a stronger measure of certainty, but may lead to an unacceptably high false-negative rate.

Due to variations in sample thickness, antibody penetration and other uncontrolled factors, these thresholds need to be set separately for every sample examined.

3. *Quantitation: how much stuff is there?* Once again, there is no easy general solution to the problem of quantifying immunofluorescence label. There are two fundamentally different types of measurements: (1) how many things are positive? and (2) how much material is being labelled?

The first question is relatively simple to answer, given a strong enough signal compared with background. The ability to count objects that show significant label above background also depends on the size of the features of interest compared with the resolution of the images. With routine fluorescence microscopy it is relatively easy counting labelled neuronal cell bodies; it is effectively impossible to count something like synaptic vesicles, even though synaptic vesicle antigens may be very well labelled. Counting can be automated to some degree if the objects of interest are labelled sufficiently strongly to be isolated from background by thresholding. Automated counting of objects that are positive for more than one marker requires adequate thresholding in each marker channel. Ideally, multiple-labelled structures should be identified using image arithmetic (i.e. identify objects that are above threshold in channel 1 AND above threshold in channel 2, where "AND" is a Boolean logical operator, equivalent to "&") prior to counting. Exactly how objects should be counted for statistically reliable sampling is a topic in its own right and will not be considered further here (see (14)).

In contrast, it is very difficult (some would say impossible) to relate the level of immunofluorescence signal directly to the amount of underlying protein or peptide of interest (5). There are several reasons for this. Overall, there are just too many variables that cannot be fully controlled from sample to sample, all of which affect the level of fluorescence signal. These include the effectiveness of the fixation, the exact concentration of antibodies within the tissue sample, the affinities of the antibodies, the quantum yield of the fluorophores, fluctuations in the intensity of the microscope illumination, the aperture of the objective, the sensitivity of the CCD camera (almost certainly different for different fluorophores), the gain of the image acquisition system, and so on. Taken together, there is little likelihood of a consistent linear relationship between the amount of protein of interest and the fluorescent signal in the final image.

Nevertheless, within an image, it may be possible to compare relative levels of labelling between different structures. For such comparisons, it is essential that images are not saturated, and that all features of interest are within the same focal plane. When relationships between levels of potentially co-localised antigens are being investigated, there are real statistical advantages in not deciding in advance the threshold for identifying structures as being positive or negative for the markers of interest. Rather, the greyscale values for each marker are recorded for each feature of interest, and any relations between their levels determined statistically by correlation analysis (10). Using this approach, the absolute levels of label have little impact on the outcome and need not be normalised.

4. Notes

1. Fixation time depends on tissue size, the fixative used and, in some cases, the sensitivity of the antigen or other markers in the tissue to fixation. With Zamboni's fixative, 24–72 h seem fine for a very wide range of epitopes. At the extremes, cultured cells can be fixed for only 1 h, whilst tissue stored for several weeks in fixative can still retain much of its antigenicity, although signal-to-noise ratios eventually get worse.
2. Fine watchmakers' forceps (e.g. Dumont® #5) are only as good as their tips. If the tips no longer meet properly (e.g. after having been dropped), they can be repaired easily enough. You need a fine grain oilstone or whetstone and old pair of mid-sized scissors. If the tips of the forceps are only slightly askew, simply use the stone, dampened with a little water or light machine oil, to grind down the ends of the forceps until they meet neatly again. When doing so, it may be helpful to use a rubber band or piece of adhesive tape to hold the tips together while they are being resharpened. If the tips are badly damaged, hold them together with a rubber band or tape, cut the tips off altogether with your old scissors (do not use good ones, since the cutting edges may be damaged in the process) and prepare to spend a patient 10–15 min using the stone to reshape the ends of the forceps until they meet at fine points again.
3. Light sources are important. Incandescent lamps produce a lot of heat and lead to the tissue surfaces drying out. Fibre optic lamps are not only cool but they allow precise direction of the light onto the preparation to maximise visibility of small structures. It is also helpful to use a dissecting microscope with a good working distance. This is rarely a problem with modern instruments, but older ones may have their otherwise close working distance extended by an auxiliary ×0.5 lens.
4. Making good whole mounts is a bit of an art, the details of which vary from tissue to tissue. Hollow organs such as gut,

urinary tract, larger blood vessels, etc. should be split longitudinally and opened out prior to stretching and pinning. The degree of stretch is determined by practise: most tissues will stretch easily to a certain point and then they will begin to tear. The aim is to get the preparation as thin as possible to aid penetration of antibodies and light alike. If you are not interested in epithelial layers (e.g. gut mucosa), you can try gently removing them once the tissue has been stretched. However, it is often better to do this after the tissue has been fixed and processed.

5. The advantage of using headless pins is that the tissue is easier to remove from the wax after it has been fixed. If they are not available consider making some either by cutting the heads off conventional pins or cutting small pieces of fine tungsten wire or stainless steel (e.g. 50–250 μm diameter).
6. If fixative is not being easily extracted from the tissue, then it is better to have more shorter washes than a smaller number of longer ones. The volume of each wash should be about 10× that of the tissue pieces. Whilst it is generally better to process the tissue straight through the protocol, it can be suspended at any step if required. Transfer the tissue to a fresh solution of the next step and tightly seal the container to prevent evaporation. If you wish to maintain predictable high quality tissue preparation do not make these variations to protocol a habit!
7. If tissues are to be embedded in PEG, they should be washed of fixative with PBS and stored in PBS–azide. Clearing through ethanol and DMSO should occur immediately prior to PEG embedding.
8. In most cases, it is much easier to separate tissue layers after the tissue has been fixed and processed. Using fine forceps and steady hands under a dissecting microscope, one can separate fine strips of innervated smooth muscle, fine branches of peripheral nerves themselves or distal components of the vasculature that are effectively impossible to isolate from fresh tissue.
9. Even though tissue samples are fixed, direct exposure to air badly affects the quality of any subsequent immunofluorescence. Any surface drying during preparation of whole mounts in particular tends to create large areas of high background and reduced signal. Even in sections, once they are in the stages of being prepared for antibody application and their subsequent processing, drying out and exposure to the air greatly inhibits the chances of obtaining strong clean labelling.
10. All secondary antibodies should be raised in the same host species, e.g. donkey or horse, which is different from all of the primary antibody host species. Pre-incubating with serum from the same species as the secondary antibody host helps prevent non-specific binding of the secondary antibodies to the tissue.

11. Using hypertonic antibody diluent helps to reduce non-specific binding of proteins, such as immunoglobins, to tissue proteins (12). Note that there is no detergent in the antibody diluent, since the tissue already has been permeabilised via its processing through xylene or DMSO. This is a marked departure from most published methods which add detergent (e.g. Tween-80) to the diluent (5). In our hands, this produces higher background and more tissue damage. The low surface tension of the mixture also requires more antibody solution to be used.
12. In general, keep small volumes of working dilutions of antisera at 4°C, typically in lots of 0.1–1.0 mL at 4–10× final working concentration. Most antibodies will eventually deteriorate if kept at working dilutions. Conversely, do not repeatedly thaw and freeze antisera, most of which will degrade after only a few freeze–thaw cycles. The best solution is to aliquot antisera into small lots (e.g. 0.5–5 μL) stored in clearly labelled containers at −70°C. Only thaw out and dilute the aliquots as required. If you have a large store of frozen antisera, make sure that your freezer is alarmed to warn of failure and that a backup unit is available for when that inevitably happens!
13. For multiple-labelling studies, it is essential to use secondary antibodies that have been re-absorbed against immunoglobins of inappropriate species. Affinity purification by itself is not sufficient to ensure adequate specificity. Several companies now provide such antibodies (e.g. Jackson Immunoresearch; Rockland). Even so, all secondary antibodies need to be tested for cross-reactivity. The best way to do this is to test mixtures of secondary antibodies against a *single* primary (e.g. donkey anti-rabbit IgG, donkey anti-mouse IgG and donkey anti-sheep IgG against a primary raised in a rabbit, Fig. 1). All secondaries should be tested against all primaries, since unexpected cross-reactivities sometimes appear.
14. There is considerable variation in the formulation of clear nail varnish. Unfortunately, some will actually lead to rapid fading of the fluorescence. Any new batch needs to be tested for this using a standard preparation.

Acknowledgments

My work in this area has been supported primarily by grants from the National Health and Medical Research Council of Australia to me and to Judy Morris. The development of multiple-labelling immunofluorescence in whole mounts was initiated by Marcello Costa and John Furness. Since then, many laboratory members and colleagues, especially Judy Morris and Sue Matthew, have added to the development and application of the methods described here, but Pat Vilimas has been a critical contributor throughout.

References

1. Rhodes KJ, Trimmer JS (2006) Antibodies as valuable neuroscience research tools versus reagents of mass distraction. J Neurosci 26:8017–8020
2. Pradidarcheep W, Labruyere WT, Dabhoiwala NF, Lamers WH (2008) Lack of specificity of commercially available antisera: better specifications needed. J Histochem Cytochem 56:1099–1111
3. Fritschy JM (2008) Is my antibody-staining specific? How to deal with pitfalls of immunohistochemistry. Eur J Neurosci 28: 2365–2370
4. Saper CB, Sawchenko PE (2003) Magic peptides, magic antibodies: guidelines for appropriate controls for immunohistochemistry. J Comp Neurol 465:161–163
5. Polak JM, Van Noorden S (2003) Introduction to immunocytochemistry, 3rd edn. BIOS Scientific Publishers, Oxford
6. Pawley J (2006) Handbook of biological confocal microscopy, 3rd edn. Springer, New York
7. Costa M, Buffa R, Furness JB, Solcia E (1980) Immunohistochemical localization of polypeptides in peripheral autonomic nerves using whole mount preparations. Histochemistry 65:157–165
8. Murphy SM, Matthew SE, Rodgers HF, Lituri DT, Gibbins IL (1998) Synaptic organisation of neuropeptide-containing preganglionic boutons in lumbar sympathetic ganglia of guinea pigs. J Comp Neurol 398:551–567
9. Gibbins IL, Teo EH, Jobling P, Morris JL (2003) Synaptic density, convergence, and dendritic complexity of prevertebral sympathetic neurons. J Comp Neurol 455:285–298
10. Gibbins IL, Jobling P, Teo EH, Matthew SE, Morris JL (2003) Heterogeneous expression of SNAP-25 and synaptic vesicle proteins by central and peripheral inputs to sympathetic neurons. J Comp Neurol 459:25–43
11. Gibbins IL (1992) Vasoconstrictor, vasodilator and pilomotor pathways in sympathetic ganglia of guinea-pigs. Neuroscience 47: 657–672
12. Grube D (1980) Immunoreactivities of gastrin (G-) cells: II. Non-specific binding of immunoglobulins to G-cells by ionic interactions. Histochemistry 66:149–167
13. Russ JC (2006) The image processing handbook, 5th edn. CRC, Boca Raton
14. Mouton PR (2002) Principles and practices of unbiased stereology: an introduction for bioscientists. Johns Hopkins University Press, Baltimore

Chapter 2

Combined In Situ Hybridization and Immunohistochemistry in Rat Brain Tissue Using Digoxigenin-Labeled Riboprobes

Natasha N. Kumar, Belinda R. Bowman, and Ann K. Goodchild

Abstract

This protocol describes a procedure for combining non-radioactive in situ hybridization (ISH) histochemistry with multi-label fluorescence immunohistochemistry (IHC) on rat brain tissue sections. This allows visualization of multiple mRNA and protein targets located within the same or different subcellular compartments. A comprehensive description of RNA probe preparation, validation and storage conditions is described. This is followed by the tissue preparation and tissue processing procedure for combined ISH/IHC on free-floating brain sections. Both the tissue and the RNA probes must be prepared and handled under strict RNase-free conditions up until the point of RNA hybridization.

Key words: In situ hybridization, Immunohistochemistry, Fluorescence microscopy, Riboprobes, Gene expression, Digoxigenin, In vitro transcription, Neurons

1. Introduction

The in situ hybridization (ISH) technique enables precise temporal and spatial localization of ribonucleic acid (RNA) targets through hybridization of tissue mRNA with complementary RNA (cRNA, also described as antisense) probes. The main advantage of this technique over other analysis methods that can identify different classes of RNA (e.g., northern blot) is that the mRNA is visualized within cells. The sensitivity and efficiency of ISH is influenced by several factors each of which must be optimized. These include probe type, probe synthesis method, tissue preparation, hybridization conditions and detection method. The theoretical basis for some of these factors is outlined below. ISH can be relatively easily combined with immunohistochemistry (IHC) in order to visualize

Emilio Badoer (ed.), *Visualization Techniques: From Immunohistochemistry to Magnetic Resonance Imaging*, Neuromethods, vol. 70, DOI 10.1007/978-1-61779-897-9_2, © Springer Science+Business Media, LLC 2012

multiple targets located within the same or different subcellular compartments (1). However, whilst coupling ISH with IHC is not uncommon, in some cases the two techniques are not compatible.

1.1. Probe Type

RNA probes (riboprobes) are used here because their hybrids are more stable, they can be produced more efficiently and they are more sensitive for detecting target mRNA than alternate probe types (2). RNA probes are also advantageous because RNA–RNA hybrids are thermostable and can thus withstand the antigen retrieval method of microwave/heat when ISH is combined with IHC. RNA–RNA hybrids are also resistant to digestion by RNases, allowing the possibility of post-hybridization RNase treatment to remove non-hybridized RNA in order to reduce non-specific staining. However, RNA probes are very sensitive to RNases so scrupulous sterile technique is required to prevent probe degradation prior to hybridization with target tissue RNA. Other probe options for ISH include single-stranded DNA probes, double-stranded DNA probes and oligonucleotide probes. Oligonucleotide probes, for example, are produced synthetically, are resistant to RNases and are usually short (40–50 base pairs) facilitating penetration into the tissue of interest; however, short probes are also more likely to bind non-specifically to tissue. Thus RNA probes, which hybridize more specifically because of their longer length, are more reliable, particularly for low abundance targets including transcripts encoding for receptors.

1.2. Radioactive vs. Non-radioactive Riboprobes

Here we describe the synthesis and use of non-radioactive digoxigenin (DIG)-labeled riboprobes. DIG is a steroid hapten not found endogenously in mammals and therefore does not bind to animal tissue (Fig. 1a). RNA probes may be labeled using nucleotides conjugated to radioactive (e.g., ^{32}P, ^{35}S) or non-radioactive (DIG, fluorochrome, biotin) tags. Radioactive labeling is an equal if not more sensitive strategy for ISH labeling; however, the material is hazardous, waste disposal is difficult, assay times are longer, the labeled probe is relatively unstable and it is not possible to discriminate between multiple mRNA targets within the same tissue, when compared with non-radioactive labeling methods. Multiple mRNA targets can be detected within the same tissue using non-radioactive probes (3, 4). Thus, non-radioactive labeling methods have largely superseded the use of isotopic probes.

1.3. Use of PCR-Generated Template

In this protocol, we describe a non-traditional method of probe synthesis. This PCR-based approach provides identical ISH products to the traditional plasmid-based approach without time-consuming cloning steps which may include plasmid propagation, preparation and linearization by restriction enzyme digest to allow transcription of the probe from the appropriate promoter (5–7). We have found PCR-based complementary DNA (cDNA) template generation for riboprobe synthesis to provide similarly efficient RNA transcription and equivalent ISH results to plasmid-generated probes.

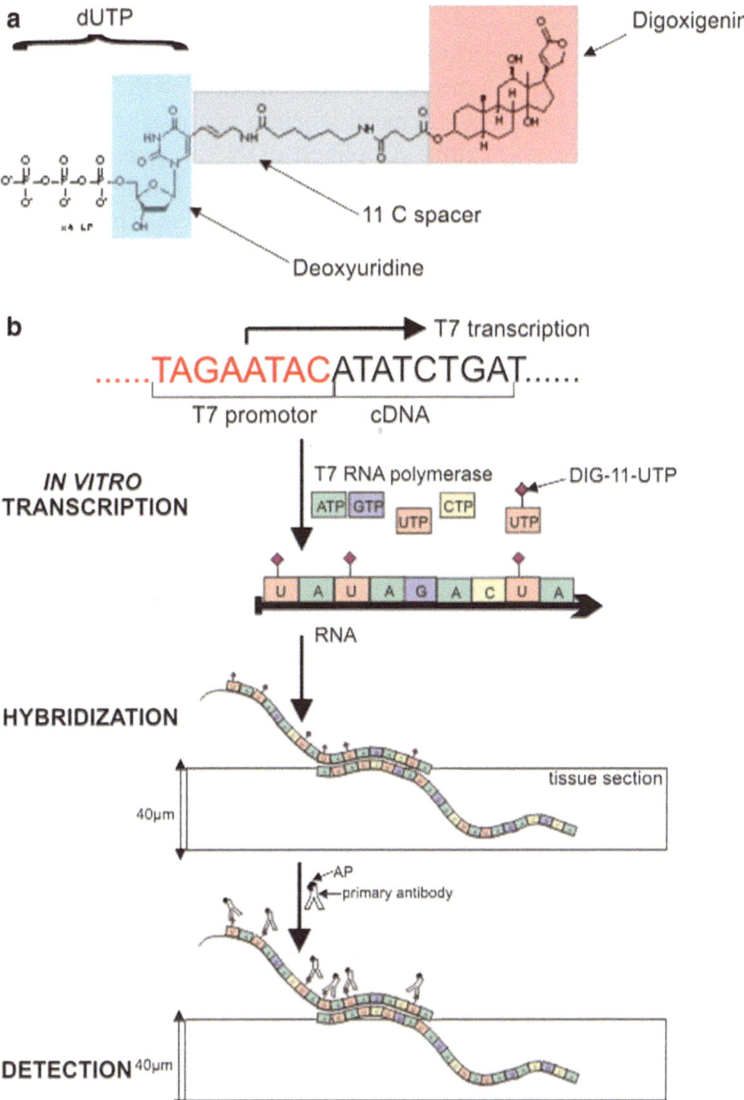

Fig. 1. (**a**) Diagram illustrating the structure of DIG-11-uridine trisphosphate (DIG-11-UTP). DIG is linked to uridine nucleotides via an 11 carbon chain spacer arm, which avoids steric hindrance and maximizes base pair matching during hybridization. The moiety attaches to nucleic acid during transcription, being incorporated every 20–25 nucleotides. The DIG moiety (aided by the spacer) protrudes from the labeled riboprobe allowing immunohistochemical detection by antibody. (**b**) Schematic representation of the mechanism of DIG-incorporated riboprobe hybridization. DIG is incorporated into the RNA sequence during in vitro transcription. During hybridization, labeled riboprobes hybridize with complementary sequences on the surface of the tissue section. Alkaline phosphatase (AP) conjugated primary antibody targets DIG-labeled hybrid. RNA is detected when AP dephosphorylates the chromogenic substrate forming an insoluble precipitate visualized using light microscopy.

We describe a method for using an antisense non-radioactive riboprobe to detect the presence of a specific complementary tissue mRNA in combination with multiple fluorescence immunohistochemical labeling at the single cell level in the same tissue section. Amplified cDNA is transcribed in vitro into strand-specific RNA probes incorporated with DIG-11-UTP by RNA polymerase-mediated transcription (Fig. 1b). DIG-labeled riboprobes are then hybridized to the target RNA in rat brain sections. The RNA hybridization signal is detected indirectly by targeting DIG by IHC (Fig. 1b). A comprehensive description of the process follows.

2. Materials

Rigorous laboratory practice should be followed to prevent RNase contamination at least until riboprobe hybridization steps are complete. This includes wearing of disposable gloves, use of RNase-free disposable plasticware, the use of Type 1 (18 megaohm "ultrapure") water or diethylpyrocarbonate (DEPC)-treated Type II (reverse osmosis) water for all buffers. To degrade RNases, glassware should be baked at 180°C for 8 h, or 300°C for 4 h or treated with 0.1% v/v DEPC water for 1 h followed by autoclaving (15 min at 15 psi), or washed with commercial RNase-degrading solution (e.g., RNase Away, Molecular Bioproducts, USA) followed by rinsing with ultrapure water. Autoclavable plastic ware may be made RNase free by DEPC treatment or washing with commercial RNase-degrading solution. It is also beneficial to store separately equipment (particularly pipettes), chemicals and buffers for RNA use only.

2.1. Riboprobe Preparation

Forward and reverse primers specific for cDNA of interest

RNA extraction kit (SV Total RNA Isolation System, Promega, Alexandria, NSW, Australia # Z3100)

Reverse transcription kit (ImProm-II™ Reverse Transcription System, Promega # A3800)

PCR reagents, including $MgCl_2$, 10× PCR Buffer, 10 mM dNTPs (Bioline, London, UK)

Taq polymerase (Bioline Immolase, London, UK, # BIO-21046)

2.2. DNA Electrophoresis

Agarose (analytical grade [Promega # V3125])

Type II water (reverse osmosis)

10× Tris-acetate-ethylenediaminetetraacetic acid (EDTA) (TAE) buffer (pH 8.3)

Ethidium bromide (Promega # H5041) (1 μg/mL)—*Hazardous substance: Irritant, toxic and suspected mutagen. Wear nitrile gloves when handling*

6× loading buffer (Promega # G1881)

DNA ladder: pGEM marker (Promega # G174A-25512001)

Gel documentation system including UV transilluminator, UV protective face mask, charge-coupled device camera and imaging software

Gel-forming equipment (tray, combs) (Biorad Laboratories, NSW)

Electrophoresis apparatus (Biorad Laboratories, NSW)

QIAquick gel extraction kit (Qiagen, Doncaster, VIC, Australia, # 28704)

QIAquick membrane-based PCR purification system (Qiagen, # 28104)

2.3. RNA Synthesis

DIG-11-UTP 10 mM (Roche Applied Science, # 11209256910)

Ampliscribe T7 FLASH kit (Epicentre Biotechnologies, Madison, WI, USA, # AS2607)

Sp6 RiboMAX large scale RNA production system (Promega, # P1280)

Ammonium acetate (7.5 M)

100% EtOH

Thermal cycler

2.4. RNA Electrophoresis

Agarose (analytical grade [Promega # V3125])

Type 1 water "ultrapure" (RNase-free water)

Ethidium bromide (1 μg/mL)—Formaldehyde (Bacto Laboratories # 809-2.5L PL)—*Hazardous substance: wear gloves when handling*

RNA denaturing gel

Gel documentation system including UV transilluminator, UV protective face mask, CCD camera and imaging software

Gel-forming equipment (tray, combs) (Biorad Laboratories, NSW)

Electrophoresis apparatus (Biorad Laboratories, NSW)

RNase Away (Molecular Bioproducts, USA, # 7002)

10× 3-[*N*-morpholino] propanesulfonic acid (MOPS) buffer

RNA marker (RNA century plus marker, Ambion, # AM7145)

RNA loading dye (NorthernMax® Formaldehyde Load Dye, Ambion, # AM8552)

2.5. Immunoblot

DIG luminescent detection kit (Roche Applied Science, Castle Hill, NSW, # 11363514910)

Nitrocellulose membrane (Boehringer Mannheim, Mannhein Germany, # 1209299)

Nitro Blue Tetrazolium (NBT, Roche Applied Science, # 11383213001)

5-Bromo-4-chloro-3-indolylphosphate (BCIP, Roche Applied Science, # 11383221001)

2.6. Tissue Preparation

Dulbecco's modified eagle medium (DMEM, Sigma, # 8900, prepare 400 mL/500 g rat)

4% Paraformaldehyde (PFA, tissue fixative solution, Sigma, # P-6148) in 0.1 M sodium phosphate buffer (PB), prepare 400 mL/500 g rat. *Hazardous substance: wear gloves, face mask and eye protection when handling*

Na_2HPO_4 (Ajax Finechem, # 478-500G)

NaH_2PO_4 (Chem Supply, # SA061-500G)

Sprague–Dawley rat (purchase from Animal Research Council, Perth, Australia)

Sodium pentobarbitone (Lethabarb, Virbac, Milperra, Australia, # LETHA450)

500–1,000 U heparin (Hospira, Mulgrave, VIC, Australia, # 12881)—anticoagulant agent

5% $NaNO_2$ (Sigma # S2252)—vasodilator agent

Perfusion instruments including large surgical scissors, small sharp scissors, large and small artery forceps and small and large haemostat. Bone rongeurs, scalpel blade and a spatula for removing the brain

70% EtOH

RNase Away

Ducted fume hood

Peristaltic perfusion pump (Masterflex, Cole Parmer, IL, USA) with 18 G draw up needle attached for transcardial perfusion

Tween-20 (Sigma # P9146)

2.7. ISH/IHC

DEPC (Sigma, # D-5758)—*Hazardous substance: handle in fume hood and wear gloves and a face mask*

25-mL Capacity screw cap pots (Labserv, Biolab, Clayton, VIC)

Phosphate buffered saline (PBS)/0.1% Tween-20 (PBT)

Pre-hybridization solution

Cryoprotectant solution

Vibrating microtome (VT1200S, Leica, Germany)

Incubation oven

Alkaline-phosphatase conjugated sheep anti-DIG antibody (Roche Applied Science, # 11093 274910)

Mouse anti-tyrosine hydroxylase antibody (Sigma, # 1299)

Donkey anti-mouse IgG conjugated to Cy3 fluorochrome (Jackson Immunoresearch, # 715-165-151)

Merthiolate (Thimerosal, Sigma-Aldrich # T5125)—*Hazardous substance, wear suitable protective clothing*

Normal horse serum (NHS) (Sigma-Aldrich # 12449C)

Fluorescence microscope (AxioImager Z1, Zeiss, Germany)

Image acquisition software (AxioVision release 4.8, Zeiss, Germany)

20× Saline-sodium citrate (SSC) buffer (pH 7.0)

10× Maleic acid (MA) buffer (pH 7.5)

Boehringer Blocking Reagent (BBR, Roche Applied Science, # 11096176001)

Levamisole (Tetramisole hydrochloride, Sigma, # 1000651675)

10× Tris buffered saline (TBS) buffer (pH 7.4)

Fluorescence mounting medium: Vectashield Hardset (Vector Labs, # H-1400)

Tris phosphate buffered saline (TPBS)

Tween-20 detergent (Sigma # P9416-100ML)

Alkaline phosphatase buffer containing NaCl, Tris–Cl, $MgCl_2$ and Tween-20 detergent (NTMT)

NBT

BCIP

STOP solution

Reagent preparation

5% $NaNO_2$: Dissolve in water. Autoclave and store at room temperature. Dilute 1:10 (0.5%) in saline for animal perfusion.

4% PFA/0.1 M PB (pH 7.4), 400 mL/500 g rat: Pre-heat 300 mL of ultrapure water to 65–70°C. In a ducted fume hood, add 16 g of PFA and mix vigorously. Add 10–15 drops of 10 N NaOH to accelerate dissolving. Mix in 4.36 g Na_2HPO_4 + 1.28 g NaH_2PO_4, cool to ~0°C and adjust to pH 7.4 with 10 N HCl. Bring volume to 400 mL with ultrapure water. Use fixative within 12 h.

10× TAE buffer (pH 8.3): 400 mM Tris-base, 10 mM EDTA, 11.4% (v/v) glacial acetic acid. Autoclave and store at room temperature.

10× TBE buffer (pH 8.3): 1.1 M Tris; 900 mM Borate; 25 mM EDTA. Store at room temperature. Working stock is 0.5× TBE.

Denaturing formaldehyde ethidium bromide stained 1.2% agarose gel (for RNA electrophoresis).

In a sterile conical flask add 0.6 g agarose, 5 mL 10× MOPS buffer, 45 mL RNase-free water. To dissolve agarose heat solution until just boiled. Cool to 65°C then add 0.9 mL formaldehyde

(37–40%), 0.5 μL EtBr (100 ng/mL) using a sterile pipette tip. Mix with gentle agitation then pour immediately into 50 mL capacity gel tray. Remove air bubbles under or between the teeth of the comb using a sterile pipette tip. *Formaldehyde is added to RNA gels to prevent intramolecular base pairing which occurs naturally in single-stranded RNA and impedes RNA gel progression. Use a gel runner dedicated for RNA use. Treat all RNA gel casting apparatus with RNase Away and then rinse with RNase-free water.*

10× MOPS buffer (pH 7.0): 200 mM (MOPS), 50 mM sodium acetate, 10 mM EDTA. Do not autoclave. Sterilize using a 0.2-μM Whatman filter. Store at room temperature, protected from light.

Pre-hybridization solution: 50% formamide, 5× SSC, pH 7.0, 250 μg/mL Herring sperm DNA, 100 μg/mL yeast tRNA (optional blocking reagent), 5% Dextran sulfate, 1× Denhardt's solution, 0.1% Tween-20. Add reagents in sequence, otherwise dextran sulfate is hard to dissolve. Mix ultrapure water, dextran sulfate and SSC first. Aliquot and store at −80°C for up to a year.

Cryoprotectant solution: 30% RNase-free sucrose, 30% ethelyene glycol, 1% polyvinylpyrrolidone (PVP-40) in 0.1 M sodium phosphate buffer, (pH 7.4). Dissolve PVP-40 in ultrapure water followed by sucrose, salts. Lastly, add ethylene glycol in a ducted fume hood. Store at −20°C.

PBT: PBS (10 mM) containing 0.1% Tween-20.

20× SSC buffer (pH 7.0): 3 M NaCl, 0.3 M Na Citrate. Autoclave and store at room temperature.

2% BBR: Dissolve BBR in 1× MABT. May be premade and stored at −20°C for up to 1 month before use.

10× MA buffer (pH 7.5): 1 M Maleic Acid, 1.5 M NaCl. Dissolve MA in ultrapure water and adjust pH with NaOH (~65 g) to ensure solubility before adding NaCl. Autoclave and store at room temperature.

Levamisole (200 mM): Store at −20°C for up to 1 month.

NTMT: 0.1 M NaCl, 0.1 M Tris–HCl (pH 9.5), 0.05 M $MgCl_2$, 0.1% Tween-20. Add 2 mM levamisole before use. Make fresh on the day of use, as buffer will acidify (due to carbon dioxide absorption) and precipitate over time.

10× TBS buffer (pH 7.4): 1 M RNase-free Tris, (pH 7.4)/1.5 M NaCl. DEPC reacts with and destroys the buffering ability of Tris. Therefore, use RNase-free Tris and use ultrapure or DEPC-treated water to make TBS buffer. Autoclave and store at room temperature.

TPBS (pH 7.4): Tris–HCl 10 mM, sodium phosphate buffer 0.1 M, 0.9% NaCl.

STOP solution: 0.1 M Tris (pH 8.5)/1 mM EDTA.

3. Methods

3.1. DIG-Labeled Riboprobe Preparation

3.1.1. Preparation of the cDNA Template for In Vitro Transcription

1. Design PCR primers specific to a 400–900 base pair sequence of the mRNA for the gene of interest (e.g., use NCBI Primer-BLAST or Primer3 software available free online or Beacon Designer software which also avoids stable secondary structures such as RNA loops). Attach SP6 universal promoter for RNA polymerase upstream of the 5′ end of the forward primer (Table 1). This will result in sense strand riboprobe synthesis. *The sense strand is identical to the tissue mRNA and will not hybridize during ISH. It is used as a negative control.* Include T7 universal promoter for RNA polymerase upstream of the 5′ end of the reverse primer (Table 1). This will transcribe in the antisense direction, resulting in synthesis of the antisense RNA strand. *The antisense strand is complementary to target mRNA in the tissue section and will hybridize during ISH.*
2. After extracting total RNA from the tissue of interest, the cDNA library is generated by reverse transcription and a scaled-up PCR reaction is performed to generate enough amplified template cDNA for *in vitro* transcription. *Since a highly concentrated cDNA template for in vitro transcription will result in a greater yield of riboprobe, a cDNA library derived from a tissue known to have an abundant expression of the gene of interest should be used. Optimization of primer annealing temperature is advisable, since the RNA promoter tags may interfere with the PCR priming process. With difficult primers, PCR may be carried out using a nested approach to ensure specificity* (8).

Table 1
Promoter sequences recognized by bacteriophage-encoded RNA polymerase

Bacteriophage	Promoter sequence
Distance from RNA polymerase initiation site	−15 −10 −5 +1 +5
	| | | | |
T7	TAATACGACTCACTATAGGGAGA
SP6	ATTTAGGTGACACTATAGAAG

The PCR mix contains cDNA derived from 30 ng of reverse transcribed total RNA, 1.5–3 mM $MgCl_2$, 0.2 mM dNTP, 0.2 μM forward and reverse primers, 12 units of heat-activated Taq polymerase and 1× reaction buffer.

PCR mix

10× PCR reaction buffer	30 μL
50 mM $MgCl_2$	12 μL
10 mM dNTPs	6 μL
5 μM Forward primer	6 μL
5 μM Reverse primer	6 μL
Immolase DNA polymerase	2.5 μL
Template cDNA	10 μL
Ultrapure water	227.5 μL
Total volume	300 μL

Distribute the PCR mix (300 μL) into 60 μL aliquots and perform PCR using a thermal cycler by initial denaturing (95°C) for 10 min, followed by 30–40 cycles of denaturation (95°C, 10 s), annealing (62°C, 30 s) and extension (72°C, 60 s), then a final extension step (72°C, 10 min). *The resulting PCR product is flanked by two different RNA polymerase initiation sites (Sp6 and T7), enabling either sense strand or antisense strand to be synthesized from the same cDNA template.*

3. Gel electrophorese (80–100 mV) 5–10 μL of the PCR product generated above with 1× loading buffer on a ethidium bromide stained 1–3% TBE agarose gel along with an DNA ladder.

   Agarose concentration of the gel should be appropriate to the size of the PCR product. The smaller the PCR product, the higher the agarose concentration required for optimal resolution. The perfect cDNA template should:

   (a) *Match the length in base pairs of the PCR product originally designed*

   (b) *Be a single, sharp, intense band on the gel*

4. Precipitate the remaining DNA fragment by adding 2.5 V 100% EtOH and 0.1 V NaOAc. Chill 30–60 min at −20°C. Centrifuge 10 min at 13,000 rpm. Remove supernatant and carefully rinse pellet with 70% EtOH. Centrifuge 5 min 13,000 rpm. Remove supernatant and dry DNA pellet then resuspend in 40–100 μL ultrapure water. *Alternatively, a vacuum concentrator may be used which may help reduce product loss.*

5. Prepare an ethidium bromide stained 1–4% agarose TAE gel using ultrapure water. Run precipitated PCR product on the gel at 80 mV for 10–20 min. Extract the cDNA band from the gel using a scalpel blade. Minimize exposure of the gel to UV light as this rapidly degrades the PCR product. Purify the DNA fragment from the gel using QIAquick gel extraction kit. In addition, RNase contamination may also be removed by extracting with phenol–chloroform. *TBE gel is used in Step iii to maximize resolution of DNA. We use a TAE gel in Step v because extraction of the cDNA from agarose gel is more efficient. Gel extraction is recommended to omit any risk of non-specific cDNA species being in vitro transcribed to riboprobe.*
6. Measure the concentration and quality of the purified cDNA on a spectrophotometer. The concentration of DNA is determined by measuring the absorbance at 260 nm. An absorbance of 1 unit at 260 nm corresponds to 50 μg of RNA/mL when dissolved in water. To ensure accuracy, readings should be greater than 0.1. A high-quality cDNA sample has 260 nm/280 nm absorbance ratio of 1.7–2.0. Purified cDNA can be frozen at −20°C and stored for years. Check the integrity of cDNA by spectrophotometry and by gel electrophoresis each time it is thawed for in vitro transcription.

3.1.2. Riboprobe Preparation by In Vitro Transcription

Purified template cDNA (double stranded) is then converted into single-stranded RNA probes which will have DIG-11-UTP incorporated. Bacteriophage DNA-dependent RNA polymerases (SP6 and T7) are employed using a commercially available kit. The kit is supplemented with the modified ribonucleotide DIG-11-UTP to make labeled RNA probes. We have found the ratio of DIG-labeled to unlabeled UTP (1:2) optimal for our purposes; however, this may vary, for example, for low abundance targets. We routinely use the SP6 kit to synthesize sense (control for ISH) riboprobe and T7 kit to synthesize the antisense riboprobe. A DIG-incorporated control RNA probe synthesized from control DNA (provided with the in vitro transcription kit) is also produced. After synthesis, the riboprobe is precipitated and purified. The control RNA probe and riboprobe(s) of interest will be tested in parallel for concentration and integrity before storage prior to hybridization.

1. Riboprobe synthesis
   A typical 20 μL reaction includes 0.5 μg purified template DNA, 3.75 mM ATP, 3.75 mM CTP, 3.75 mM GTP, 2.5 mM UTP, 1.25 mM DIG-11-UTP, 1.25 mM DTT and 1× enzyme mix. Thaw all reagents on ice. If a precipitate is present the transcription buffer requires incubation at 37°C for 5 min.

Do not take enzyme out of freezer until the moment it is required. Mix the following at room temperature, in order specified:

Ultrapure water	20 μL
10× transcription buffer	2 μL
Purified cDNA fragment	0.25–1 μg
100 mM rATP	0.75 μL
100 mM rGTP	0.75 μL
100 mM rCTP	0.75 μL
100 mM rUTP	0.5 μL
10 mM DIG-11-UTP	2.5 μL
100 mM DTT	2 μL
T7 or Sp6 enzyme	2 μL

Incubate at 37–40°C for 2 h.

Longer incubations (>1 h) result in higher yield. Do not use DEPC-treated water as this inhibits RNA polymerase activity.

2. Precipitation and purification of the riboprobe
   Make the volume of the solution up to 100 μL by adding RNase-free water, then add 34 μL of 7.5 M $(NH_4)_2OAc$ and 340 μL of 100% EtOH, mix by inversion and chill at –80°C for 30 min. Centrifuge at 11,000 rpm, 4°C for 10 min. Discard the supernatant and carefully rinse the pellet once with RNase-free 70% EtOH. Centrifuge at 11,000 rpm, 4°C for 2 min. Discard the supernatant and air dry pellet until it is a little clear around the edges (*over drying RNA makes it difficult to dissolve*). Resuspend the pellet in 100 μL of ultrapure water. Repeat the entire precipitation process as above. We find greater RNA yield when one of the 100% EtOH steps is performed overnight at –20°C instead of –80°C for 30 min. Resuspend the pellet in 100 μL ultrapure water. Make 5 μL aliquots of riboprobe in 0.2-mL sterile tubes and store at –80°C. Riboprobe may be stored at –80°C in ultrapure water for a year without degradation. *For long-term storage, resuspension in ethanol or formamide is recommended as these both protect RNA from RNases to a greater extent than ultrapure water. For archival storage, chelation by addition of RNase-free EDTA is recommended to prevent the action of Mg^{2+} and other metals which over time catalyze non-specific cleavages in RNA.*
3. Testing the probe
   (a) Determine riboprobe *concentration* by measuring the absorbance at 260 nm using a spectrophotometer. An absorbance

of 1 unit at 260 nm corresponds to 40 μg of RNA/mL when dissolved in water. To ensure accuracy, A260nm readings should be greater than 0.1. A *pure* RNA preparation has a 260/280 nm absorbance reading of 2.0. We yield 30–40 μg of riboprobe from 500 ng of purified cDNA template when sourcing highly abundant mRNA species.

(b) Determine *the integrity and size* of the riboprobe by running 5 μL (~100 ng) on a 1.2% formaldehyde denatured agarose gel alongside a commercially available RNA marker. *Denaturing conditions remove secondary structure from the RNA and therefore the RNA runs on the gel as single-stranded product. This is essential to determine accurate length of the RNA species. The size of the transcribed RNA should be equal to the size of the template cDNA in base pairs.* The RNA should appear as a sharp band on the EtBr stained gel. Add 3-V RNA loading dye to 1-V RNA (~5 μL). Incubate at 65°C for 5 min to remove secondary structure, followed by immediate chilling on ice. Electrophorese RNA on denaturing formaldehyde agarose gel under RNase-free conditions in 1× MOPS buffer. Visualize gel with UV transillumination.

(c) Assess *DIG-11-UTP incorporation* of the riboprobe by immunoblot

Label and orient the edges of a strip of nitrocellulose membrane (Fig. 2). Handle the membrane with a pair of tweezers and perform washes in an RNase-free glass petri dish. Wash the strip of membrane in 100% EtOH, followed by 2× SSC for 2 min each. Air dry the membrane. Spot the membrane in rows, using 1 μL dilution series of control DIG-labeled RNA (1/1, 1/50, 1/50, 1/100 dilutions). Similarly, spot 1 μL serial dilutions of the probe to be tested (1/1, 1/50, 1/50, 1/100) onto the membrane. Bake the membrane at 120°C for 30 min, then rinse twice with PBT. After this, incubate the membrane with sheep anti-DIG Fab primary antibody conjugated to alkaline phosphatase (1:1,000) in PBT for 30 min with agitation. Quench the antibody with three 5-min PBT washes then incubate the membrane in NTMT for 5 min. Develop the membrane with colourization solution (NBT: 4.5 μL/mL NTMT; BCIP: 3.5 μL/mL NTMT) until purple dots become apparent at the lowest concentration. Quench by rinsing several times with ultrapure water. *Incorporation of DIG-11-UTP into the riboprobe of interest is estimated from the intensity of the purple dots relative to the dots of the control labeled RNA* (Fig. 2).

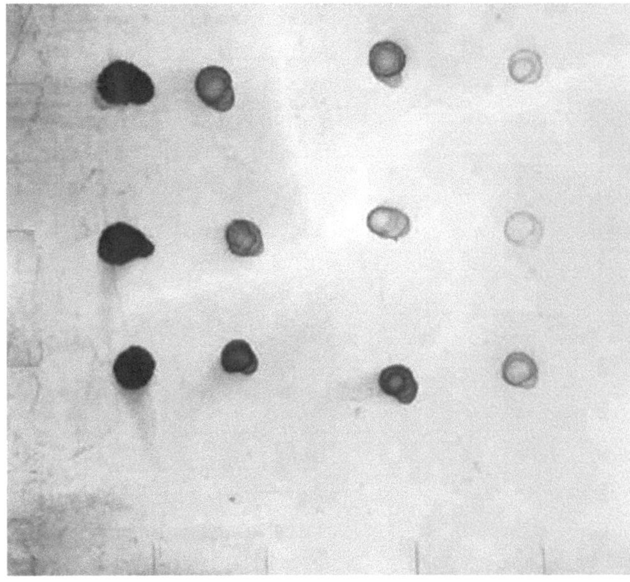

Fig. 2. Immunoblot. Dot blot comparison of DIG incorporation efficiency into antisense (*top row*), sense (*middle row*) and control RNA probe of known concentration (*bottom row*). The amount of DIG incorporated into the riboprobe was determined by spotting serial dilutions of each probe onto a nitrocellulose membrane and performing detection steps for the presence of DIG. The labeled dilutions were 1/1, 1/10, 1/50, 1/100 (*left* to *right*). In this case, antisense blots at 1/10 and 1/50 are appropriate.

3.2. Tissue Preparation

Retrograde labeling of sympathetic neurons in the spinal cord or bulbospinal neurons in the rostral ventolateral medulla was performed as previously described (9, 10). On the day of perfusion, clean surgical instruments and surfaces with 70% EtOH followed by RNase Away. Deeply anesthetized (Lethabarb, 80 mg/kg i.p.) animals are transcardially perfused, using a peristaltic pump, with 150 mL ice-cold heparinized (1,000 U) DMEM (pH 7.4), over 4 min followed by 250 mL ice-cold 4% PFA/0.1 M PB (pH 7.4) over 15 min under RNase-free conditions. Fixative should be made fresh on the day of sacrifice to prevent the formation of degraded aldehyde by-products that may cause an increase in background labeling. Tissue is extracted and blocked to no more than 1 cm^3 maximizing even fixative penetration. Drop-fix, using the same fixative, the tissue block for 12–24 h at 4°C. *Small tissue blocks will be fixed to a much greater extent than large tissue blocks. The aim of the fixation step is to harden the tissue sufficiently to allow for sectioning and withstand processing of free floating sections and also to preserve the tissue morphology. However overfixation can inhibit ISH signal most likely due to prevention of riboprobe access to cellular mRNAs. Overfixation may also impede the antigen retrieval process required for IHC* (11). *Other methods of tissue preparation include fresh frozen tissue without fixation using on-slide ISH processing or the use of paraffin embedded sections.*

3.3. Free-Floating Tissue Sections

Cut 30–50 μm brain tissue sections with a vibrating microtome into 25-mL capacity screw cap pots placed on ice and filled with cold PBT. RNase-free conditions should be used.

The use of free-floating tissue sections permits access of histochemical agents to both surfaces of the tissue section. A small plastic sieve is used to strain sections during buffer changes. Tissue sections are either processed for combined ISH and IHC or may be stored free floating in cryoprotectant solution for up to 6 months.

3.4. In Situ Hybridization and Fluorescence Immunohistochemistry

Pre-hybridization

1. Wash freshly sectioned tissue in PBT for 10 min on a shaker with gentle agitation at 4°C (if the sections were previously stored in cryoprotectant solution, they should be washed 3× 10 min with PBT).
2. Remove PBT with a glass pasteur pipette. Add enough thawed pre-hybridization solution to just cover brain sections (~1 mL/30 brain sections). Incubate sections at 37°C for 30 min and 55–60°C for 1+ h with gentle agitation. *Pre-hybridization takes place under high stringency conditions that improve riboprobe access, specific binding and reduce background staining. Pre-hybridization solution is composed of all the elements of the hybridization solution minus the probe.*

Hybridization

1. Thaw DIG-labeled RNA probe on ice then add to the pre-hybridization solution (at a final concentration of 0.2–1 μg/mL). Avoid touching tissue sections with pipette tip in order to prevent damage. Tightly seal pots and incubate at 55–60°C overnight with agitation. *Occasionally, pre-denaturing the probe by incubation at 75°C for 3 min or alkaline hydrolysis of long riboprobes (>1 kb) to enhance tissue penetration can assist the hybridization process. However, alkaline hydrolysis fragmentation of the riboprobe confers the disadvantages associated with short probes* (see Sect. 1).

Post-hybridization washes

2. Wash sections twice for 30 min in 2× SSC/0.1% Tween-20 at 55–60°C, with agitation.
3. Wash sections twice for 30 min in 0.1× SSC/0.1% Tween-20 at 55–60°C, with agitation.

Tissue is washed to remove unbound riboprobe or imperfectly matched RNA sequences that are not tightly bound. RNA–RNA hybrids are now extremely stable and resistant to RNases. Henceforth,

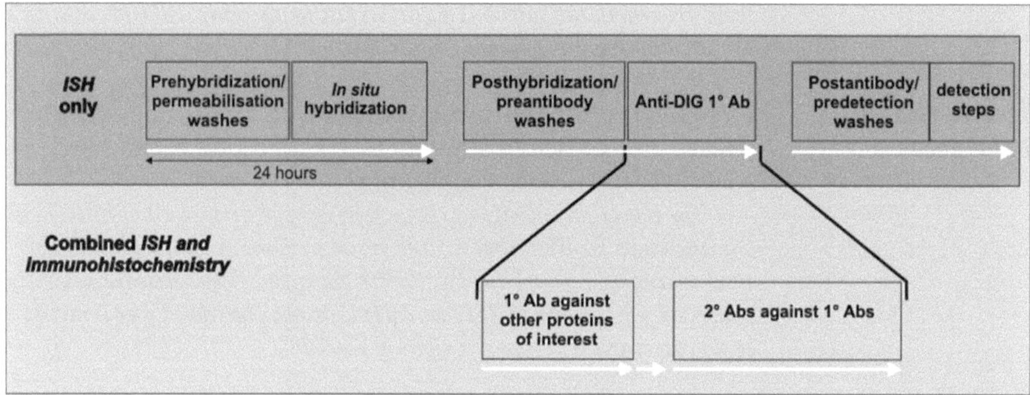

Fig. 3. Timeline of the combined in situ hybridization (ISH) and immunohistochemical procedure. The combined procedure (*light gray box*) includes all of the steps involved for ISH (*dark gray box*), with addition of primary (1°) and secondary (2°) antibodies (Abs) against proteins of interest. These additional steps add 36–48 h to the ISH-only protocol. Each *white arrow* represents a 24-h period.

there is no requirement for RNase-free conditions. Gradually decreasing stringency conditions (stepwise reduction in salt concentration) avoids large changes in osmotic pressure.

Pre-antibody washes and primary antibody incubation

1. Wash sections twice for 15 min with 1× MABT at room temperature, with agitation.
2. Incubate the sections with 2% BBR in 1× MABT for 1–3 h at room temperature with gentle agitation.
3. Add NHS to a final concentration of 10%. Incubate for 1.5–2 h at room temperature.
4. Add alkaline-phosphatase conjugated sheep anti-DIG antibody (1:1,000) to reveal the DIG-labeled riboprobe and other primary antibodies of interest (e.g., mouse anti-tyrosine hydroxylase antibody 1:5000) that will be detected with immunofluorescence to the above solution. Incubate antibodies with gentle agitation for 20 min at room temperature, then 4°C for 48 h.

Longer blocking steps reduce background staining. Where IHC is not to be combined with ISH, proceed straight to the riboprobe predetection step. Figure 3 illustrates the timeline for the combined ISH/IHC procedure compared to ISH alone.

Post-antibody washes and secondary antibody incubation

1. Wash sections 3× 30 min with TPBS.
2. Incubate sections in donkey anti-mouse IgG conjugated to Cy3 fluorochrome (Jackson Immunoresearch) and 2% NHS in TPBS/0.05% merthiolate. Cover pots with aluminum foil to

prevent photobleaching. The riboprobe detection substrates also require protection from light. Incubate pots with mild agitation for 20 min at room temperature, then overnight at 4°C.

Pre-detection and post-antibody washes

1. Wash sections 3× 30 min with MABT (with 2 mM levamisole).
2. Wash sections 2× 10 min with NTMT buffer.

Addition of levamisole inhibits endogenous tissue alkaline phosphatase but does not diminish the intensity of specific labeling for alkaline phosphatase (12).

Riboprobe detection

1. Make a mixture of colourization solution (NBT: 4.5 μL/mL NTMT; BCIP: 3.5 μL/mL NTMT). Filter with 0.45-μM syringe filter and add 1 mL to each pot.
2. Incubate sections at room temperature with mild agitation for 3–24 h. Incubation at 4°C will reduce the rate of detection whilst incubation at 30–37°C will increase the rate of reaction.
3. When the signal is strongest without producing background staining, stop the reaction by washing 3× 15 min in 0.1 M Tris, pH 8.5/1 mM EDTA. Leave in TBS at 4°C until mounting.

After detection, avoid using buffers containing phosphates (e.g., TPBS) as they may react with alkaline phosphatase therefore generating false positive ISH labeling.

Mounting and preservation

Sections are mounted onto clean glass slides and coverslipped when almost dry with fluorescence mounting medium. Drying slides are protected from light and then sealed with clear nail polish. *The NBT/BCIP stain is sensitive to photobleaching and tends to diffuse out with time and with thicker tissue sections. When viewed under a light microscope, the alkaline phosphatase catalyzed blue/black precipitant product is visible in the cytoplasm of each labeled cell (Fig.* 4*A1–E1*)*. Fluorescence IHC is viewed in darkfield using a Zeiss Z1 microscope. The data is analyzed using filter sets with wavelength excitation filters that differentiate between different fluorophores (FITC, Cy3 or AMCA; Figs.* 4 *and* 5*). Images are captured using a CCD camera and stored as Axiovision software files before being converted to TIFF files. All fluorescent images are pseudocoloured. Where double- and triple-labeled (Fig.* 4*) neurons are observed, the signal for each fluorophore must be clearly separate, but overlap completely when merged.*

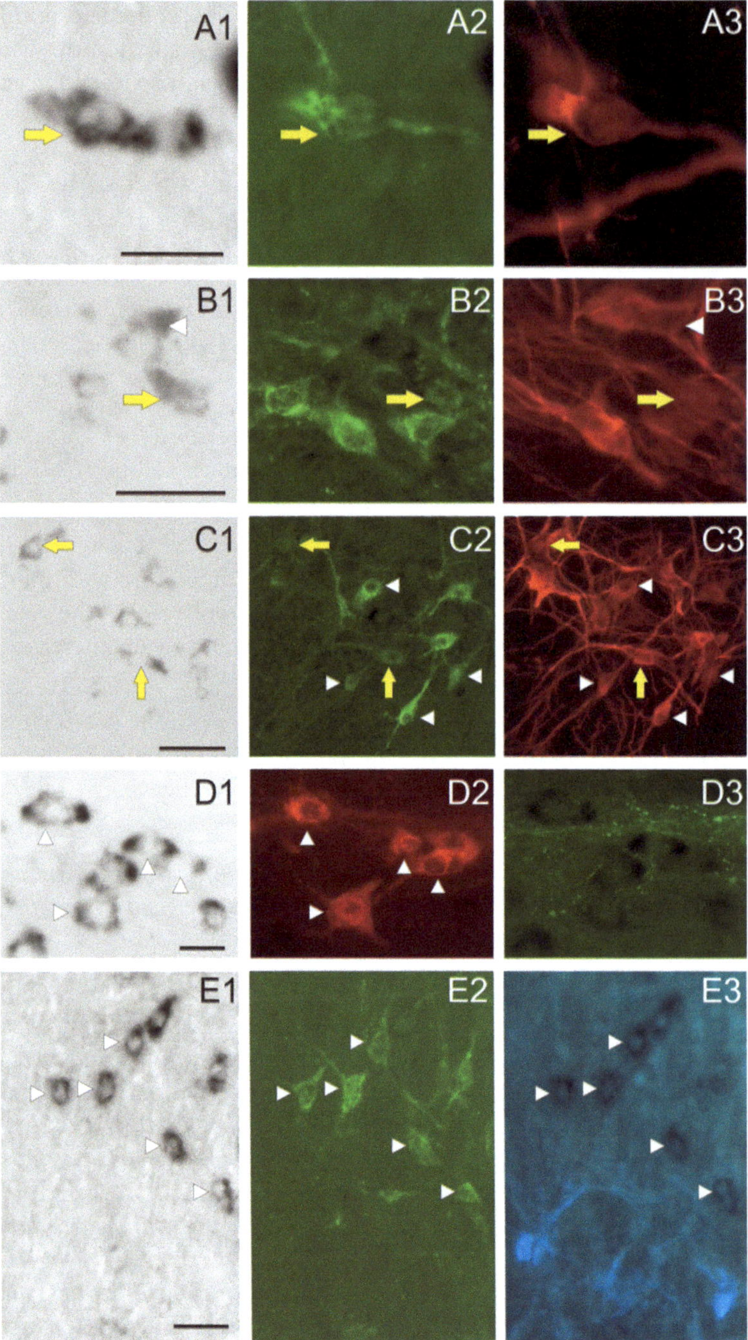

Fig. 4. Combined ISH and immunohistochemistry (IHC) labeling. **A–C:** Highly abundant preproenkephalin (PPE, A1, B1, C1) mRNA is colocalized with retrogradely labeled (CTB-ir, A2, B2, C2) neurons projecting from the brainstem to the spinal cord. Many PPE+ neurons are also catecholaminergic (TH-ir, A3, B3, C3). **D:** PPE mRNA (D1) colocalizes with retrogradely labeled neurons (CTB-ir, D2) projecting from the spinal cord to the adrenal gland that closely appose TH-ir terminals (D3). **E:** Less abundant muscarinic 2 receptor (M2R, E1) mRNA is colocalized with retrogradely labeled neurons (CTB-ir, E2) projecting from the brainstem to the spinal cord. M2R does not colocalize with catecholaminergic neurons (TH-ir, E3) in the vicinity. *Arrow heads* indicate double-labeled neurons; *arrows* indicate triple-labeled neurons. *Scale bars* = 25 μm (A, D) or 50 μm (B, C, E).

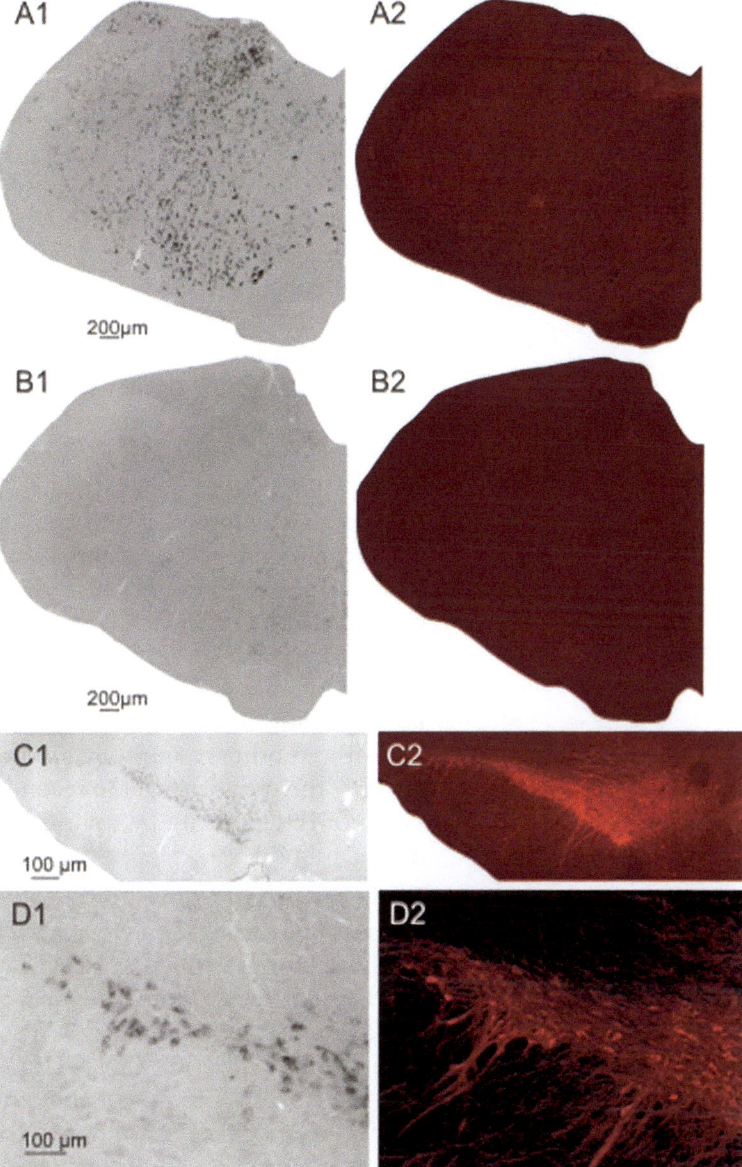

Fig. 5. Combined ISH and IHC labeling of abundant and rare mRNAs. **A:** Low power image of rat brainstem coronal hemisection with many neurons containing preproenkephalin (PPE, *blue/black*) mRNA in high abundance (A1). The same tissue section is labeled with immunoreactivity for vesicular acetylcholine transporter (vAChT, A2). **B:** When heat is removed from the pre-hybridization process, ISH labeling for PPE mRNA is almost non-existent (B1), whereas immunohistochemical labeling (B2) remains unaffected. **C–D:** Low (C) and medium (D) power images of less abundant mRNA species (inositol trisphosphate receptor subtype 1, C1, D1) combined with immunofluorescence (tyrosine hydroxylase, C2, D2).

4. Notes on Compatibility of ISH and IHC

1. The most important aspect is the inclusion of appropriate controls. These include synthesis of a control RNA probe alongside the target sense and antisense riboprobes, during RNA synthesis steps. The control RNA probe is analyzed (immunoblot, gel electrophoresis) in parallel with the sense and antisense riboprobes to ensure that the riboprobes are specific for a single mRNA. The sense probe is used during the ISH procedure to ensure antisense probe specificity. When testing a new antisense probe or new primary antibody, perform in parallel ISH/IHC using a known antisense riboprobe/antibody targeting a highly abundant mRNA/protein species in the tissue of interest, to ensure that the ISH process is correct and that poor tissue preparation is not an issue (positive control). To preserve mRNA, it is important to perform riboprobe hybridization before other histochemical procedures. In some cases, the entire ISH histochemistry protocol can be performed prior to fluorescence IHC (13). All immunohistochemical primary antibodies were initially titrated and appropriate controls used. Negative controls for IHC include (1) incubation of tissue without primary antibody followed by secondary antibody to verify the specificity of the secondary antibody and (2) inhibition of IHC staining by a purified antigen absorbing the primary antibody.
2. During analysis, observed differences in signal may reflect differences in hybridization efficiency (ISH) or antibody binding (IHC) at targets in phenotypically different cells. On the other hand, it may reflect transcripts that are less abundant in a particular tissue (e.g., receptor mRNA, Fig. 5C, D). Therefore, it is important to perform ISH/IHC using tissue known to contain the mRNA/protein of interest (positive control tissue) and where possible, known not to contain the mRNA/protein of interest (negative control tissue).
3. One aspect in which ISH and IHC are complementary is heating of tissue sections. The use of heat during pre-hybridization is crucial for the penetration of riboprobes during the hybridization process (Fig. 5A vs. B). The absence of the heating step may not have deleterious effects on IHC (Fig. 5B2) but heat can enhance antigenicity (14). In our experience, however, some primary antibodies do not combine well with ISH under the conditions described here, e.g., PNMT, pMAPK.
4. Overstaining during the ISH detection step, despite lack of background labeling, can cause quenching of the fluorescence label (Fig. 6). This is particularly common with highly abundant

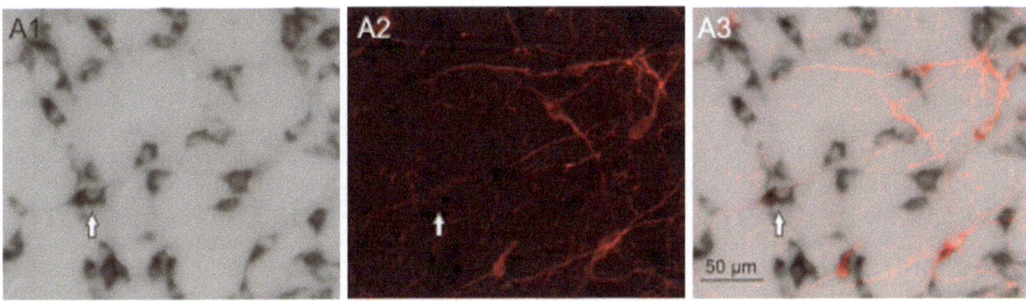

Fig. 6. Immunofluorescent quenching due to the presence of colocalization with ISH. Overdetection of highly abundant mRNAs often results in quenching of the immunofluorescent signal of combined antibodies. Despite very low signal to noise, G_s alpha mRNA (A1) staining is so *dark* that it has quenched tyrosine hydroxylase-ir (A2), making it difficult to visualize double-labeled neurons (see *arrow* in A3).

mRNAs. We suggest wet mounting tissue sections intermittently during the ISH detection process to gauge fluorescence labeling.

5. Previous methods, including those described here, have shown that ISH combined with multiple immunofluorescence labeling does not compromise the detection of target mRNAs, however ISH combined with enzymatic IHC detection methods (e.g., immunoperoxidase) causes a decrease in mRNA signal that could significantly compromise the detection, particularly of low abundance target mRNAs (1).

Acknowledgments

Some aspects of the combined ISH and IHC protocol described here are based on the method originally described by Dr Ruth Stornetta (University of Virginia Health System, VA, USA).

References

1. Pineau I, Barrette B, Vallires N et al (2006) A novel method for multiple labeling combining in situ hybridization with immunofluorescence. J Histochem Cytochem 54:1303–1313
2. Cox KH, DeLeon DV, Angerer LM et al (1984) Detection of mRNAs in sea urchin embryos by in situ hybridization using asymmetric RNA probes. Dev Biol 101:485–502
3. Stornetta RL, Spirovski D, Moreira TS et al (2009) Galanin is a selective marker of the retrotrapezoid nucleus in rats. J Comp Neurol 512:373–383
4. Barroso-Chinea P, Aymerich MS, Castle MM et al (2007) Detection of two different mRNAs in a single section by dual in situ hybridization: a comparison between colorimetric and fluorescent detection. J Neurosci Methods 162:119–128
5. Angerer LM, Angerer RC, Barbara AH et al (1991) Localization of mRNAs by in situ hybridization. Methods Cell Biol 35:37–71
6. Higuchi R, Krummel B, Saiki RK (1988) A general method of in vitro preparation and specific mutagenesis of DNA fragments: study

of protein and DNA interactions. Nucleic Acids Res 16:7351–7367

7. Young ID, Ailles L, Deugau K et al (1991) Transcription of cRNA for in situ hybridization from polymerase chain reaction-amplified DNA. Lab Invest 64:709–712
8. Haff LA (1994) Improved quantitative PCR using nested primers. Genome Res 3:332–337
9. Springell DA, Powers-Martin K, Phillips JK et al (2005) Phosphorylated extracellular signal-regulated kinase 1/2 immunoreactivity identifies a novel subpopulation of sympathetic preganglionic neurons. Neuroscience 133: 583–590
10. Li Q, Goodchild AK, Seyedabadi M et al (2005) Preprotachykinin A mRNA is colocalized with tyrosine hydroxylase-immunoreactivity in bulbospinal neurons. Neuroscience 136:205–216
11. Armstrong E, Partanen J, Alitalo K et al (1992) Localization of the fibroblast growth factor receptor-4 gene to chromosome region 5q33-qter. Genes Chromosomes Cancer 4:94–98
12. Ponder BA, Wilkinson MM (1981) Inhibition of endogenous-tissue alkaline-phosphatase with the use of alkaline-phosphatase conjugates in immunohistochemistry. J Histochem Cytochem 29:981–984
13. Stornetta RL, Schreihofer AM, Pelaez NM et al (2001) Preproenkephalin mRNA is expressed by C1 and non-C1 barosensitive bulbospinal neurons in the rostral ventrolateral medulla of the rat. J Comp Neurol 435: 111–126
14. Shi S, Key M, Kalra K (1991) Antigen retrieval in formalin-fixed, paraffin-embedded tissues: an enhancement method for immunohistochemical staining based on microwave oven heating of tissue sections. J Histochem Cytochem 39:741–748

Chapter 3

In Situ Hybridization Within the CNS Tissue: Combining In Situ Hybridization with Immunofluorescence

Dominic Bastien and Steve Lacroix

Abstract

In situ hybridization (ISH) is a useful method to investigate de novo mRNA expression in tissue sections. The high specificity and sensitivity of this technique combined with the great preservation of tissue and cellular morphology conferred by fixatives such as 4% paraformaldehyde make ISH a tool of choice for detecting genes of interest in individual cells in the central nervous system (CNS). Here, we provide a step-by-step protocol for a novel method that combines radioactive ISH with immunofluorescence on the same tissue section to identify cell populations expressing selected mRNA transcripts. This novel method has several major advantages over previously described double-labeling light microscopic methods combining enzymatic immunohistochemistry and ISH, including: (1) complete protection against loss of hybridization signal that normally occurs during the immunoenzymatic reaction, (2) improved immunolabeling sensitivity due to the proteinase K (PK) digestion step during ISH, (3) detection of several proteins specific for different cell populations on the same tissue section, and (4) counterstaining of tissue sections without affecting visualization of immunolabeling. This new method will be particularly useful for investigators looking to identify cell populations producing mRNAs expressed in low abundance, such as cytokines, chemokines, and growth factors, in the intact and injured/diseased mammalian CNS.

Key words: Brain, Double labeling, Immunohistochemistry, Method, Protocol, Riboprobe, Spinal cord

1. Introduction

One way of investigating de novo gene expression in central nervous system (CNS) and other tissues is to use in situ hybridization (ISH). The ISH technique is based on the concept of hybridization between a target RNA in tissue and a labeled single-stranded nucleic acid probe, which can either be a complementary RNA

Emilio Badoer (ed.), *Visualization Techniques: From Immunohistochemistry to Magnetic Resonance Imaging*, Neuromethods, vol. 70, DOI 10.1007/978-1-61779-897-9_3, © Springer Science+Business Media, LLC 2012

(cRNA) probe (also called riboprobe) or DNA oligonucleotide probe. The result of such hybridization is a stable double-stranded hybrid molecule, which is protected from single-stranded RNA digestion with RNase A and tolerates very high stringency washes. The ISH signal can later be visualized by detecting radiolabeled or conjugated nucleotides (e.g., radioisotope, fluorophore, biotin, digoxigenin) incorporated into the "protected" probe. As reviewed elsewhere (1), ISH offers many advantages over most other techniques used for detecting target mRNAs in tissue samples (e.g., RT-PCR, RNase protection assays, and Northern blots): (1) visualization of the spatial distribution of target mRNAs at the single cell level, (2) high specificity compared to RT-PCR, and (3) the possibility to identify the cellular populations that express target mRNAs by combining on the same tissue section, ISH with histochemistry, immunohistochemistry, neuronal tract tracing, or even ISH (e.g., combination of radioactive and nonradioactive probes).

Over the last 2 decades, many studies have combined ISH with immunohistochemistry to identify cellular populations that synthesize a gene of interest. In these studies, cell-specific antigenic markers were first detected using the immunoperoxidase technique, followed by detection of target mRNAs using radioactive or nonradioactive probes. We recently found, however, that peroxidase reaction products cause a decrease in mRNA signal, and this can seriously compromise the detection of mRNA transcripts (Fig. 1; see also (1)). This situation is even more problematic if one has to study tissues in which cells enriched in endogenous peroxidase are abundant. This is notably the case following CNS insult or disease, in which leukocytes, rich in endogenous peroxidase, are often recruited in large numbers and remain inside CNS tissues for protracted periods (2, 3). We further found that immunoperoxidase reaction products cause the loss of signal from target mRNAs already hybridized with riboprobes, making the possibility to perform ISH before immunoperoxidase staining at least equally challenging (Fig. 1). This latter finding emphasizes even more the destructive effects of free radicals generated during the immunoperoxidase reaction on mRNA signal integrity, and highlights the importance of developing novel multilabeling methods that do not involve the generation of highly reactive products such as the immunoperoxidase method.

The apparent incompatibility between ISH signal and certain enzymatic reaction products has prompted us to develop a new methodology that allows for the combination of ISH with immunofluorescence to identify CNS cell types expressing selected mRNA transcripts. This novel method has several advantages over the more traditional double-labeling technique combining immunoperoxidase labeling with ISH: (1) immunofluorescence labeling

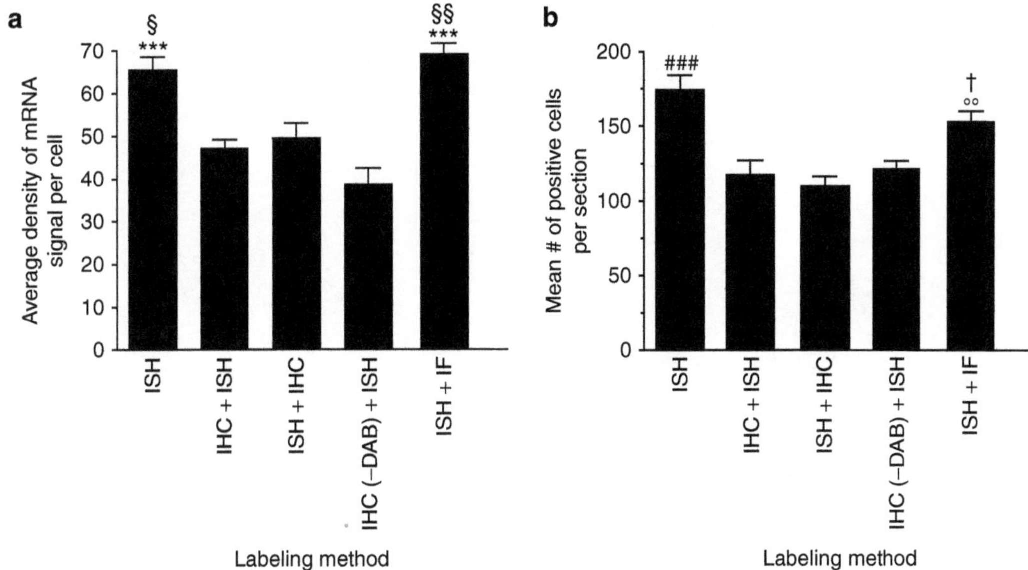

Fig. 1. Peroxidase reaction products cause a decrease in mRNA signal. (**a**, **b**), Quantification of MIP-1α mRNA signal in adjacent spinal cord sections from spinal cord injured mice using different labeling methods. MIP-1α mRNA expression was quantified by measuring the average density of MIP-1α mRNA signal per cell (**a**) and by counting the number of cells expressing MIP-1α mRNA (**b**) in adjacent coronal sections located between 1,260 and 1,470 μm caudal to the lesion epicenter. The following labeling methods were compared: (1) in situ hybridization (ISH) alone, (2) immunoperoxidase labeling (IHC) followed by ISH, (3) ISH followed by IHC, (4) IHC performed without DAB (–DAB) followed by ISH, and (5) ISH followed by multiple immunofluorescence labeling (IF). Note that hybridization signal is significantly reduced when immunoperoxidase labeling is involved, whereas signals detected using the new multiple labeling method are quantitatively equal to those obtained using ISH alone. Statistical significance was assessed by a one-way ANOVA test. Post hoc comparisons were made using the Tukey test. $^{***}P<0.001$ compared with IHC (–DAB) + ISH. $^{§§}P<0.01$, $^{§}P<0.05$ compared with IHC + ISH or ISH + IHC. $^{\#\#\#}P<0.001$ compared with IHC + ISH, ISH + IHC, or IHC (–DAB) + ISH. $P<0.01$ compared with ISH + IHC. $^{†}P<0.05$ compared with IHC + ISH or IHC (–DAB) + ISH. The final, definitive version of this figure has been published in Pineau et al. (1), http://jhc.sagepub.com/content/54/11/1303.full by SAGE Publications Ltd. Copyright © 2006 by SAGE Publications, Inc. All rights reserved.

does not compromise the detection of target mRNAs, (2) the simultaneous detection of several antigens corresponding to as many cell populations on the same tissue section is possible, and (3) tissue sections can be counterstained with fluorescent reagents that specifically bind nucleic acids (e.g., DAPI), therefore making sure that all double-labeled cells are associated with a single nucleus. Without performing nuclear counterstaining, we found in our previous work that it was basically impossible to ascribe with certainty the co-localization of a specific target mRNA with a given cell type (1). This was mostly due to the presence of multiple cellular layers in tissue sections and to the accumulation of cell clusters in the injured/diseased CNS (Fig. 2).

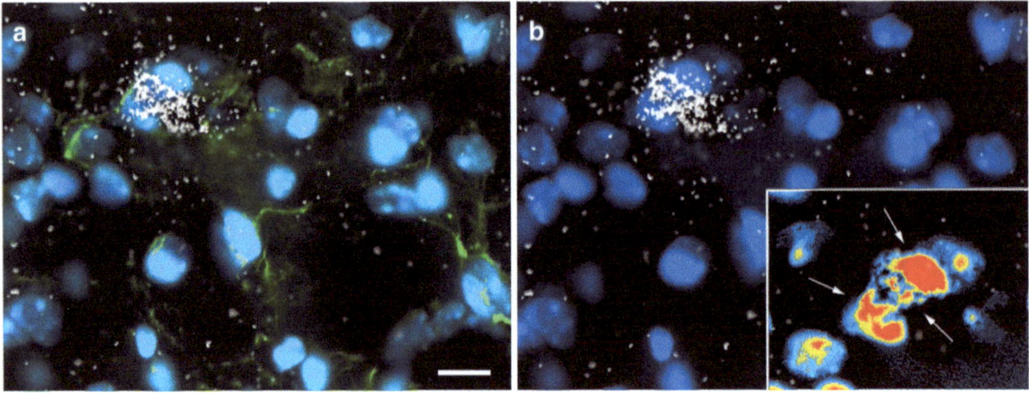

Fig. 2. Difficulty to ascribe with certainty the co-localization of a specific target mRNA within individual cells in certain conditions. (**a**) Photomicrographs showing ISH signal for IL-6 mRNA that appears to be co-localized within an astrocyte, as visualized using the astrocytic marker GFAP (*green*; the reader is referred to the online version for the color reproduction of this figure). However, counterstaining with the nuclear dye DAPI ((**b**), *blue*) reveals the presence of a small agglomeration of cells containing at least three distinct nuclear profiles. Note that the nucleus (DAPI, *blue*) of the IL-6 mRNA+cell is not co-localized with the astrocytic marker GFAP but rather found in between two GFAP-ir astrocytes. *Arrows* indicate nuclear profiles (see *inset* in (**b**)). Scale bar = (in (**a**)) (**a**, **b**) 10 μm. The final, definitive version of this figure has been published in Pineau et al. (1), http://jhc.sagepub.com/content/54/11/1303.full by SAGE Publications Ltd. Copyright © 2006 by SAGE Publications, Inc. All rights reserved.

2. Materials

The products we used are listed below. Comparable products from other suppliers may also be effective.

2.1. In Situ Hybridization

Presented here is a detailed protocol for the detection of mRNAs encoding specific proteins in brain and spinal cord tissue sections. The ISH protocol described below has been adapted from Simmons et al. (4) and optimized for labeling studies involving the simultaneous detection of multiple antigens (e.g., cell-specific markers or any particular protein of interest) using immunofluorescence. Our ISH protocol uses ^{35}S-labeled single-stranded cRNA riboprobes to detect target mRNA because of their greater sensitivity compared to DNA oligonucleotide probes.

2.1.1. Perfusion Fixation and Tissue Preparation

1. Prepare the following solutions the day before animal perfusion: a 0.9% (w/v) saline solution, 4% paraformaldehyde (PFA) in 0.1 M borax (sodium tetraborate decahydrate, $Na_2B_4O_7 \cdot 10H_2O$; pH 9.5) (see Note 1), and 4% PFA–borax containing 10% sucrose. Keep solutions at 4°C.
2. Perfusion instruments: fine forceps, scissors, hemostatic forceps (Fine Science Tools).
3. A variable-speed, planetary gear-driven, digital peristaltic pump (Model 78023-00) equipped with four channels, Click'N'Go cartridges, connectors, and Tygon® LFL tubing resistant to fixatives (all from Cole-Palmer) (see Note 2).

4. Microdissection instruments: fine forceps, scalpel, scissors, microrongeurs with extra-fine tips, spring scissors, microdissection spatula (Fine Science Tools).
5. Cryostat or sliding microtome (Leica Microsystems Inc.).
6. Positively charged microslides (Surgipath Canada Inc.).
7. A cryoprotectant solution made of 0.05 M sodium phosphate buffer, pH 7.3, 30% (v/v) ethylene glycol, and 20% (v/v) glycerol. All solutions must be RNase free.

2.1.2. Preparation of cRNA Riboprobes

1. [α-^{35}S]UTP (Perkin Elmer, store at –20°C) and a Savant SpeedVac Plus concentrator (model SC110A).
2. DNA template prepared from a transcription vector (e.g., pCRII-TOPO, pBluescript II) containing common bacteriophage promoters (e.g., SP6, T7, and/or T3) and in which the cDNA corresponding to the target mRNA (or any specific portion of it) has been subcloned. The coding sequence chosen for riboprobe synthesis needs to be carefully selected to match only the intended gene, as verified by using the BLAST algorithm in Genbank.
3. Promega Riboprobe® Combination System kit designed for in vitro preparation of highly specific single-stranded RNA probes. The kit contains Riboprobe® System Transcription Buffers (Transcription Optimized 5× Buffer, nuclease-free water, dithiothreitol [DTT, 1 M]), ribonucleotide triphosphates (rNTPs: rATP, rCTP, rGTP, rUTP, each at 10 mM in separate tubes) (see Note 3), recombinant RNasin® ribonuclease inhibitor, RNA polymerases (SP6, T7, and/or T3), and RNase-free DNase I. Store at –20°C.
4. Transfer RNA (tRNA; 10 mg/mL), diethylpyrocarbonate (DEPC)-treated H_2O, 10 mM Tris solution (pH 8.0) containing 10 mM $MgCl_2$, phenol:chloroform:isoamyl alcohol 25:24:1 saturated with 10 mM Tris/1 mM EDTA (TE; pH 8.0), 5 M ammonium acetate (pH 8.0), 70 and 100% ethanol (EtOH), TE (pH 8.0).
5. Liquid scintillation counter (model WinSpectra 1414, Wallac).

2.1.3. Prehybridization and Hybridization

Precautions must be taken to prevent RNase contamination and all materials and solutions used must be RNase-free grade. To do so, use autoclaved plastic and glassware devoted exclusively to the prehybridization and hybridization steps of the ISH protocol. Prepare all solutions for prehybridization and hybridization using DEPC-treated H_2O. We typically prepare DEPC-H_2O by stirring overnight 1 L of Milli-Q H_2O in the presence of 0.01% DEPC under a fume hood at room temperature (RT) followed by autoclaving (see Note 4).

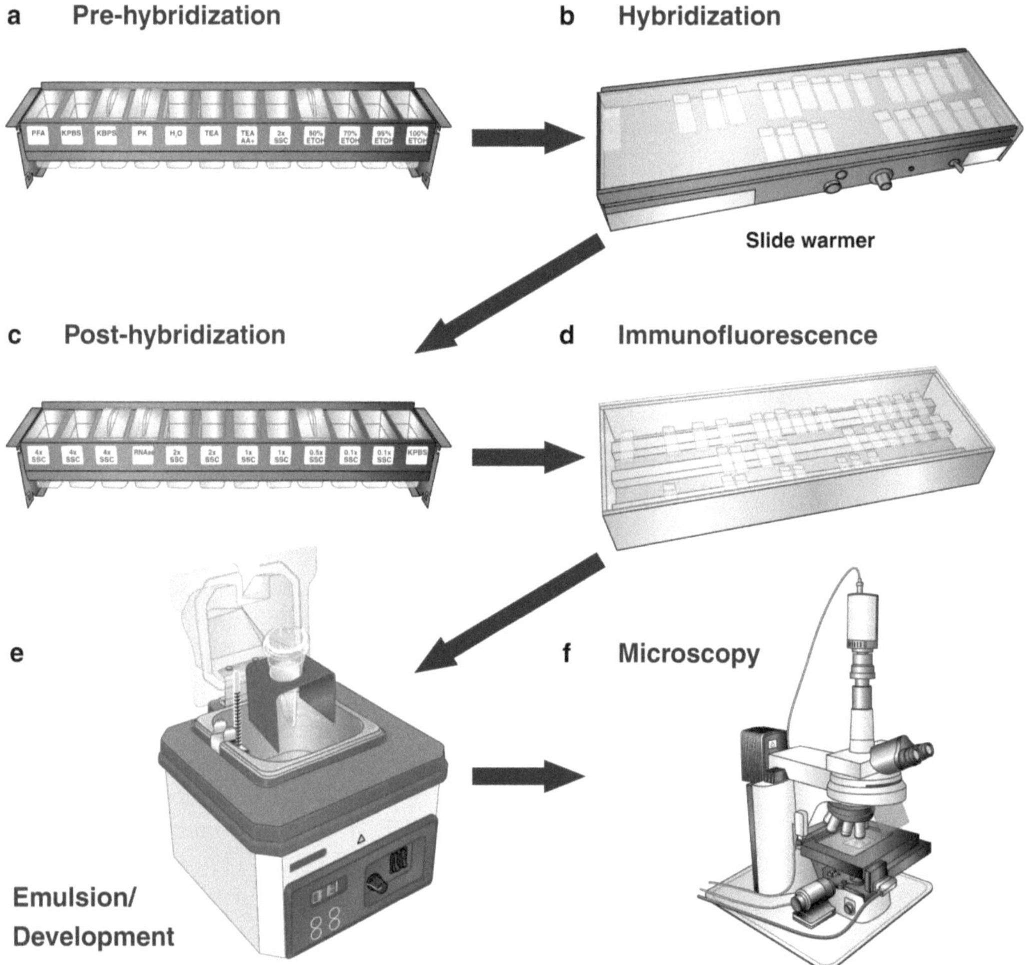

Fig. 3. Schematic diagram summarizing the major steps of the protocol combining ISH with multiple immunofluorescence labeling. (**a–f**) The protocol consists of six major steps, namely: prehybridization (**a**), hybridization (**b**), posthybridization (**c**), immunofluorescence (**d**), emulsion and development (**e**), and microscopy (**f**). Refer to the Sects. 2 and 3 for a more complete description of the protocol.

1. Poly-L-lysine-coated slides and a small artist's brush with a fine point.
2. Tissue-Tek Slide Staining Set (model 4451, Sakura Finetek) including staining dishes and slide holders (24-place). We note that each Tissue-Tek staining dish has a full capacity for 250 mL of solution (see Fig. 3). Prepare the following prehybridization solutions and reagents:
   - Four percent PFA in 0.1 M borax, pH 9.5 (see Sect. 2.1.1)
   - KPBS:
     - 3.81 g potassium phosphate, dibasic, anhydrous (K_2HPO_4)
     - 0.45 g potassium phosphate, monobasic, anhydrous (KH_2PO_4)

 - 8.1 g NaCl
 - Complete to 1 L using DEPC-H_2O
- Proteinase K (PK) buffer:
 - 100 mL 1 M Tris–HCl, pH 8.0
 - 100 mL 0.5 M EDTA, pH 8.0
 - Complete to 1 L using DEPC-H_2O
- Saline-sodium citrate (SSC) 20×, pH 7.5:
 - 175.3 g sodium chloride
 - 88.2 g sodium citrate, dihydrate, $C_6H_5Na_3O_7{\cdot}2H_2O$
 - Complete to 1 L using DEPC-H_2O
- Proteinase K (PK, 20 mg/mL), 0.1 M triethanolamine (TEA, pH 8.0), acetic anhydride (minimum 98%), 50–100% EtOH

3. Vacuum desiccator including transparent polycarbonate cover, desiccator plate, and vacuum tubing (VWR).
4. Lab-Line Slide Warmer (a thermostat-controlled surface for slide support) (model 26025, TekniScience Inc.). Digital water bath (model Isotemp 205, Fisher Scientific).
5. Prepare the following hybridization solutions and reagents:
 - Denhardt's solution 50×:
   - 0.5 g Ficoll
   - 0.5 g polyvinylpyrrolidone
   - 0.5 g bovine serum albumin, Fraction V, ≥96%
   - Complete to 50 mL using DEPC-H_2O, filter, and store at −20°C in 1-mL aliquots
 - Dextran sulfate 50%:
   - 20 g of dextran sulfate in a 50-mL Falcon conical tube
   - Add 40 mL of DEPC-H_2O, vortex, and leave to rest for 2 days at 4°C. Readjust the volume to 40 mL after it has completely dissolved. Keep protected from light at 4°C
 - Solution I (for 50 mL):
   - 31.25 mL of deionized formamide
   - 3.75 mL of 5 M NaCl
   - 625 μL of 1 M Tris, pH 8.0
   - 125 μL of 0.5 M EDTA, pH 8.0
   - 1.25 mL of Denhardt's solution 50×
   - 12.5 mL of 50% Dextran sulfate
   - Protect from light and store at −20°C
 - tRNA (10 mg/mL), 1 M DTT, DEPC-H_2O

2.1.4. Posthybridization

Starting at this point, it is not necessary to be RNase-free anymore. However, it remains essential to wear gloves due to handling material contaminated by radioactivity.

1. Tissue-Tek Slide Staining Set (model 4451, Sakura Finetek) including staining dishes and slide holders (24-place) (see Note 5). Prepare the following posthybridization solutions and reagents:
   - RNase buffer:
     - 100 mL of 5 M NaCl
     - 10 mL of 1 M Tris–HCl, pH 8.0
     - 1 mL of 0.5 M EDTA, pH 8.0
     - Complete to 1 L using DEPC-H_2O
   - 20× SSC, pH 7.5 (see Sect. 2.1.3)
   - RNase A (10 mg/mL), 1 M DTT, 50–100% EtOH

2.2. Immunofluorescence (IF)

To identify cell populations expressing the target mRNA transcript coding for a gene of interest, ISH can be combined with immunofluorescence. For this, tissue sections are first prehybridized, hybridized, and posthybridized, as described earlier, and then processed for multiple-immunofluorescence labeling using the materials listed below.

1. CoverWell incubation chambers (Invitrogen Canada Inc.) and a humidity chamber with a number of ridges on which slides can be deposited. Also, prepare/purchase the following solutions and reagents:
   - KPBS (see Sect. 2.1.3)
   - Triton X-100, normal serum, primary antibodies, secondary antibodies conjugated to Alexa fluorophores (Invitrogen Canada Inc.), 4′,6-diamidino-2-phenylindole dilactate (DAPI) (Invitrogen Canada Inc.)
2. Vacuum desiccator including transparent polycarbonate cover, desiccator plate, and vacuum tubing (VWR). Use the same as for posthybridization.

2.3. Detection of ISH Signal

2.3.1. X-Ray Film Autoradiography

1. Darkroom, X-ray films (model Kodak BioMax-MR, Amersham Biosciences Inc.), and autoradiography cassettes (model FBXC, Fisher)
2. X-ray film processor (model Mini-Med/90, AFP Imaging Corp.)

2.3.2. Liquid Emulsion Autoradiography

1. Humidifier and digital water bath
2. Dip Miser set (Electron Microscopy Sciences) which consists of a glass slide coating cup, a stand for immersing the cup in water, and a cleaning brush

3. Kodak NTB autoradiography emulsion (Mandel Scientific Company Inc.)
4. Slide boxes and desiccant
5. Kodak D-19 Developer and Rapid Fixer (Sigma-Aldrich)
6. Ethanol 50–100%, Hemo-D xylene substitute (Fisher Scientific), and DPX Mounting Medium (Cedarlane Laboratories Ltd.)

3. Methods

3.1. In Situ Hybridization

3.1.1. Perfusion Fixation and Tissue Preparation

1. Inject the animal with an overdose of an injectable anesthetic (e.g., a mixture of ketamine and xylazine). Fill the perfusion pump tubing with cold 0.9% saline solution.
2. Once the animal is deeply anesthetized (i.e., no response to hard pinch), pin the rodent onto a dissection board (ventral side up). Working quickly, use the fine forceps and scissors to open the chest cavity. Grab the flap of the xiphoid process with the hemostatic forceps and clamp back the entire chest to create an adequate working field, thereby exposing the heart.
3. Use the fine forceps to hold the heart and introduce the perfusion needle into the left ventricular cavity (see Note 6). Perfuse with cold 0.9% saline solution followed by 4% PFA–borax (see Note 2). Once the flow of perfusion saline has started circulating in blood vessels, use scissors to make a cut in the right atrium to allow blood and perfusion solutions to drain.
4. After perfusion with the fixative, use the forceps, scalpel, and scissors to expose the skull and vertebral column. Remove the bone overlaying the brain and spinal cord using microrongeurs with extra-fine tips. Dissect out the brain and spinal cord using fine forceps, spring scissors, and a microdissection spatula. Before brain removal, make sure to carefully remove the dura mater. Postfix the tissue for 2 days in 4% PFA–borax. The day before sectioning, place the brain and spinal cord overnight in a 4% PFA–borax/10% sucrose solution.
5. We routinely cut the brain and the spinal cord using either a cryostat or a microtome, depending on the fragility of the tissue (we use the cryostat to cut delicate or fragile tissue samples, e.g., injured spinal cord tissue, mouse peripheral nerves).
6. Collect 30-μm-thick cryostat sections directly onto positively charged microslides. Separate into different series of adjacent sections. Store slides at –20°C until ISH.
7. Collect 30-μm-thick microtome sections into tissue culture plate wells containing a cold cryoprotectant solution and store at –20°C. Free-floating sections (and slides) can be stored for months at –20°C without any apparent loss of mRNA signal.

3.1.2. Preparation of cRNA Riboprobes

We normally synthesize riboprobes ahead of time, i.e., before thawing/mounting tissue sections.

1. Dry down the radioactivity using the SpeedVac. We typically use 200 μCi of [α-^{35}S]UTP per riboprobe.
2. In order to transcribe the cDNA and not the plasmid, linearize the transcription vector with an enzyme that cleaves within the multiple cloning site downstream of the cDNA insert.
3. After linearization and purification of the DNA template, synthesize the radiolabeled cRNA probe using the Riboprobe® Combination System kit. In more detail, the riboprobe is synthesized by adding the following reagents to the dried radiolabeled UTP: 250 ng of the linearized DNA template resuspended in 1 μL of TE, 2 μL of Transcription Optimized 5× buffer, 3 μL of nuclease-free water, 1 μL of 0.1 M DTT, 1 μL of rNTP mixture containing 2 mM of each rNTP except rUTP (see Note 3) (i.e., rATP, rGTP, rCTP), 1 μL of RNase inhibitor, and 1 μL of RNA polymerase. Incubate for 60 min at 37°C.
4. Remove unincorporated nucleotides using the ammonium acetate method (see Note 7). First, add 100 μL of DNase solution (1 μL of DNase, 2.5 μL of tRNA, 2.5 μL of DEPC-treated H_2O, 94 μL of 10 mM Tris/10 mM $MgCl_2$) and incubate 10 min at RT to remove the DNA template. Extract once with 100 μL of phenol:chloroform:isoamyl alcohol, centrifuge at >14,000 × g for 5 min and save the top phase. Precipitate the cRNA with 80 μL of 5 M ammonium acetate and 500 μL of 100% cold (–20°C) EtOH. Leave for 20 min on dry ice. Then, centrifuge for 15 min at 4°C and save the pellet. Wash with 200 μL of 70% cold EtOH, centrifuge again for 10 min at 4°C and save the pellet. Air-dry the pellet and resuspend in 100 μL of TE.
5. Measure the incorporated radioactivity in 3 × 1 μL aliquots of the riboprobe by liquid scintillation spectrometry. Radioactivity in a good riboprobe normally yields 2–5 × 10^6 cpm/μL (we do not use probes yielding <1 × 10^6 cpm/μL as they typically lead to poor signal-to-noise ratios). Store the riboprobe at –20°C until use.

3.1.3. Prehybridization and Hybridization

To normalize background intensity, we strongly recommend to prehybridize, hybridize, and posthybridize all sections in parallel (to see how to do this, refer to Fig. 3).

1. First, remove slides from the freezer and thaw overnight at RT under vacuum. If free-floating sections are being used, mount them onto poly-L-lysine-treated slides using a small artist's brush and a dish containing KPBS (see Note 8). Let sections dry under vacuum for at least 2 h before proceeding to step 2.

2. Immerse slide-mounted sections in the following solutions (this procedure must be done in the order stated): 4% PFA–borax for 20 min, KPBS for 5 min (2×), PK buffer plus 125 μL of the PK enzyme (final concentration, 10 μg/mL) for 10–25 min at 37°C (see Note 9), briefly rinse in DEPC-treated H_2O, rinse in 0.1 M TEA, 0.1 M TEA plus 625 μL of acetic anhydride for 10 min, 2× SSC for 5 min, and ascending concentrations of EtOH (3 min in each of 50, 70, 95, 2 × 100%).
3. Air-dry the sections and then place them with desiccant in a desiccator under vacuum at RT for at least 1 h until hybridization.
4. Turn on the slide warmer to 60°C and preheat a water bath at 65°C.
5. Prepare the hybridization mixture (10^7 cpm/mL) by combining (for making 1 mL, which is enough to hybridize nine slides): 822 μL of Solution I, 50 μL of tRNA, 10 μL of 1 M DTT, 118 μL of DEPC-treated H_2O minus the riboprobe volume, and 1×10^7 cpm of the riboprobe. Mix well and store at −20°C. The hybridization mixture can be made ahead of time.
6. Ten minutes before the hybridization step, place the Falcon tube containing the hybridization mixture in the water bath at 65°C to melt any annealed strands. Using a micropipette, spot 90 μL of the hybridization mixture onto each slide and seal immediately under a coverslip (see Note 10). Remove air bubbles that may prevent the hybridization solution from contacting the tissue (resting on a dark surface for best visibility, use a spatula to carefully remove all air bubbles). Place the slides on the preheated slide warmer and incubate for ~15–18 h.

3.1.4. Posthybridization

1. The next day, gently remove the coverslips from the slides by agitating in 4× SSC at RT for 30–60 min (see Note 11). Insert slides immediately into another slide holder immersed in 4× SSC. After four additional 5-min washes in 4× SSC plus 250 μL of 1 M DTT (see Note 12), digest single-stranded RNA (i.e., nonhybridized RNA) by incubating slides into RNase buffer to which 500 μL of the RNase A enzyme are added (final concentration, 20 μg/mL) for 30 min at 37°C (see Note 13). RNase A treatment is the most critical step of the posthybridization procedure. Omitting this step will result in a very intense background due to the presence of riboprobes that have bound nonspecifically to the CNS tissue. It is important to note, however, that RNase A treatment will result in the digestion of ribosomal RNA, therefore interfering with subsequent cytoplasmic staining of Nissl substance. Following RNase A treatment, submerge slide-mounted sections in the following solutions to which 250 μL of 1 M DTT are added, in the stated order: 2× SSC for 5 min (2×); 1× SSC for 5 min; 0.5× SSC for

10 min; 0.1× SSC for 30 min at 60°C; and briefly rinse in 0.1× SCC (see Note 14). Rinse in KPBS for 5 min and then immediately process for immunofluorescence (see Note 15).

3.2. Immunofluorescence (IF)

1. Block for 1 h in KPBS + 0.25% Triton X-100 + 5% normal serum (the normal serum must be from the same species as the secondary antibodies). Incubate sections in primary antibodies overnight at 4°C. Use CoverWell incubation chambers to reduce the amount of primary antibodies needed and prevent evaporation. Store slides in a humidity chamber to facilitate transport to the 4°C room. The next day, leave slide-mounted sections for 1 h at RT and then carefully remove CoverWell incubation chambers in KPBS using fine-tipped forceps. Rinse sections in KPBS, 3×5 min. Use CoverWell incubation chambers to incubate sections in secondary antibodies conjugated to Alexa fluorophores (1:200 dilution) for 2.5 h. Rinse in KPBS 3×5 min, counterstain in DAPI (1:5,000 dilution) for 20 min, and rinse successively in KPBS (3×5 min) and H_2O (1×5 min).
2. Air-dry slide-mounted sections and place them with desiccant under a vacuum desiccator at RT. Protect from light.

3.3. Detection of ISH Signal

At this point, you can choose to skip the X-ray film autoradiography detection method and proceed directly with the liquid emulsion autoradiography method. Note that dehydration and defatting steps that normally precede dipping into liquid NTB emulsion are voluntarily omitted to avoid fading of the fluorescence.

3.3.1. X-Ray Film Autoradiography

1. Tape slides onto cardboard sheets cut to fit perfectly inside autoradiography cassettes. Expose tissue sections to X-ray films in a darkroom. Store cassettes for 1–2 days at 4°C (see Note 16).
2. Leave cassettes for 1 h at RT. Process X-ray films in the darkroom using the automatic X-ray film processor. Automatic processing is faster than manual processing and produces autoradiograms of uniform quality, which is obviously better for quantification.
3. Remove slides from cassettes and put them back in the vacuum desiccator.

3.3.2. Liquid Emulsion Autoradiography

1. Turn on the humidifier and set the water bath to 42°C in the darkroom.
2. Dilute the autoradiography emulsion 1:1 in Milli-Q H_2O in a 50-mL Falcon conical tube. Warm in water bath for 30 min at 42°C. Gently mix the tube by inversion until the emulsion

solution becomes homogeneous. Slowly pour 40 mL of liquid emulsion into the Dip Miser and let rest in the water bath for an additional 30 min. Remove air bubbles by dipping blank slides that were previously cleaned in Hemo-D followed by EtOH.

3. Dip slide-mounted sections (one by one) in liquid NTB emulsion in the darkroom. Dip slowly and one time only to avoid streaks. Place slides in a vertical position in a rack and allow them to dry for at least 2–3 h.
4. Once slides are completely dry, place them in slide boxes containing desiccant bags. Cover with foil (to protect from light) and store at 4°C. In general, 1 day of exposure on X-ray film translates into 1 week in emulsion.
5. At time of developing, insert slides into a slide holder at RT in the darkroom. Develop slide-mounted sections in D-19 Developer for 3.5 min at 14–16°C. Stop the reaction by briefly rinsing (15 s) slides in Milli-Q H_2O at 14–16°C and then fix in freshly made Rapid Fixer for 5 min also at 14–16°C. Rinse slides in running H_2O for 60 min.
6. Quickly dehydrate in graded concentrations of EtOH (50, 70, 95, 2 × 100%), rapidly clear in Hemo-D, and mount coverslip with DPX medium. The presence of target mRNA can be visualized by the agglomeration of grains within the cell cytoplasm using either brightfield or darkfield microscopy. We note that water-based mounting media that are commonly used to reduce fluorescence quenching (e.g., PVA-DABCO, Fluoromount G) are incompatible with darkfield microscopy. For quantification of ISH signal and examples of multiple labeling combining ISH with immunofluorescence, please refer to Fig. 4 and our previously published studies (1, 5, 6).

3.4. Conclusion

Our previous work has shown that immunoperoxidase reaction products cause mRNA signal loss and can therefore compromise double-labeling studies combining immunohistochemistry with ISH. To resolve this problem, we recently developed a new method that allows for the combination of ISH and immunofluorescence without causing alteration of mRNA signal integrity. This new method, described herein, yielded stronger ISH signal when compared to previous double-labeling methods and allowed for simultaneous detection of multiple antigenic markers on the same tissue section. Using this technique, we successfully identified the cells that synthesize specific cytokines and chemokines that are known to be expressed at relatively low levels in the injured/diseased rodent nervous system.

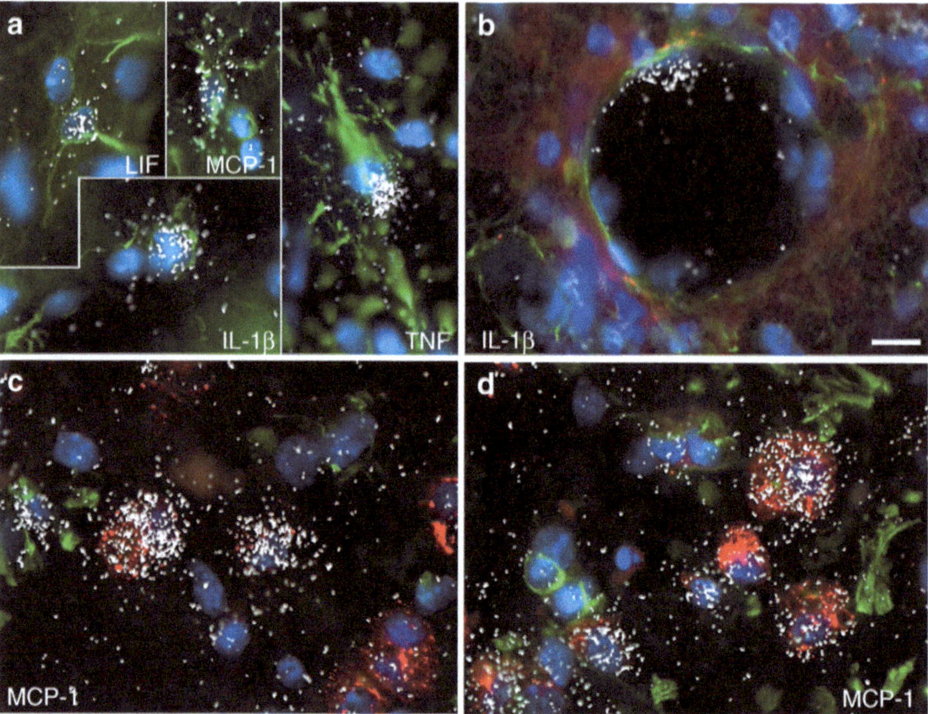

Fig. 4. Examples of multiple labeling combining ISH with immunofluorescence in the injured mouse spinal cord. (**a–d**) Photomicrographs taken from the mouse spinal cord showing co-localization of LIF (**a**), MCP-1 (**a**, **c–d**), IL-1β (**a**, **b**), and TNF (**a**) mRNAs within GFAP-ir astrocytes ((**a**), *green*), an endothelial cell (**b**), and Gal-3-ir activated microglia/macrophages ((**c**, **d**), *red*). Nuclear counterstaining with DAPI is shown in *blue*. Scale bar = (in (**b**)) (**a**, **d**) 10 μm. The final, definitive version of this figure has been published in Pineau et al. (1), http://jhc.sagepub.com/content/54/11/1303.full by SAGE Publications Ltd. Copyright © 2006 by SAGE Publications, Inc. All rights reserved.

4. Notes

1. Heat up to 65°C (but do not exceed) and stir to help dissolve the PFA. After the fixative solution has cooled to RT, adjust pH to 9.5 by adding about 5 mL of a 10 N NaOH solution to each liter of fixative. Lastly, filter the fixative through No. 5 Whatman paper and cool to 4°C.

2. We set the flow rate of the perfusion pump at 12.5 mL/min for mice and 25 mL/min for rats. Mice are routinely perfused with 50 mL of cold 0.9% saline solution followed by 50 mL of 4% PFA–borax. Rats are perfused with 100 mL of 0.9% saline solution followed by 250 mL of 4% PFA–borax.

3. Because rNTPs are very sensitive to freeze–thaw cycles, we found that the best strategy is to combine them (resulting in 2 mM for each rNTP) and prepare aliquots to be stored at −20°C. However, make sure that you do not add rUTP to the

mix if you are to label riboprobes with radioactive UTP (we typically do not incorporate cold rUTP in the in vitro transcription reaction).

4. DEPC is carcinogenic. It should therefore be handled with extreme care (wear gloves and mask and mix under a fume hood) and completely destroyed by autoclaving (45 min at 123°C on liquid cycle) before using DEPC-treated water.
5. Use different slide rack, dishes, and slide holders from those used for the prehybridization.
6. The perfusion needle is connected to the perfusion tubing via a Luer lock connector. We typically use a 20-gauge needle to perfuse mice and a blunt 18-gauge needle to perfuse rats. Note that for rat perfusion, we normally make a small incision in the left ventricle and introduce the blunt 18-gauge needle into the aorta, thus passing through the aortic valve. To secure the needle in place, clamp the aorta with the needle in it using hemostatic forceps. For mice, we do not recommend to try inserting the perfusion needle into the aorta. Instead, hold the 20-gauge needle in place inside the left ventricle.
7. Some groups prefer to use Quick Spin Columns to remove unincorporated nucleotides (4, 7).
8. Make sure to remove the cryoprotectant from the tissue by washing sections thoroughly with KPBS (3×10 min) before proceeding to mounting.
9. Pretreatment with PK is the most important step of the prehybridization protocol. This step is necessary to permeabilize cell membranes, and therefore to expose target mRNAs to probes. In a previous study, we demonstrated that proteolytic digestion of tissue sections is absolutely critical to obtain maximal ISH signal intensity, but that excessive PK pretreatment results in the loss of target mRNAs (1). Interestingly, we also found that PK digestion facilitates the penetration and access of antibodies to their antigenic sites in tissue fixed with 4% PFA–borax (pH 9.5) and sections mounted onto slides (1). Based on data published in that study, we recommend for double-labeling studies that combine ISH with immunohistochemistry, to digest 30-μm-thick CNS tissue sections in PK (10 μg/mL) for 10 min to detect cell surface and membrane antigens and 20–25 min for cytosolic and nuclear antigens (see Table 1 for permeabilization conditions for immunofluorescence labeling of particular antigens). Ideally, time of enzymatic digestion should be optimized for each antigen of interest.
10. Some groups prefer to seal the edges of the coverslip with a bead of liquid rubber cement or DPX mounting medium to prevent the riboprobe from drying (4, 7, 8).

Table 1
Permeabilization conditions that have proved to be optimal for immunofluorescence detection of specific antigens in slide-mounted brain and spinal cord tissue sections (30-μm thick) fixed with 4% PFA–borax, pH 9.5

Antibody	Specificity	Immunogen/clone	Dilution	PK digestion time (min)	Source	Catalog/lot #
Polyclonal						
Carbonic anhydrase II (CAII)	CAII found in oligodendrocytes	Purified protein from rat hemolysates	1:2,000	25	Dr. Said Ghandour	N/A
Glial fibrillary acid protein (GFAP)	GFAP present in astrocytes	GFAP isolated from cow spinal cord	1:750	10	Dako	Z 0334/5193
Ionized calcium-binding adaptor molecule 1 (Iba1)	17 kDa EF hand protein expressed in macrophages and microglia	Synthetic peptide of C-terminus Iba1 N′-PTGPPAKKAISELP-C′	1:750	10	Wako	019-19741/ HNH3687
Monoclonal						
7/4	40 kDa antigen expressed by neutrophils and Type I inflammatory monocytes	Clone 7/4	1:800	15	AbD Serotec	MCA771GA
Galactose-specific lectin-3 (Gal-3)	32 kDa Mac-2 glycoprotein in macrophages/activated microglia	Spleen cells fused with NS-1 myeloma cells/ Clone M3/38	1:500	10–15	ATCC	TIB-166/211921
Glial fibrillary acid protein (GFAP)	GFAP present in astrocytes	Clone 2.2B10	1:200	10	Invitrogen	13-0300/509384A
HuC/HuD (human neuronal proteins)	Elav family members HuC, HuD, and Hel-N1	Clone 16A11	1:80	20	Invitrogen	A-21271

11. If needed, use fine-tipped forceps to remove the coverslips. Make sure not to cause any damage to the tissue.
12. According to Simmons et al. (4), DTT plays two key roles during ISH: (1) it stabilizes ^{35}S attachment to the riboprobe and (2) it reduces the formation of disulfide bonds between the ^{35}S-labeled riboprobe and sulfhydryl groups in the tissue, which obviously would lead to nonspecific binding of the riboprobe and increase background levels.
13. Discard the 4× SSC washing solutions and RNase buffer into the liquid radioactive waste container.
14. Washes under conditions of high stringency (i.e., low salt concentration and high temperature) ensure dissociation of noncomplementary hybrids. Make sure, however, that you do not start losing tissue sections during the 30-min wash in 0.1× SSC at 60°C. If so, stop immediately and proceed to the next wash.
15. In the event that ISH alone would be performed (i.e., no immunohistochemistry), skip the 5-min wash in KPBS and proceed to dehydrating the sections in ascending concentrations of EtOH (3 min in each of 50, 70, 95, 2×100%). Add 1.25 μL of 20× SSC to the first two EtOH rinses. Air-dry the sections and store them in a vacuum desiccator at RT. Use a different desiccator from the one used for the prehybridization.
16. The time of exposure will depend on the abundance of the target mRNA within tissue and the desired signal intensity needed for imaging and quantification.

Acknowledgments

This study was supported by the Canadian Institutes of Health Research (CIHR). Steve Lacroix is supported by a Career Award from the Rx&D Health Research Foundation and CIHR. We thank Nicolas Vallières, Nadia Fortin, and Isabelle Pineau for their technical assistance.

References

1. Pineau I, Barrette B, Vallières N, Lacroix S (2006) A novel method for multiple labeling combining in situ hybridization with immunofluorescence. J Histochem Cytochem 54:1303–1313
2. Bohatschek M, Werner A, Raivich G (2001) Systemic LPS injection leads to granulocyte influx into normal and injured brain: effects of ICAM-1 deficiency. Exp Neurol 172:137–152
3. Fleming JC, Norenberg MD, Ramsay DA, Dekaban GA, Marcillo AE, Saenz AD, Pasquale-Styles M, Dietrich WD, Weaver LC (2006) The cellular inflammatory response in human spinal cords after injury. Brain 129:3249–3269

4. Simmons DM, Arriza JF, Swanson LW (1989) A complete protocol for in situ hybridization of messenger RNAs in brain and other tissues with radio-labeled single-stranded RNA probes. J Histotechnol 12:169–181
5. Pineau I, Lacroix S (2007) Proinflammatory cytokine synthesis in the injured mouse spinal cord: multiphasic expression pattern and identification of the cell types involved. J Comp Neurol 500:267–285
6. Pineau I, Libo S, Bastien D, Lacroix S (2010) Astrocytes initiate inflammation in the injured mouse spinal cord by promoting the entry of neutrophils and inflammatory monocytes in an IL-1 receptor/MyD88-dependent fashion. Brain Behav Immun 24:540–553
7. Winzer-Serhan UH, Broide RS, Chen Y, Leslie FM (1999) Highly sensitive radioactive in situ hybridization using full length hydrolyzed riboprobes to detect alpha 2 adrenoceptor subtype mRNAs in adult and developing rat brain. Brain Res Brain Res Protoc 3:229–241
8. Broide RS, Trembleau A, Ellison JA, Cooper J, Lo D, Young WG, Morrison JH, Bloom FE (2004) Standardized quantitative in situ hybridization using radioactive oligonucleotide probes for detecting relative levels of mRNA transcripts verified by real-time PCR. Brain Res 1000:211–222

Chapter 4

Visualizing $GABA_B$ Receptor Internalization and Intracellular Trafficking

Paola Ramoino, Paolo Bianchini, Alberto Diaspro, and Cesare Usai

Abstract

The number of neurotransmitter receptors on the plasma membrane is regulated by the traffic of intracellular vesicles. Golgi-derived vesicles provide newly synthesized receptors to the cell surface, whereas clathrin-coated vesicles are the initial vehicles for sequestration of surface receptors, which are ultimately degraded or recycled. Here we use confocal laser scanning microscopy and multiple immunofluorescence analysis to study constitutive $GABA_B$ receptor internalization and intracellular trafficking in the single-celled organism *Paramecium primaurelia*. $GABA_B$ receptors display a dotted vesicular pattern dispersed on the cell surface and throughout the cytoplasm and are internalized via clathrin-dependent and -independent endocytosis. Indeed, $GABA_B$ receptors colocalize with the adaptin complex AP2, which is implicated in the selective recruitment of integral membrane proteins to clathrin-coated vesicles, and with caveolin 1, which is associated with uncoated membrane invaginations. After internalization, receptors are targeted to the early endosomes, characterized by the molecular markers EEA1 and rab5. Some of these receptors, addressed to recycling back to the plasma membrane, move from the early endosomes to the endosomal recycling compartment that is characterized by the presence of rab4 immunoreactivity. Receptors that are addressed to degradation exit the endosomal pathway at the early endosomes and move to the late endosome–lysosome pathway. In fact, some of the $GABA_B$-positive compartments were identified as lysosomal structures by double staining with the lysosomal marker LAMP1. $GABA_B$ vesicle structures also colocalize with TGN38- and rab11-immunoreactivity. TGN38 and rab11 proteins are associated with post-Golgi and recycling endosomes, respectively.

Key words: $GABA_B$ receptor, Endocytosis, Clathrin, Adaptin complex, Caveolin, Rab proteins, Immunofluorescence, Confocal microscopy, Ciliated protozoa

1. Introduction

1.1. $GABA_B$ Receptor Internalization

$GABA_B$ receptors, the metabotrophic receptors for GABA, are members of the G-protein coupled superfamily of receptors (GPCR) and are made up of a dimer of seven-transmembrane spanning subunits. $GABA_B$ receptors are widely distributed throughout

Emilio Badoer (ed.), *Visualization Techniques: From Immunohistochemistry to Magnetic Resonance Imaging*, Neuromethods, vol. 70, DOI 10.1007/978-1-61779-897-9_4, © Springer Science+Business Media, LLC 2012

the central nervous system where they act postsynaptically to cause a long-lasting hyperpolarization through the activation of a potassium conductance. They are also present presynaptically where they act as auto and heteroreceptors to inhibit neurotransmitter release by modulation of Ca^{2+} channels. $GABA_B$ receptors play a complex role in the regulation of excitatory transmission and their activation can have both inhibitory and disinhibitory effects (1).

The classical paradigm of receptor function assumes that receptors localize on the cell surface and are activated by the binding of agonist ligands. After activation, most receptors are endocytosed from cell surface and travel to low pH endosomes, allowing the ligand to detach before the receptor is recycled back to the cell surface or sent through late endosomes to lysosomes for degradation (2). Increasing evidence shows that some GPCRs are not totally inactive in the absence of ligands but exhibit a constitutive activity, too, with elevated basal levels of intracellular signaling (3, 4). It was found that receptor internalization from the neuronal surface occurring both constitutively and in response to agonist exposure is mediated by clathrin-dependent endocytosis (5–7). Clathrin-coated vesicles are the initial vehicles for sequestration of surface receptors, which are ultimately degraded or recycled. Endocytosis of such membrane proteins involves a series of steps beginning with the clustering of receptors at specific sites of the plasma membrane, regions that later turn into clathrin-coated pits (Fig. 1). Receptors do this by recruiting cytosolic AP2 adaptor complexes through their cytoplasmic tails.

AP2 is a key component of the endocytotic machinery that links cargo membrane proteins to the clathrin lattice, selects molecules for sorting into clathrin-coated vesicles and recruits clathrin to the plasma membrane (for reviews, see (8–12)). It is composed of subunits: α, $\beta 2$, $\mu 2$, and σ (Fig. 1). The $\mu 2$ subunit (AP50) binds the endocytic sequence motif of cargo proteins, whereas the $\beta 2$ subunit binds to clathrin and the α region interacts via distinct domains with amphiphysin, AP180 and eps15. In addition to the AP2 adaptor complex, amphiphysin interacts with dynamin and the disruption of dynamin–amphiphysin interaction by recombinant amphiphysin src homology 3 (SH3) domain in vivo leads to a potent block in clathrin-mediated endocytosis (13). Dynamin, a large GTP-binding protein, pinches off vesicles at constricted clathrin-coated pits by forming a ring-like structure collaring the neck of the vesicle that is thought to drive vesicle separation. Eps15 binds the C-terminal domain of the AP2 adaptor α-subunit and mediates the interaction of AP2 with proteins such as epsin, CALM/AP180 and synaptojanin, implicated in regulation of receptor-mediated endocytosis. Eps15 function in clathrin-dependent endocytosis seems to be restricted to the early events leading to clathrin-coated pit formation: indeed eps15 is not present in clathrin-coated vesicles (14).

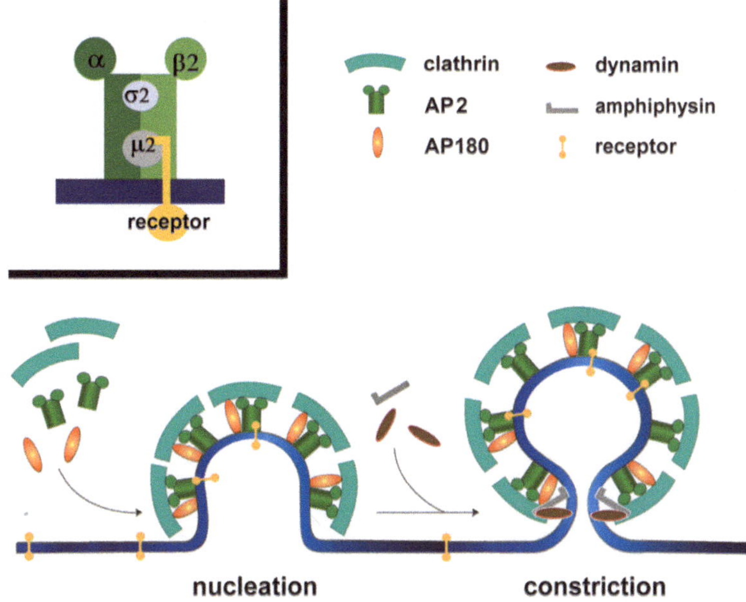

Fig. 1. Internalization of receptors by the clathrin-mediated pathway. Clathrin-mediated vesicle endocytosis is divided into four steps: coated pit nucleation (recruitment of AP2, AP180, and clathrin to form a curved membrane), invagination (deep invagination before the formation of the characteristic dynamin neck structure), fission (constriction and closure of the neck), and uncoating. Schematic representation of the AP2 complex (*inset*).

It has been shown that endocytosis of receptors may also occur through other membrane structures, including noncoated membrane invaginations and caveolae (15–18). The β_2-adrenergic receptor, which is endocytosed by clathrin-coated pits in several cell types (19, 20), is endocytosed by membrane invaginations resembling to caveolae in other cells (21, 22). Cholecystokinin receptors have been observed in both clathrin-coated pits and caveolae in the same cells (23). Caveolae are cholesterol- and sphingolipid-rich smooth invaginations of the plasma membrane that partition into raft fractions and the expression of which is associated with caveolin 1.

Incubation of mammalian cells in media containing either sucrose or chlorpromazine has been shown to inhibit clathrin-mediated endocytosis by interfering with the clathrin-adaptor interaction or by altering the structure of the clathrin itself (24–26). Finally, caveolae-mediated internalization is blocked by nystatin and filipin, sterol-binding agents that disrupt caveolar structure and function (27, 28).

1.2. GABA$_B$ Receptor Intracellular Trafficking After Internalization

Endocytosed receptors can be dephosphorylated and then recycled to the plasma membrane, contributing to functional resensitization of signal transduction. Alternatively, they are degraded into lysosomes, reducing the number of receptors present in the plasma membrane and promoting the down-regulation of receptors, a process that leads to functional desensitization of signal transduction (2).

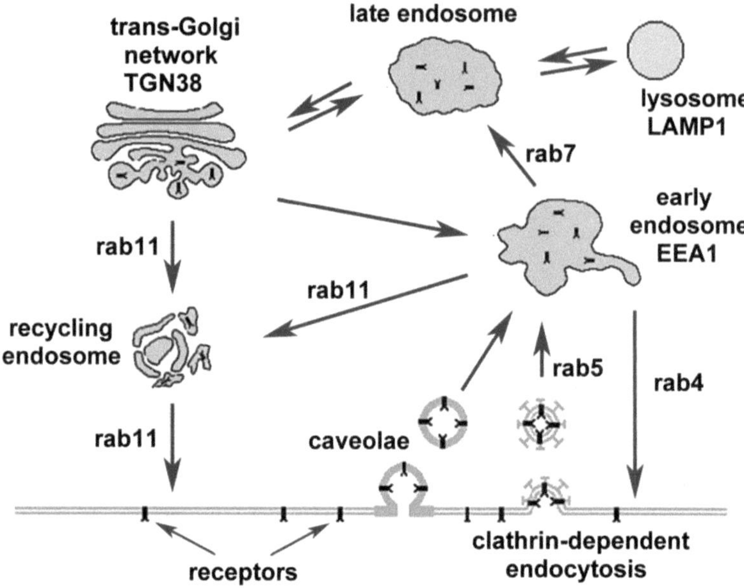

Fig. 2. Intracellular trafficking of receptors and role of Rab GTPases. Receptors on the cell surface can be selectively recruited into clathrin-coated pits that can invaginate inward and pinch off vesicles into the cytoplasm. Clathrin is the main structural component of the coat. Coated pits and vesicles contain adaptor-protein complexes AP2 and several accessory proteins. Upon shedding of the clathrin coat, the receptor-containing vesicle fuses with the sorting endosome by action of Rab5 and the aid of accessory endosomal proteins such as EEA1. Receptors can then be recycled via the recycling endosome, an event mediated by Rab4 or Rab11, or transported to late endosomes. From late endosomes receptors can be recycled via the Golgi or degraded into the lysosome. The diagram also shows how receptors internalized in a caveolin-dependent manner can join the clathrin endocytic pathway at the level of the sorting endosome and are also dependent on dynamin activity (modified from (40)).

These processes of receptor regulation are mediated by a continuous traffic of vesicular and tubular intermediates, which need to be coordinated to ensure proper progression of cargo through the different compartments (Fig. 2). Several members of the rab proteins family have been localized in distinct compartments of the endocytic pathway and play different roles in endocytosis and recycling (for reviews, (29–32)). Rab5 and rab4 are both localized to early endosomes, but exert opposite effects on the uptake of membrane-bound proteins. Rab5 plays a role in the formation of clathrin-coated vesicles at the plasma membrane, in their subsequent fusion with early endosomes, in the homotypic fusion between early endosomes and in the interaction of early endosomes with microtubules. Rab4 has been implicated in the regulation of membrane recycling from the early endosomes to the recycling endosomes or directly to the plasma membrane.

In accordance with its functional diversity, rab5 lies at the centre of a complex machinery made up of several effector proteins.

Among these proteins, EEA1 was identified as a core component of the homotypic endosome docking and fusion machinery and was shown to play a role in the docking/tethering of the endosome membranes (29–32). EEA1 is predominantly localized to early endosomes and is regarded as a specific marker of this compartment. Because of this localization and given its function in endosome membrane docking it has been proposed that EEA1 may confer directionality to rab5-dependent vesicular transport to the early endosomes. Another effector protein for rab5 is rabaptin-5. Rabaptin-5 binds directly to the GTP-bound form of rab5 and is recruited to early endosomes by rab5 in a GTP-dependent manner (29–32). It stabilizes rab5 in the GTP-bound active form by down-regulating the GTP hydrolysis and it is required for the homotypic fusion between early endosomes as well as for the heterotypic fusion of clathrin-coated vesicles with early endosomes in vitro. Rabaptin-5 also interacts with GTP-bound rab4 (via a distinct structural unrelated N-terminal rab-binding domain), but does not appear to interact with rab11, a GTPase that is highly present on the recycling endosome and whose activity is required for receptor recycling through this compartment (29–32). Thus, the same effector interacts with the two rab proteins, which act sequentially in transport through the early endosomes.

1.3. Paramecium as a Model to Study Receptor Internalization

The excitable ciliate *Paramecium* is easily cultured and manipulated and has provided an excellent model for studying the complexities of membrane trafficking (sorting, retrieval, transport, and fusion) in a single-celled organism by both morphological and immunocytochemical techniques (33, 34).

The internalization of material in ciliated protozoa occurs via different mechanisms, even if most nutrients, particulate or not, are taken up by phagosomes formed at the bottom of the oral cavity. The endocytosis of small-sized molecules occurs at the parasomal sacs located next to the ciliar basal bodies. Indeed, in ciliates only defined areas on cell surface are potential sites for endocytic uptake since most of the surface is covered internally by an extensive system of alveoli and an underlying fibrous epiplasm (35). This system is interrupted only at the cytopharynx, the cytoproct, contractile vacuole pores and along the junctions of the abutting units of alveolar membrane sacs. Only the punctuate indentations of the plasma membrane, called parasomal sacs, and pellicular pores are potential endocytic entry ports of all fluid phase and putative receptor-mediated endocytosis (36, 37).

Detailed morphological and tracer studies on endocytosis carried out by electron microscopy showed that in *Paramecium multimicronucleatum* fluid phase markers such as horseradish peroxidase (HRP) and in *Tetrahymena pyriformis* receptor-mediated markers such as cationized ferritin are internalized via coated pits and are found in coated vesicles (36, 37). Both coated pits and vesicles are

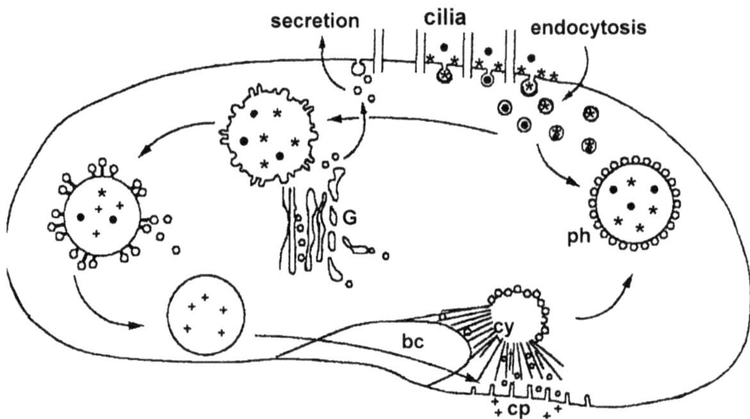

Fig. 3. Schematic drawing of the endocytic pathway internalization in *Paramecium*. Without phagocytosis inhibition, the ingested material is directed by the oral membranelle beating into the buccal cavity (bc), the cytopharynx (cy) and, at last, into the nascent phagosome. As lysosomes fuse with the phagosome (ph), the cargo is digested and pinocytic vesicles containing digestion products are pinched off. Finally, the indigestible material is excreted at the cytoproct (cp). Receptor and fluid-phase endocytosis occurs at the parasomal sacs located next to the ciliar basal bodies. Exogenous fluid and plasma membrane components are internalized by vesicles which fuse with food vacuoles, and the denaturation and digestion of endosome and food vacuole contents are completed together. (*G*) Golgi apparatus; (*asterisk*) receptor and (*filled circle*) fluid-phase endocytosis; (*plus*) degraded material (modified from (33)).

also labeled in fixed cells when a monoclonal antibody against the plasma membrane of *P. multimicronucleatum* (C6 antigen) is applied to cryosections, suggesting that both membrane bound and fluid phase markers are internalized at the coated pits (34). However, there is increasing evidence for clathrin-independent pathways, mediated by caveolae or noncoated vesicles (38, 39). It was shown that exogenous fluid and plasma membrane components are internalized by vesicles which are first localized in the cortical region of the cell and then migrate in the cytoplasm and fuse with other endosomal compartments, until their content is transferred to the phagosomes (39, 40). In the end, denaturation and digestion of both endosome and phagosome contents are completed together (Fig. 3).

2. Materials

2.1. Antibodies

Antibodies against the protein of interest; here:

1. Monoclonal antibodies anti-clathrin (clone 23; 1:400), -adaptin β (clone 74; 1:2,000), -AP50 (clone 31; 1:100), -AP180 (clone 34; 1:100), -eps15 (clone 17; 1:100), -dynamin II (clone27; 1:100), -amphiphysin (clone 15; 1:4,000), -caveolin

1 (clone 2297; 1:1,000), -rab4 (clone 7; 1:400), -rab5 (clone 15; 1:100), -rab11 (clone 47; 1:400), -rabaptin-5 (clone 42; 1:200), -EEA1 (clone 14; 1:1,000), -LAMP1 (clone 25; 1:100), and -TGN38 (clone 2; 1:100) were purchased from Transduction Laboratory (BD Biosciences, San Jose, CA, USA).

2. The polyclonal guinea pig anti-$GABA_B$ R1 (1:1,000) receptor from Sigma Chemical Co. (St. Louis, MO, USA).
3. The secondary antibodies anti-guinea pig Alexa Fluor 594 (1:300) and anti-mouse Alexa Fluor 488 (1:300) were obtained from Molecular Probes (Invitrogen, Carlsbad, CA, USA).

2.2. Solution

2.2.1. Dextran-Coupled Texas-Red (DXT-TXR, Molecular Probes, Invitrogen, Carlsbad, CA, USA)

1. Dilute in distilled water 2 mg/mL 10,000 Mw DXT-TXR. The stock solutions may be stored at 2–6°C for several weeks, with the addition of sodium azide to a final concentration of 2 mM to inhibit bacterial growth. For long-term storage, divide the solution into aliquots and freeze at ≤–20°C.
2. Working solution: dilute 1:100 in culture medium.

2.2.2. Alexa Fluor 594 Transferrin (Molecular Probes, Invitrogen, Carlsbad, CA, USA)

1. Reconstitute in 1 mL of deionized water to obtain a 5 mg/mL solution in phosphate buffer saline (PBS).
2. Add sodium azide at a final concentration of 2 mM.
3. Store solutions at 2–6°C, protected from light. Do not freeze.

2.2.3. Sucrose (Sigma Chemical Co., St Louis, MO, USA)

1. Dilute 51.36 g sucrose in 10 mL distilled H_2O.
2. Working solution: dilute 1:100 in culture medium.

2.2.4. Nystatin (Sigma Chemical Co., St Louis, MO, USA)

1. Dilute 2 mg nystatin in 1 mL DMSO. Divide the solution into aliquots and frozen at ≤–20°C. They have a shelf life of 24 months when stored frozen.
2. Working solution: dilute 1:1,000 in culture medium.
3. Solutions and aqueous suspensions are stable for 2–3 days.

2.2.5. Filipin (Sigma Chemical Co., St Louis, MO, USA)

1. Dilute 0.1 mg filipin in 1 mL DMSO.
2. Working solution: dilute 1:1,000 in culture medium.
3. Solutions of filipin in aqueous or organic solvents are not considered to be stable and it is recommended that they be prepared fresh and not stored as stock solutions.

2.2.6. Chlorpromazine (Sigma Chemical Co., St Louis, MO, USA)

1. Dilute 0.2 mg chlorpromazine in 1 mL distilled water. Divide the solution into aliquots and frozen at ≤–20°C.
2. Working solution: dilute 1:1,000 in culture medium.

2.2.7. Phosphate Buffer Saline, pH 7.4

Stock solutions:

(a) 0.2 M monobasic NaH_2PO_4:

- Dissolve by stirring at room temperature (RT) 27.6 g in distilled H_2O to make 1 L.

(b) 0.2 M dibasic Na_2HPO_4:

- Dissolve by stirring at RT 28.4 g in distilled H_2O to make 1 L.

Note: Be sure the monobasic and dibasic are anhydrous and not hydrated; the formula weights will be different.

PBS working solution: 0.1 M, pH 7.4

1. Mix 95 mL of 0.2 M monobasic (A solution) with 405 mL dibasic (B solution).
2. Add 500 mL water containing 90 g NaCl.
3. Adjust pH to 7.4 by adding dilute 1 N HCl if necessary.
4. Filter through Millipore filter, store at 4°C.

Note: PBS buffer is used for routine immunohistochemical staining. PBS is often used for diluting secondary antibodies. PBS-Triton X-100 is often used for washing steps.

2.2.8. 4% Paraformaldehyde in 0.1 M PBS

1. Dissolve 8 g paraformaldehyde (PFA) in 100 mL of distilled H_2O.
2. Heat to water (on heated stirrer and stir using magnetic bead) to 55–60°C to partially dissolve PFA.
3. Add 1 M NaOH dropwise (one or two drops) until solution clears.
4. Add 100 mL of 0.2 M PBS, mix.
5. Store at 4°C. Prolonged storage (>1 week) may require that pH is checked. pH should be around 7.4. For long-term storage, divide the solution into aliquots and freeze at ≤–20°C.

Do not autoclave! Caution: Formaldehyde is toxic, and the preparation process involving heating results in considerable vaporization, which increases the hazard. It is essential to use appropriate safety procedures, such as working in a fume hood.

2.2.9. 3% Bovine Serum Albumin in PBS Plus 1% Triton X-100

1. Dissolve 3 g of bovine serum albumin (BSA) in 50 mL PBS.
2. Add 10 mL of 10% Triton X-100.
3. Bring volume up to 100 mL PBS.
4. Transfer into 10-mL centrifuge tubes. Keep the working aliquot at 2–8°C and freeze the other aliquots for up to 12 months at –20°C. Record the date prepared, expiration date, and the technician's initials on each tube.

2.2.10. 1% BSA Plus 0.1% Triton X-100

1. Dissolve 1 g of BSA in 50 mL PBS.
2. Add 1 mL of 10% Triton X-100.
3. Bring volume up to 100 mL PBS.
4. Transfer into 10-mL centrifuge tubes. Keep the working aliquot at 2–8°C and freeze the other aliquots for up to 12 months at –20°C. Record the date prepared, expiration date, and the technician's initials on each tube.

2.2.11. 1 N HCl

1. Take 10 mL of concentrated HCl.
2. Add enough distilled water to make 120 mL.

Store at 4°C.

2.2.12. 1 M NaOH

1. Dissolve 4 g of NaOH in distilled water.
2. Make up to 100 mL gives 1 N (or 1 M).

Store at 4°C.

3. Methods

3.1. In Vivo Experiments

3.1.1. Inhibition of $GABA_B$ Receptor Internalization

1. To block the phagocytic activity, pretreat the cells for 15 min with trifluoperazine (2.5–5 μM) in the culture medium.
2. Incubate the cells in a culture medium containing a clathrin (150 mM sucrose or 0.2 μg/mL chlorpromazine) or a caveolin (0.1 μg/mL filipin or 2 μg/mL nystatin) inhibitor for 15–30 min.
3. Add the polyclonal antibody anti-$GABA_B$-receptor R1 (1:100, final concentration) and incubate at 25°C for 15–30 min.
4. Fix the cells with PAF 4% in PBS 0.2 M for 30 min.
5. Wash in PBS.
6. Process for $GABA_B$ immunolabeling.

Controls are cells without inhibitor.

3.1.2. Inhibition of $GABA_B$ Receptor Internalization After Receptor Accumulation on the Cell Membrane

1. Incubate the cells at 4°C for 30 min.
2. Add the prewarmed anti-$GABA_B$ receptor antibody (1:100, final concentration) and incubate at 4°C for other 30 min.
3. Wash in a cold culture medium.
4. Add the inhibitor.
5. Chase for 20 min at 25°C.
6. Fix the cells with PAF 4% in PBS 0.2 M for 30 min.

7. Wash in PBS.
8. Process for $GABA_B$ immunolabeling.

Controls are cells without inhibitor.

3.1.3. $GABA_B$ Receptor Internalization Visualized by a Double Labeling

To further assess whether $GABA_B$ receptors are internalized by clathrin-dependent and/or -independent mechanisms, cells are incubated for various times both with the antibody anti-$GABA_B$ receptors and with a fluid-phase or a receptor-mediated endocytosis marker.

1. Pretreat the cells for 15 min with trifluoperazine (2.5–5 μM) in the culture medium.
2. Add the anti-$GABA_B$ receptor antibody (1:100, final concentration) and the fluid-phase endocytosis marker dextran (0.02% w/v, final concentration) or the receptor-mediated endocytosis marker transferrin (25 μg/mL, final concentration).
3. Incubate at 25°C for 5–10 min.
4. Fix the cells with PAF 4% in PBS 0.2 M for 30 min.
5. Wash in PBS.
6. Process for $GABA_B$ immunolabeling.

3.2. Indirect Immunofluorescence

3.2.1. Immunolabeling for Confocal Microscopy

All steps are carried out in a humidity chamber.

1. Fix the cells in 4% PFA in PBS buffer (0.1 M, pH 7.4) for 30 min.
2. Wash three times with PBS.
3. Incubate for 60 min with 3% BSA and 1% Triton X-100 in PBS.
4. Remove the blocking permeabilizing buffer.
5. Incubate the cells overnight at 4°C with both the polyclonal antibody against the $GABA_B$ receptor (1:1,000 dilution) and one of the monoclonal antibodies.
6. Wash three times in 1% BSA in PBS plus 0.1% Triton X-100 for 10 min each.
7. Apply a cocktail of the secondary antibodies anti-guinea pig IgG conjugated to Alexa Fluor 594 (dilution 1:300) and anti-mouse IgG conjugated to Alexa Fluor 488 (dilution 1:300) for 2 h at 37°C. Cover to protect from light. Always include a control with secondary antibodies alone.
8. Wash three times with PBS.
9. Mount the cells in glycerol buffer.

Note. The absence of cross-reactivity between the secondary antibodies was verified by omitting one of the primary antibodies during incubation. Moreover, for every combination of double labeling, single-labeled vesicles can be always observed in the cells.

3.2.2. Immunofluorescence Staining with Signal Amplification

Immunofluorescence can be amplified by using the Alexa Fluor 488 signal amplification kit for mouse antibody (Molecular Probes).

1. Incubate the cells removed from blocking buffer overnight at 4°C in primary antibodies (anti-caveolin 1 and anti-$GABA_B$ receptor).
2. Wash for three times in 1% BSA in PBS plus 0.1% Triton X-100 for 10 min each.
3. Apply 10 μg/mL Alexa Fluor 488 rabbit anti-mouse (component A) and Alexa Fluor 594 goat anti-guinea pig (dilution 1:300) for 30 min at 37°C.
4. Wash for three times with 1% BSA in PBS plus 0.1% Triton X-100.
5. Incubate for 30 min at 37°C with 10 μg/mL Alexa Fluor 488 goat anti-rabbit (component B) and Alexa Fluor 594 goat antiguinea pig.
6. Wash three times with PBS.
7. Mount the cells in glycerol buffer.

3.3. Confocal Image Acquisition

- Acquire images by laser confocal scanning microscopy (CLSM) (for example, Leica TCS SP5 AOBS) equipped with an Ar (457-476-488-514 nm) 100-mW laser and a HeNe (543 nm) 1.5-mW laser using a one Airy disk unit pinhole diameter and an oil immersion HCX PL APO 63×/1.4 objective.
- Excite Alexa Fluor 488 with the 488-nm line of the Ar laser and collect its fluorescence in a spectral window of 500–530 nm.
- Excite Alexa 594 fluorescence using the 543-nm excitation wavelength and collect its fluorescence in a spectral window of 590–620 nm.
- Excite Texas-Red fluorescence using the 543-nm excitation wavelength and collect its fluorescence in a spectral window of 620–680 nm.
- Take serial optical sections through the cell at a *z*-step of 50–75 nm with a laser power of 1.5 mW and illumination attenuated by a 50% transmission neutral-density filter to reduce photobleaching, a 1 Airy unit pinhole diameter.
- Acquire the double-stained images through green and red channels according to a time-sequential protocol to reject possible cross-talk artifacts.
- Use the software program of CLSM for image acquisition, storage and analysis.

3.4. Image Analysis

Image analysis can be performed using different software including those from CLSM or other commercial products such as Bitplane's Imaris, Molecular Devices'MetaMorph and MatWork's Matlab softwares, or public domain programs such as NIH's MIPAV and ImageJ (and its distributors, i.e., Fiji). We have used the ImageJ 1.34f software (Wayne Rasband, National Institute of Health, USA).

3.4.1. Background Correction

Background correction can be done in several ways.

Rolling-Ball Background Correction

1. Open the red and green images (File/Open); both images must be 8-bit or 16-bit grayscale (Image/Type/8 bit).
2. Choose Process/Subtract background. This menu command removes uneven background from images using a "rolling ball" algorithm.
3. Set the radius to at least the size of the largest object that is *not* part of the background. Running the command several times may produce better results.

Region of Interest Background Correction

1. Open the red and green images (File/Open); both images must be 8-bit or 16-bit grayscale.
2. Use the selection tools to select an area of background.
3. Run the menu command Plugins/ROI/BG Subtraction from region of interest (ROI). This macro will subtract the mean of the ROI from the image plus an additional value equal to the standard deviation of the ROI multiplied by the scaling factor you enter (3 by default).

3.4.2. 2D Cytofluorogram

We generated cytofluorograms by plotting the gray values of pixels in red and green channel images against each other (ICA plugin of ImageJ).

1. Open the red and green images (File/Open); both images must be 8-bit or 16-bit grayscale.
2. Subtract background (see Sect. 3.4.1).
3. Run the ICA plugin (Plugins/Colocalization analysis/Intensity Correlation analysis).
   - Select red fluorescent image (Channel 1) and green fluorescence image (Channel 2).
   - Select the combination of channel color (Red:Green).
   - Choose "one" in the ROI box.
   - Select "Keep merged ROI" to obtain the merged image.
   - Select "Display color Scattered plot" to obtain a 2D cyofluorogram.

3.4.3. z-Profile of the Fluorescence Intensity of Double-Stained Vesicles

1. Import image sequence of the two stacks.
2. Subtract background.
3. Open Image/Color/RGB merge and select "Keep Source Image" to obtain a double-labeled image.

4. Select a ROI defining a yellow vesicle on the double-labeled image.
5. Open Analyse/Tools/ROI Manager and select "Add."
6. Copy the selected ROI separately on green and red stacks.
7. Open Image/Stacks/Plot *z*-axis profile for each stack. This generates a plot window with mean intensity. The raw data are available by selecting "list." This can be easily copied or saved/imported to Excel.
8. Plot the mean intensity vs. the slice number.

3.4.4. Quantitative Estimation of Colocalized Proteins

The literature is full of different methods for colocalization analysis, which probably reflects the fact that one approach does not necessarily fit all circumstances. Here we report one of the analyses present in ImageJ.

The quantitative estimation of colocalized proteins was performed calculating the "colocalization coefficients" (41). They express the fraction of colocalizing molecular species in each component of dual-color image and are based on the Pearson's correlation coefficient, a standard procedure for matching one image with another in pattern recognition (42). If two molecular species are colocalized, the overlay of their spatial distribution has a correlation value higher than that expected by chance alone. Costes et al. developed an automated procedure to evaluate the correlation between the green and red channels with a significance level >95% (43). The same procedure automatically determines an intensity threshold for each color channel based on a linear least-square fit of the green and red intensities in the image's 2D correlation cytofluorogram. Costes' approach was accomplished by macroroutines integrated as plugins (WCIF Colocalization Plugins, Wright Cell Imaging Facility, Toronto Western Research Institute, Canada).

1. Open the red and green images (File/Open); both images must be 8-bit or 16-bit grayscale.
2. Subtract background (see Sect. 3.4.1).
3. Run Plugins/Colocalization Analysis/Colocalization Threshold to obtain the result window.
4. Select red fluorescent image (Channel 1) and green fluorescence image (Channel 2).
5. Choose "none" in the ROI box.
6. Select the combination of channel color (Red:Green).
7. Click set options to activate the Results Options dialog and to choose the results to be reported. By default, all results are reported. In results tM1 represents colocalization of red vs. green and tM2 colocalization of green vs. red (tM = Manders colocalization coefficient corrected for threshold).
8. Carry out the experiments on at least 15 cells and repeat them on different occasions.

3.4.5. Quantification of Cell Membrane Fluorescence

1. Open the image (File/Open).
2. Remove background (see Sect. 3.4.1).
3. (a) When a single ROI is used:
   - Select a ROI defining the area to be analyzed.
   - Select Analyze/Measure. This generates a window with the mean fluorescence intensity value.

   (b) When more ROIs are used:
   - Open ROI Manager plugin (Analyze/Tools/Roi Manager) on toolbar.
   - Select ROIs and "Add" to ROI manager (hotkey: t). Clicking "Show All" helps avoid analyzing the same cell twice.
   - Once finished selecting ROIs to be analyzed, click the "More>>" button and select "Combine"
   - Click "Measure." This generates a window with the mean fluorescence intensity value.

3.5. Statistical Analysis

Value the significance of differences between means by Student's *t*-test (using, for example, GraphPad Prism, GraphPad, San Diego, CA).

3.6. Conclusions

In conclusion, using immunolabeling in confocal microscopy and image analysis we have shown the presence in *Paramecium* of proteins involved in the initiatory steps of endocytosis in mammalian cells.

According to our results, we assume that in *Paramecium* $GABA_B$ receptors undergo constitutive endocytosis by clathrin- and caveolin-dependent mechanisms, then are partly recycled to cell plasma membrane and partly degraded into lysosomes. Furthermore, we have shown that rab-like proteins are involved in the vesicle transport from one compartment to another. Comparing our data with the internalization and recycling mechanisms of other GPCRs observed in mammalian cells, we may deduce that a system typical of mammalian neuronal cells is already present in the single-celled organism *Paramecium*. Interestingly, this general mechanism has been maintained through evolution.

4. Notes

1. $GABA_B$ receptor internalization is mediated by clathrin-coated pits.
   To show the involvement of clathrin-coated vesicles in constitutive endocytosis of $GABA_B$ receptors, cells are double labeled with a guinea pig anti-$GABA_B$ receptor R1-subunit antibody

(red fluorescence) and with a monoclonal anti-clathrin antibody (green fluorescence). Staining with an anti-clathrin antibody led to a punctuate pattern both on cell membrane and inside the cytoplasm, representing endocytic vesicles and exocytic vesicles from the trans-Golgi network (TGN) (Fig. 4a, b). Importantly, $GABA_B$ receptor and clathrin-coated vesicle clusters are partly colocalized (yellow fluorescence). Colocalized pixels are shown in blue in the 2D cytofluorogram reported in Fig. 4c, d. The colocalization percentage of $GABA_B$ receptors with proteins involved in the endocytosis is given in Table 1.

Colocalization along the *z*-axes of double-stained vesicles is also demonstrated by the similarity of green and red profiles of their fluorescence intensity. In Fig. 5, the first (a), middle (b), and last (c) images of a 40-plane 3-μm-thick *z*-stack (left side) and the *z*-profile of the fluorescence intensity of three different double-stained vesicles of each plane (right side) are shown.

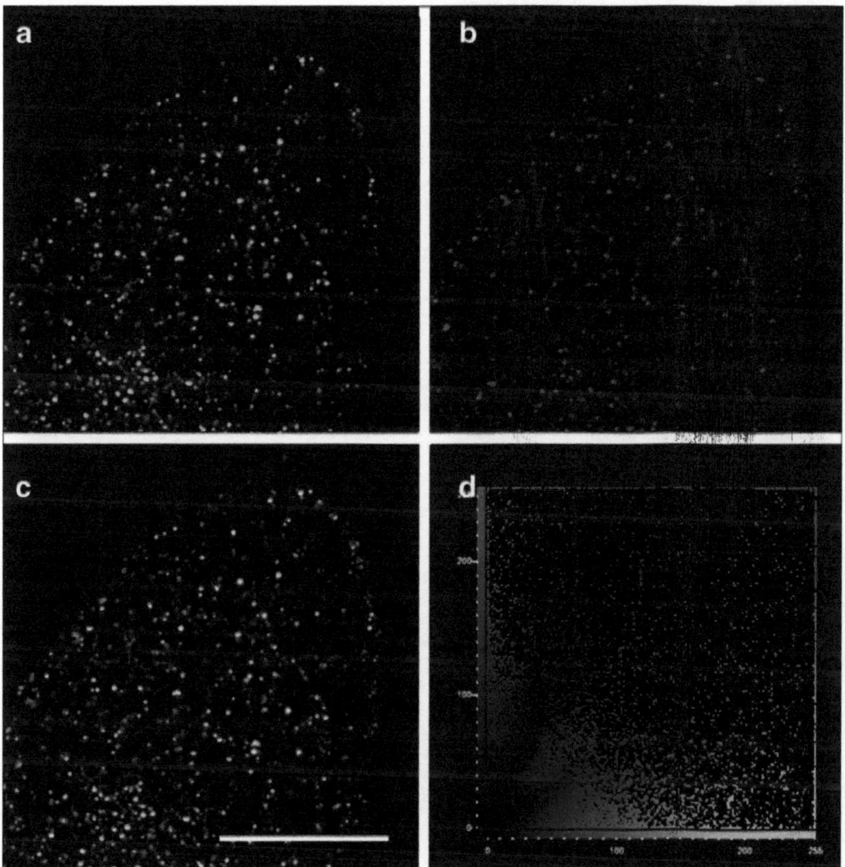

Fig. 4. Colocalization of $GABA_B$ receptors and clathrin. In cells labeled with a polyclonal antibody against $GABA_B$ receptor (**b**) and a monoclonal antibody against clathrin HC (**a**), a clustered distribution of fluorescence is detected on the plasma membrane and inside the cytoplasm. $GABA_B$ receptors and clathrin vesicles are partly colocalized ((**c**), *yellow* fluorescence). Bar, 20 μm. (**d**) 2D cytofluorogram: colocalized pixels are clustered along the diagonal line (visualized in *blue*).

Table 1
Colocalization of $GABA_B$ receptor labeling with various endocytic protein labeling

Red	Green	Red/green (%)	Green/red (%)
$GABA_B$ receptor	Clathrin	18 ± 3	19 ± 3
$GABA_B$ receptor	Caveolin 1	21 ± 6	25 ± 4

Every colocalization value is the average from four optical sections of ten cells. Data were calculated as the mean ± s.e.m. and are given in percent

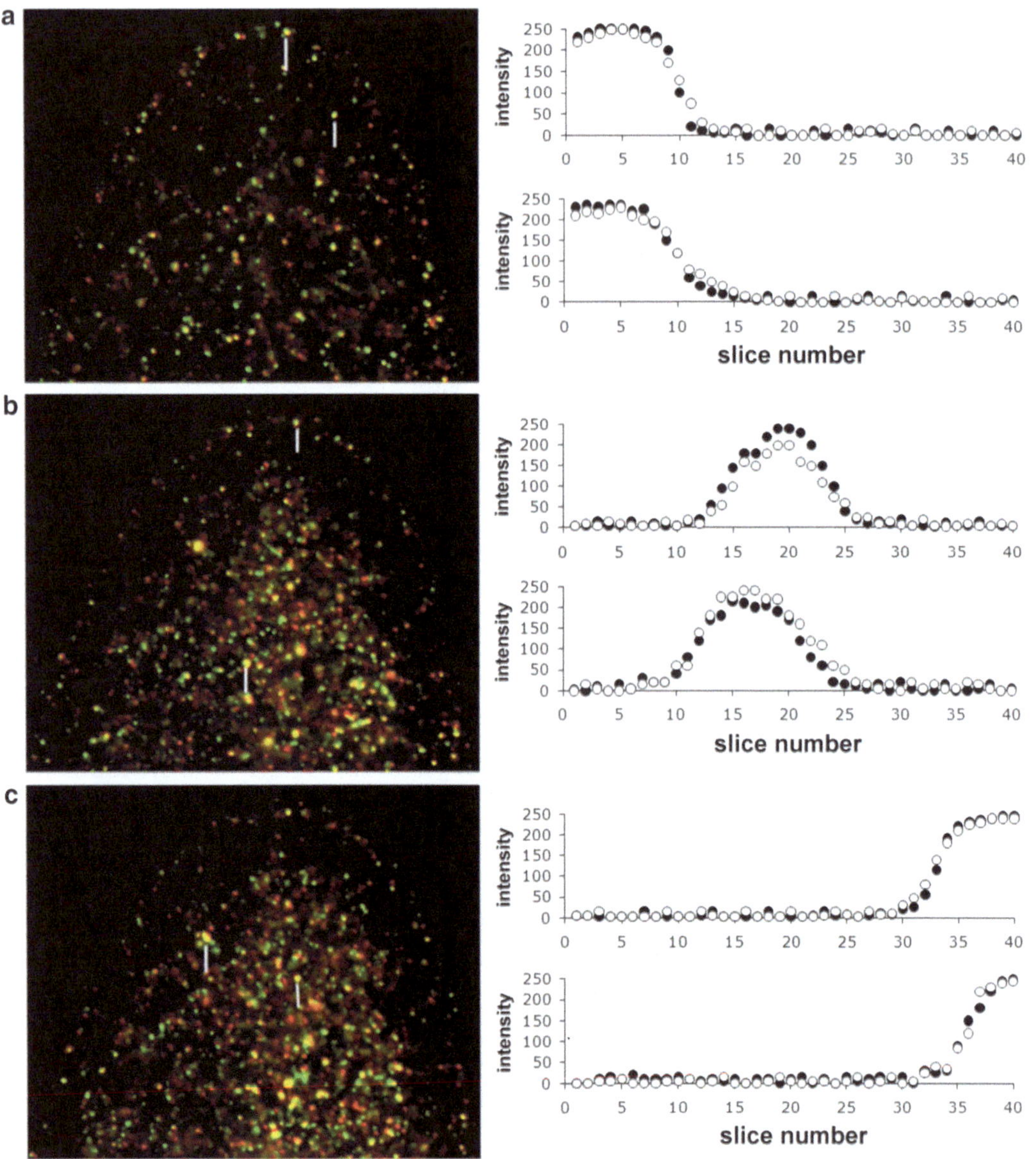

Fig. 5. *z*-Stack profile of fluorescence intensity of double-labeled vesicles. The *left side* of the figure shows three optical planes ((**a**), *first*; (**b**), *middle*; (**c**), *last*) from a stack of 40 images (total thickness 3 mm). For each focal plane a sample of three vesicles that show colocalization (*yellow*) of $GABA_B$ receptor (*red*) and clathrin (*green*) is selected (*white bars*). The *right side* of the figure shows fluorescence-intensity distribution along the *z*-axis for each selected vesicle (*green*, *open circle*) (*red*, *filled circle*).

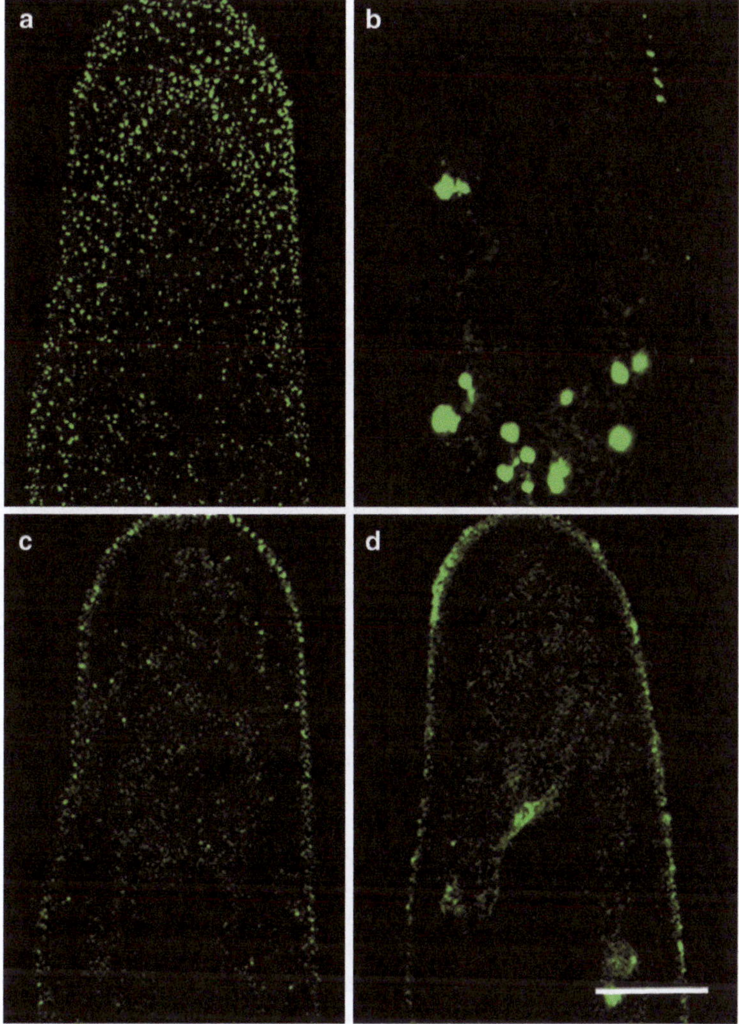

Fig. 6. GABA$_B$-receptor internalization is mediated by clathrin-coated vesicles. In cells whose phagocytic activity is blocked by trifluoperazine, 20-min treatment with 150 mM sucrose (**c**) or 0.2 μg chlorpromazine (**d**) inhibits receptor internalization, which can be seen by receptor accumulation on the cell membrane and receptor reduction inside the cytoplasm. Controls are cells incubated with the anti-GABA$_B$ receptor antibody for 1 (**a**) and 20 min (**b**) in the absence of inhibitors; the antibody is localized in endosomes and phagosomes. Incubation temperature, 25°C. Bar, 20 μm.

Treatment of cells with 150 mM sucrose or 0.2 μg/mL chlorpromazine significantly inhibits the internalization of receptors, as shown by the considerable reduction in receptors inside the cytoplasm (Fig. 6c, d) when compared with the control (Fig. 6a, b). This observation strongly suggests that GABA$_B$ receptor internalization is mediated by clathrin-dependent endocytosis. In these in vivo experiments phagocytosis was blocked by trifluoperazine, a calmodulin antagonist that inhibits the formation of phagosomes and stimulates endocytosis in *Paramecium* (44).

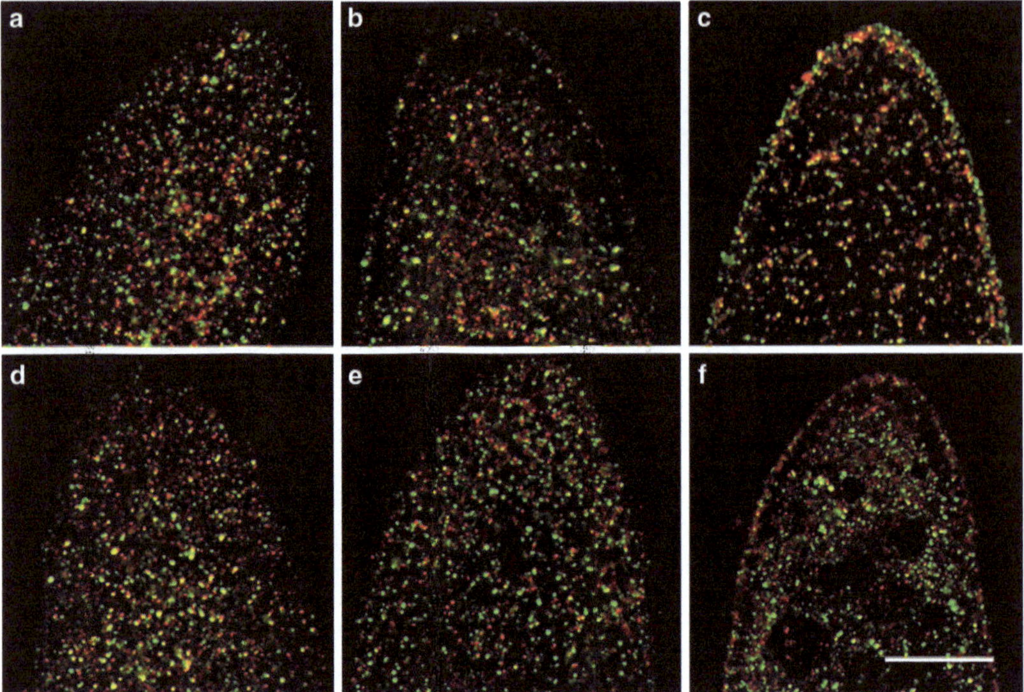

Fig. 7. Colocalization of $GABA_B$ receptors (*red*) with proteins involved in clathrin-mediated endocytosis (*green*): (**a**) adaptin β, (**b**) AP50, (**c**) AP180, (**d**) eps15, (**e**) dynamin, and (**f**) amphiphysin. Colocalization is seen as a *yellow* fluorescence. Bar, 20 μm.

2. The clathrin adaptor protein AP2 colocalizes with the $GABA_B$ receptor.

   After observation of a significant clathrin-mediated endocytosis of $GABA_B$ receptors, we analyzed the colocalization between the receptor subunits and the endocytic adaptor complex AP2 by using antibodies against the β2 (adaptin β) and μ2 (AP50) subunits. All antibodies revealed clusters of fluorescence on the cell surface and inside the cytoplasm (Fig. 7) that represent AP2 in clathrin-coated pits on the plasma membrane and in coated vesicles in their endocytic pathway. Colocalization of some $GABA_B$ receptor clusters with AP2 complexes, seen as yellow staining in the merged panel, could be detected on the plasma membrane (Fig. 7a, b).

   $GABA_B$ receptors also partly colocalized with AP180 (Fig. 7c), a monomeric assembly protein involved in the organization and assembly of clathrin triskelia in clathrin-coated vesicles. In clathrin-mediated endocytosis, AP180 interacts with membrane phospholipids and AP2 complex.

3. Other proteins involved in the $GABA_B$ receptor internalization in *Paramecium*.

   We analyzed the subcellular localization of amphiphysin, dynamin and eps15. Staining the cell with anti-eps15 antibody results in a clear punctate staining pattern throughout the

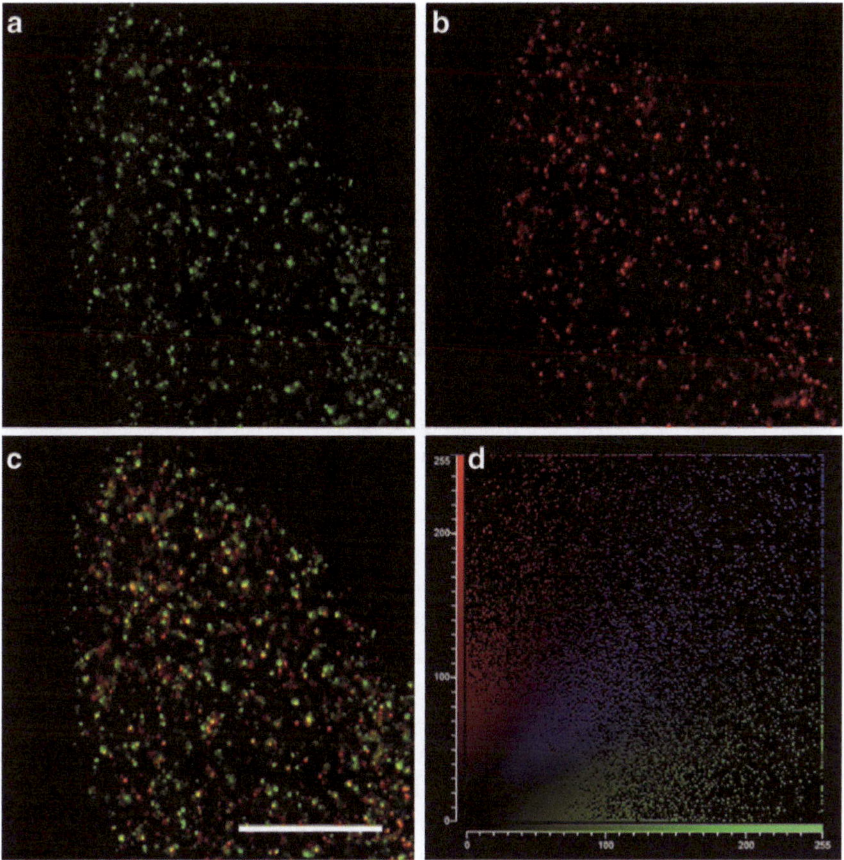

Fig. 8. $GABA_B$ receptors interact with caveolin 1. Cells labeled with an anti-$GABA_B$ receptor antibody (**b**) and an anti-caveolin 1 antibody (**a**). (**c**) Colocalization is seen as a *yellow* fluorescence. Bar, 20 μm. (**d**) 2D cytofluorogram: colocalized pixels are clustered along the diagonal line (visualized in *blue*).

whole cell (Fig. 7d). Staining of the same cells with an anti-dynamin (Fig. 7e) or with an anti-amphiphysin antibody (Fig. 7f) revealed a similar punctate pattern representing coated pits, endocytic vesicles, and exocytic vesicles from the TGN. A partial colocalization of eps15, dynamin and amphiphysin (green stain) with $GABA_B$ receptors (red stain) was found, visible as a yellow stain (Fig. 7d–f).

4. $GABA_B$ receptor clathrin-independent endocytosis.
   Constitutive endocytosis of $GABA_B$ receptors in *Paramecium* also involves a clathrin-independent mechanism. When observed by confocal microscopy, caveolin 1 was revealed in distinct classes of intracellular endosomes. Colocalization of $GABA_B$ receptors and caveolin 1 was observed by indirect immunofluorescence of endogenous caveolin 1 and $GABA_B$ receptors both on cell surface and throughout the cytoplasm (Fig. 8).

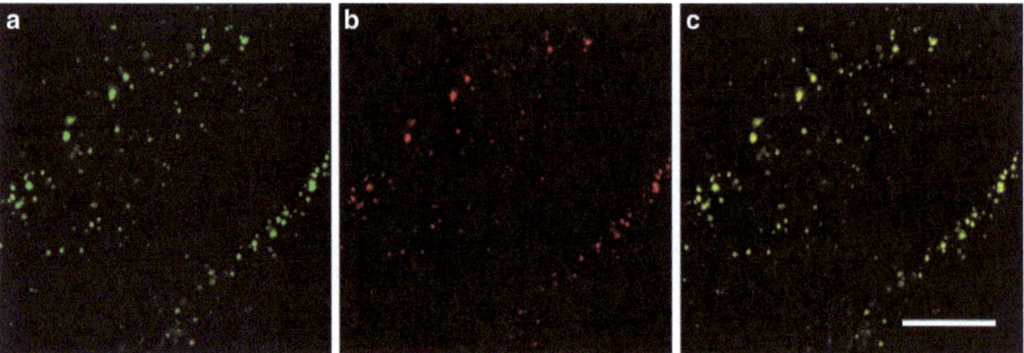

Fig. 9. $GABA_B$ receptors are internalized by a clathrin-independent endocytosis. Cells blocked in their phagocytic activity by trifluoperazine were incubated for 5 min at 25°C with anti-$GABA_B$ receptor antibody (**a**) and dextran-coupled Texas-Red (**b**). Colocalization of $GABA_B$ receptors and dextran (**c**) is seen inside the vesicles located on the cell membrane (*yellow*). Bar, 20 μm.

To further assess whether $GABA_B$ receptors are internalized by clathrin-independent mechanisms, cells were incubated for various times both with the antibody anti-$GABA_B$ receptors and with dextran, a fluid-phase endocytosis marker. This revealed a colocalization of $GABA_B$ receptors and dextran in *Paramecium* cells (Fig. 9).

Moreover, when endocytosis was blocked by filipin (0.1 μg/mL) or by nystatin (2 μg/mL), sterol-binding agents that disrupt caveolar structure and function, the receptor internalization decreased (Fig. 10c, d). In these experiments, cells were incubated in the anti-$GABA_B$ receptor antibody for 30 min at 4°C (a temperature inhibiting phagosome and endosome formation) (45), so that receptors were accumulated on the cell membrane (Fig. 10a). After removal of the excess of antibody, cells were incubated at 25°C. Eighty-four percent receptors were internalized in untreated cells after 20-min incubation at 25°C (Fig. 11), as shown both by the reduction of cell membrane fluorescence intensity and by the fluorescence localization into endosomes and phagosomes (Fig. 10b). Only 37 and 46% fluorescence was internalized in filipin ($P < 0.01$) and nystatin-treated cells ($P < 0.01$), respectively (Fig. 11). These results suggest that both endocytic processes (clathrin-dependent and/or -independent) occur in *Paramecium* and that the amount of receptors internalized through the two pathways is similar (Table 1).

5. Intracellular $GABA_B$ receptors trafficking after internalization: colocalization with endosomal markers.

   $GABA_B$ receptor antibody strongly labeled early endosomes, identified by antibody immunostaining against the molecular markers EEA1 and rab5. Both $GABA_B$-like receptor and rab5-like immunoreactivities were localized in vesicles showing a

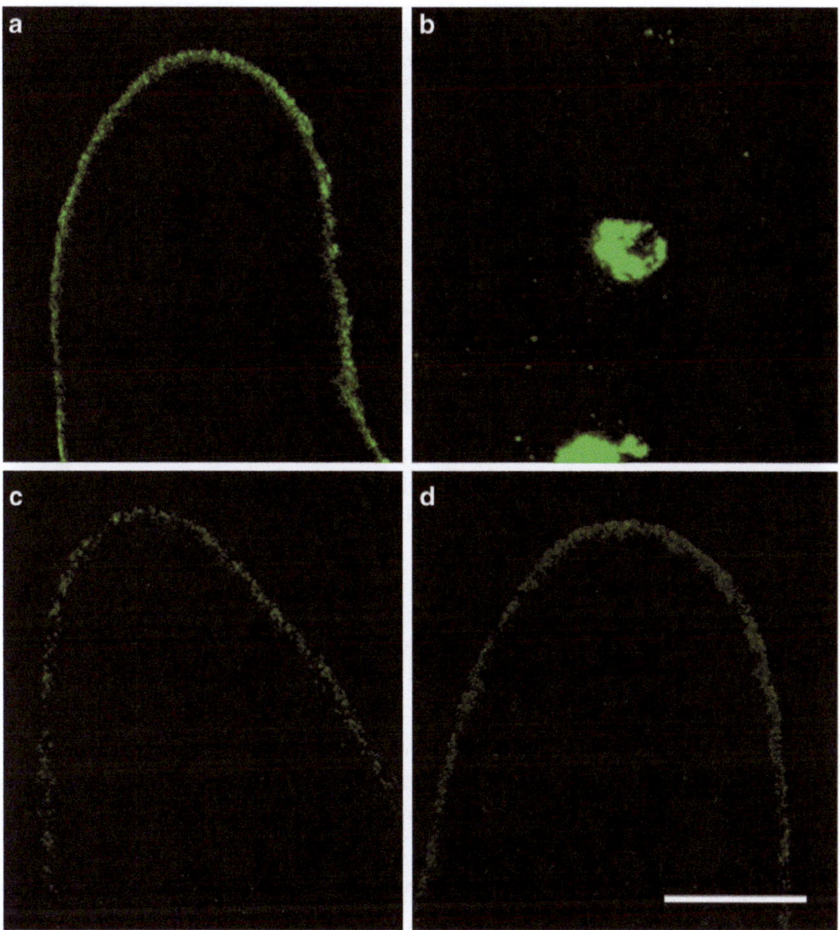

Fig. 10. GABA$_B$ receptors are internalized by noncoated endocytosis. Cells preincubated at 4°C for 30 min and labeled with anti-GABA$_B$ receptor antibody (dilution 1:100) for 30 min (**a**) are fixed after a 20-min chase at 25°C in the absence (**b**) or in the presence of non-coated-pit endocytosis inhibitors filipin (0.1 μg/mL) (**c**) and nystatin (2 μg/mL) (**d**). Bar, 20 μm.

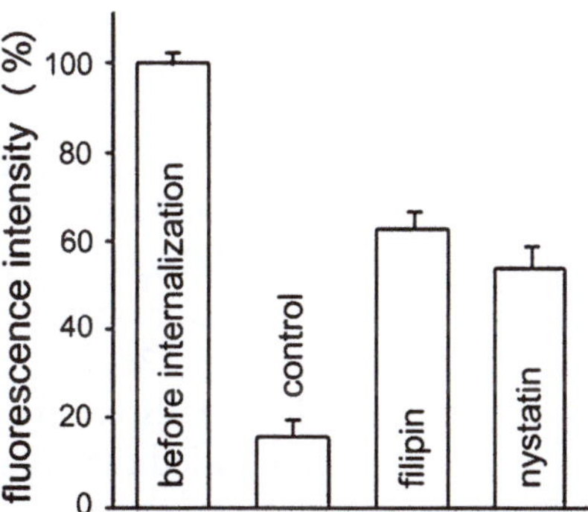

Fig. 11. Measurement of the internalization shown in Fig. 10. Constitutive receptor internalization is partially inhibited by filipin (0.1 μg/mL) and nystatin (2 μg/mL) (38% and 47%, respectively); ($P < 0.01$, Student's *t*-test). Data were normalized to cells before internalization at 25°C (shown in Fig. 10a).

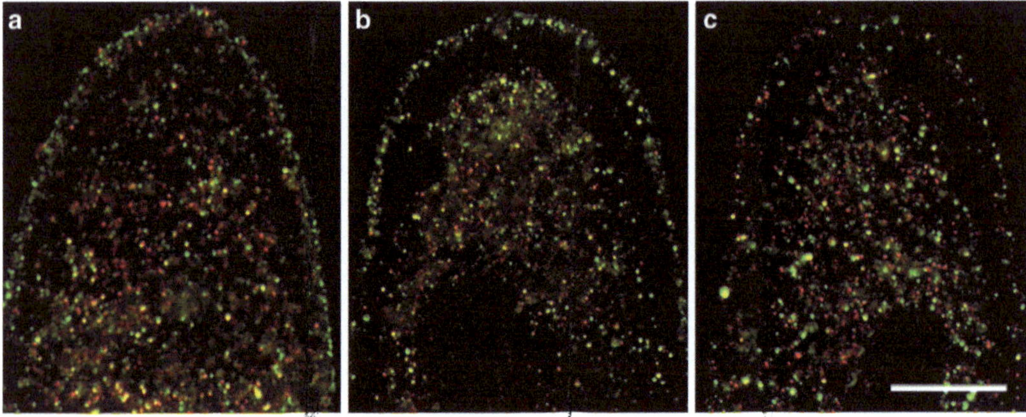

Fig. 12. Colocalization of $GABA_B$ receptors (*red*) with proteins involved in the intracellular trafficking (*green*): (**a**) rab5, (**b**) EEA1, (**c**) rabaptin-5. In all cases examined, most receptors aggregate in clusters and their fluorescence is localized in spots distributed on the plasma membrane and throughout the cytoplasm. Colocalization is seen as a *yellow fluorescence*. Bar, 20 μm.

clustered distribution near the cell membrane and in the most inner portion of the cytoplasm; with the double immunolabeling method most of the peripheral vesicle clusters appeared yellow because $GABA_B$-like receptors and rab5-like immunoreactivities partly colocalized (Fig. 12a). A similar colocalization pattern was obtained using, together with the anti-$GABA_B$R1 antibody, either the monoclonal antibodies anti-EEA1 (Fig. 12b) or anti-rabaptin-5 (Fig. 12c).

6. Retrieval of $GABA_B$ receptors by recycling endosomes.

Internalized receptors, destined for recycling back to the plasma membrane, that traffic from the early endosomes to the endosomal-recycling compartment were characterized by the presence of rab4 and rab11 proteins. To visualize $GABA_B$ receptor retrieval, the cells were double labeled with anti-$GABA_B$R1 and, either anti-rab4 or anti-rab11 as primary antibodies, visualized with anti-guinea pig Alexa Fluor 594 and anti-mouse Alexa Fluor 488 secondary antibodies, respectively. All immunostained vesicles showed a clustered distribution near the cell membrane and inside the cytoplasm; the yellow immunofluorescent vesicles, due to the colocalization of the $GABA_B$-like receptor with either rab4-like or rab11-like immunofluorescence, distributed in the peripheral zone is visible in Fig. 13a, b.

Rab11 also controls the traffic to the TGN, which is generally identified through the marker TGN38. Moreover, TGN38 protein was shown to undergo constitutive cycling through the cell surface by passing late endosomes as it moves back to the TGN (46). Evidence was given for the presence of $GABA_B$ receptors in the TGN by the colocalization of $GABA_B$-like with TGN38-like immunoreactivity; the yellow fluorescence was

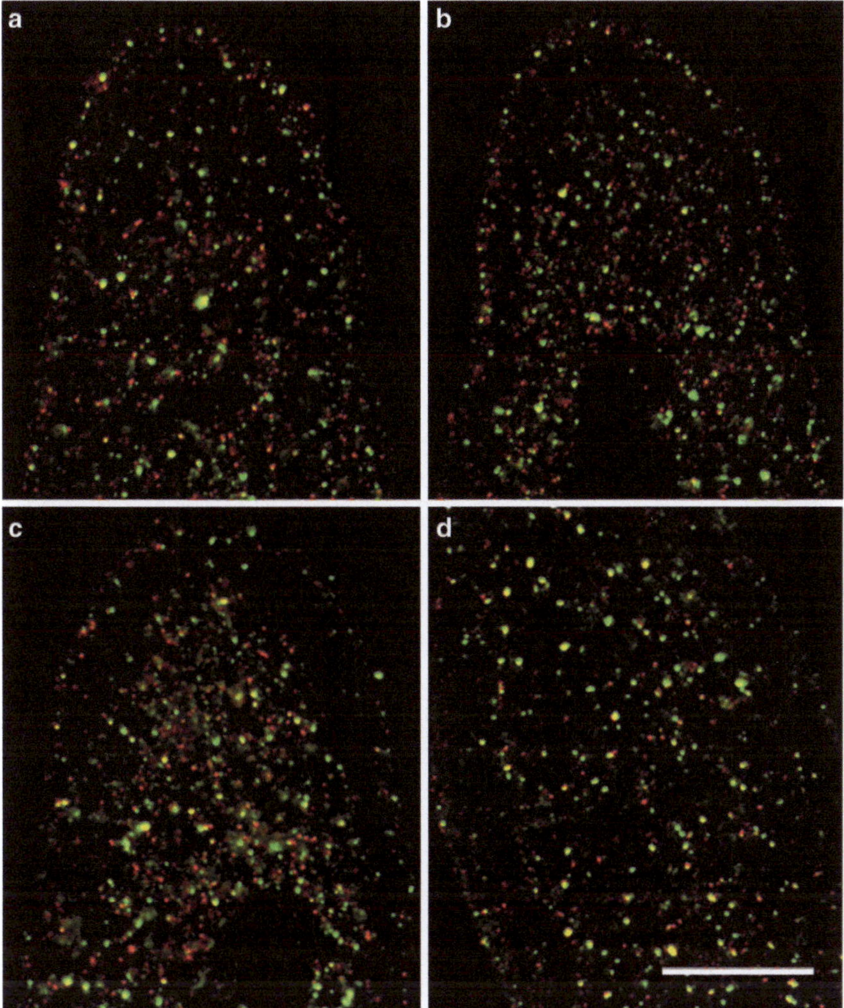

Fig. 13. Internalized $GABA_B$ receptors are retrieved to cell membrane or degraded into lysosomes. Colocalization of $GABA_B$ receptors (*red*) with (**a**) rab4, (**b**) rab11, (**c**) TGN38, and (**d**) LAMP1 proteins (*green*) is seen as a *yellow fluorescence*. Bar, 20 μm.

detected in a small number of vesicles located in the cytoplasm (Fig. 13c). Indeed, the Golgi apparatus of *Paramecium* is decentralized and occurs as hundreds of small stacks of three or so highly cisternae scattered throughout the cytoplasm (47).

7. $GABA_B$ receptors are degraded into lysosomes.
   In addition to receptor recycling, another possible fate for sequestered receptors is their degradation. Receptors destined for degradation exit the endosomal pathway at the early endosomes and traffic to the late endosome–lysosome pathway. We have shown that in *Paramecium* most of the $GABA_B$-like receptor immunostained vesicles were also immunolabeled with the lysosomal marker LAMP1 antibody (Fig. 13d).

References

1. Bowery NG, Bettler B, Froestl W, Gallagher JP, Marshall F, Raiteri M et al (2002) International Union of Pharmacology. XXXIII. Mammalian gamma-aminobutyric acid(B) receptors: structure and function. Pharmacol Rev 54:247–264
2. Ferguson SS (2001) Evolving concepts in G protein-coupled receptor endocytosis: the role in receptor desensitization and signaling. Pharmacol Rev 53:1–24
3. Seifert R, Wenzel-Seifert K (2002) Constitutive activity of G-protein-coupled receptors: cause of disease and common property of wild-type receptors. Naunyn Schmiedebergs Arch Pharmacol 366:381–416
4. Royle SJ, Murrell-Lagnado RD (2003) Constitutive cycling: a general mechanism to regulate cell surface proteins. Bioessays 25:39–46
5. Tehrani MHJ, Barnes EM Jr (1997) Sequestration of gamma-aminobutyric acid A receptors on clathrin-coated vesicles during chronic benzodiazepine administration in vivo. J Pharmacol Exp Ther 283:384–390
6. Grampp T, Notz V, Broll I, Fischer N, Benke D (2008) Constitutive, agonist-accelerated, recycling and lysosomal degradation of $GABA_B$ receptors in cortical neurons. Mol Cell Neurosci 39:628–637
7. Traub LM (2010) Tickets to ride: selecting cargo for clathrin-regulated internalization. Nat Rev Mol Cell Biol 10:583–596
8. Kirchhausen T (1999) Adaptors for clathrin-mediated traffic. Annu Rev Cell Dev Biol 15:705–732
9. Takei K, Hauke V (2001) Clathrin-mediated endocytosis: membrane factors pull the trigger. Trends Cell Biol 11:385–391
10. Traub LM (2003) Sorting it out: AP-2 and alternate clathrin adaptors in endocytic cargo selection. J Cell Biol 163:203–208
11. Owen DJ, Collins BM, Evans PR (2004) Adaptors for clathrin coats: structure and function. Annu Rev Cell Dev Biol 20:153–191
12. Rodemer C, Haucke V (2008) Clathrin/AP-2-dependent endocytosis: a novel playground for the pharmacological toolbox? Handb Exp Pharmacol 186:105–122
13. Shupliakov O, Low P, Grabs D, Gad H, Chen H, David C et al (1997) Synaptic vesicle endocytosis impaired by disruption of dynamin-SH3 domain interactions. Science 276:259–263
14. Cupers P, Jadhav AP, Kirchhausen T (1998) Assembly of clathrin coats disrupts the association between Eps15 and AP-2 adaptors. J Biol Chem 23:1847–1850
15. Nichols BJ, Lippincott-Schwartz J (2001) Endocytosis without clathrin coats. Trends Cell Biol 11:406–412
16. Tsao PI, von Zastrow M (2001) Diversity and specificity in the regulated endocytic membrane trafficking of G-protein-coupled receptors. Pharmacol Ther 89:39–147
17. Johannes L, Lamaze C (2002) Clathrin-dependent or not: is it still the question? Traffic 3:443–451
18. Chini B, Parenti M (2004) G-protein coupled receptors in lipid rafts and caveolae: how, when and why do they go there? J Mol Endocrinol 32:325–338
19. von Zastrow M, Kobilka BK (1994) Antagonist-dependent and -independent steps in the mechanism of andrenergic receptor internalization. J Biol Chem 269:18448–18452
20. Zhang J, Ferguson SS, Barak LS, Menard L, Caron MG (1996) Dynamin and beta-arrestin reveal distinct mechanisms for G protein-coupled receptor internalization. J Biol Chem 271:18302–18305
21. Raposo G, Dunia I, Delavier-Klutchko C, Kaveri S, Strosberg AD, Benedetti EL (1989) Internalization of beta–adrenergic receptor in A431 cells involves non-coated vesicles. Eur J Cell Biol 50:340–352
22. Dupree P, Parton RG, Raposo G, Kurzchalia TV, Simons K (1993) Caveolae and sorting in the trans-Golgi network of epithelial cells. EMBO J 12:1597–1605
23. Roettger BF, Rentsch RU, Pinon D, Holicky E, Hadac E, Larkin JM et al (1995) Dual pathways of internalization of the cholecystokinin receptor. J Cell Biol 128:1029–1041
24. Daukas J, Zigmond SH (1985) Inhibition of receptor-mediated but not fluid-phase endocytosis in polymorphonuclear leukocytes. J Cell Biol 101:1673–1679
25. Heuser JE, Anderson RG (1989) Hypertonic media inhibit receptor-mediated endocytosis by blocking clathrin-coated pit formation. J Cell Biol 108:389–400
26. Wang L-H, Rothberg KG, Anderson RGW (1993) Mis-assembly of clathrin lattices on endosomes reveals regulatory switch for coated pit formation. J Cell Biol 123: 1107–1117
27. Schnitzer JE, Oh P, Pinney E, Allard J (1994) Filipin-sensitive caveolae-mediated transport in endothelium: reduced transcytosis, scavenger endocytosis, and capillary permeability of selected macromolecules. J Cell Biol 127: 1217–1232

28. Lamaze C, Schmid SL (1995) The emergence of clathrin-independent pinocytic pathways. Curr Opin Cell Biol 7:573–580
29. Seabra MC, Mules EH, Hume AN (2002) Rab GTPases, intracellular traffic and disease. Trends Mol Med 8:23–30
30. Seachrist JL, Ferguson SSG (2003) Regulation of G protein-coupled receptor endocytosis and trafficking by Rab GTPases. Life Sci 74:225–235
31. Grosshans BL, Ortiz D, Novick P (2006) Rabs and their effectors: achieving specificity in membrane traffic. Proc Natl Acad Sci USA 103:11821–11827
32. Stenmark H (2009) Rab GTPases as coordinators of vesicle traffic. Nat Rev Mol Cell Biol 10:513–525
33. Allen RD, Fok AK (2000) Membrane trafficking and processing in *Paramecium*. Int Rev Cytol 198:277–318
34. Plattner H, Kissmehl R (2003) Molecular aspects of membrane trafficking in *Paramecium*. Int Rev Cytol 232:185–216
35. Allen RD (1988) Cytology. In: Görtz H-D (ed) *Paramecium*. Springer, Berlin, pp 4–40
36. Nilsson JR, van Deurs P (1983) Coated pits and pinocytosis in *Tetrahymena*. J Cell Sci 63:209–222
37. Allen RD, Fok AK (1993) Endosomal membrane traffic of ciliates. In: Plattner H (ed) Advances in cell and molecular biology of membranes, membrane traffic in protozoa. JAI Press, Greenwich, CT, pp 283–309
38. Ramoino P, Fronte P, Fato M, Beltrame F, Robello M, Diaspro A (2001) Fluid phase and receptor mediated endocytosis in *Paramecium*. Eur Biophys J 30:305–312
39. Ramoino P, Gallus L, Beltrame F, Diaspro A, Fato M, Rubini P et al (2006) Endocytosis of $GABA_B$ receptors modulates membrane excitability in the single-celled organism *Paramecium*. J Cell Sci 119:2056–2064
40. Ramoino P, Usai C, Beltrame F, Fato M, Gallus L, Tagliaferro G et al (2005) $GABA_B$ receptor intracellular trafficking after internalization in *Paramecium*. Microsc Res Tech 68:290–295
41. Manders EM, Verbeek FJ, Aten JA (1993) Measurement of co-localization of objects in dual colour confocal images. J Microsc 169:375–382
42. Gonzalez RC, Wintz P (1987) Digital image processing, 2nd edn. Addison Wesley, Massachusetts
43. Costes SV, Daelemans D, Cho EH, Dobbin Z, Pavlakis G, Lockett S (2004) Automatic and quantitative measurement of protein-protein colocalization in live cells. Biophys J 86: 3993–4003
44. Fok AK, Leung SS-K, Chun DP, Allen RD (1985) Modulation of the digestive lysosomal system in *Paramecium caudatum*. II. Physiological effects of cytochalasin B, colchicine and trifluoperazine. Eur J Cell Biol 37:27–34
45. Fok AK, Leung SS-K, Allen RD (1984) Modulation of the digestive lysosomal system in *Paramecium caudatum*. I. Effects of temperature. Eur J Cell Biol 34:265–270
46. Mallet WG, Maxfield FR (1999) Chimeric forms of furin and TGN38 are transported with the plasma membrane in the trans-Golgi network via distinct endosomal pathways. J Cell Biol 146:345–359
47. Estève J-C (1972) L'appareil de Golgi des Ciliés. Ultrastructure, particulièrement chez *Paramecium*. J Protozool 19:609–618

Chapter 5

Using Total Internal Reflection Fluorescence Microscopy (TIRFM) to Visualise Insulin Action

James G. Burchfield, Jamie A. Lopez, and William E. Hughes

Abstract

Total Internal Reflection Fluorescence Microscopy is a powerful application of fluorescence microscopy that allows selective imaging of fluorescent molecules that are either in or close to the plasma membrane of a cell. Thus, it is ideally suited to imaging the trafficking of molecules to and from the plasma membrane. Here we describe its use to study the trafficking and exocytosis of the glucose transporter (GLUT4) in response to insulin stimulation.

Key words: TIRF, TIRFM, Microscopy, GLUT4, IRAP, Insulin, Trafficking, Exocytosis, Fluorescence, eGFP, pHluorin

1. Introduction

The plasma membrane defines the extracellular and intracellular environments of a cell. Information must traverse this barrier for a cell to be able to interact with the extracellular environment. Thus, many fundamental cellular processes occur at the plasma membrane. Specific receptors or transporters in the plasma membrane can bind to a range of molecules such as nutrients, hormones, growth factors, cytokines or neurotransmitters, triggering a cascade of events signalling a particular cellular response. Regulated exocytosis largely relies on the engagement of extracellular cues at the plasma membrane, which trigger the export of intracellular cargo. This process of regulated exocytosis involves membranous tubulovesicular or granule structures enriched in cargo being transported to the plasma membrane. At the plasma membrane the exocytotic vesicles undergo attachment processes to ensure fidelity,

Emilio Badoer (ed.), *Visualization Techniques: From Immunohistochemistry to Magnetic Resonance Imaging*, Neuromethods, vol. 70, DOI 10.1007/978-1-61779-897-9_5, © Springer Science+Business Media, LLC 2012

and to facilitate the final process where the membranes of the vesicle and plasma membrane coalesce, resulting in the cargo being delivered either into or through the plasma membrane. Whilst the size of exocytotic vesicles and the precise regulatory mechanisms controlling exocytosis may vary (synaptic vesicles are ~50 nm whilst insulin granules can be 450 nm; calcium flux is an important trigger for neurotransmission but not for exocytosis of GLUT4 glucose transporter), many of the behaviours associated with exocytotic vesicles are common. Exocytotic vesicles are firstly transported to the plasma membrane and attach to the subplasmalemmal surface. Attachment of vesicles is typified by a transition from a freely mobile state to an immobile/static state. Finally, vesicles fuse with the plasma membrane to deliver cargo, a process with a characteristic behaviour (discussed later). Thus, techniques and methods devised to image these processes in one cell type and/or exocytotic cargo can generally be applied to other systems. Total Internal Reflection Fluorescence Microscopy (TIRFM) is one such technique that is ideal for imaging the trafficking events described earlier.

TIRFM or evanescent wave microscopy is based on the principle that a beam of light will be totally internally reflected when travelling to a medium of lower refractive index (e.g., from glass [*n1*] into media [*n2*]) and the angle of incidence (θi) is greater than the critical angle. Total internal reflection creates an evanescent field or standing wave of electromagnetic radiation in the lower refractive index medium that is capable of exciting fluorescent molecules (Fig. 1a). The intensity of excitation, however, decays exponentially from the interface and thus only fluorescent molecules close to the interface will be excited. In the case of cells grown on a glass coverslip, only molecules on or close to the plasma membrane are excited and imaged. In comparison to standard epifluorescence microscopy (Fig. 1b), TIRFM yields greatly improved resolution of PM and sub-PM structures (Fig. 1c) due to the elimination of image-degrading noise resulting from the excitation of out-of-focus fluorescent molecules (see (1) for review).

We have recently used TIRFM to better understand the regulation of exocytosis associated with the physiological process of blood glucose homeostasis. Tissues such as the brain need glucose constantly and low concentrations of glucose in the blood can cause seizures, loss of consciousness and death. In contrast, prolonged elevation of blood glucose concentrations, as in poorly controlled diabetes, can result in blindness, renal failure, cardiac and peripheral vascular disease and neuropathy. Therefore, blood glucose concentrations need to be maintained within narrow limits. This homeostasis is accomplished by the finely tuned hormonal regulation of peripheral glucose uptake and hepatic glucose production. After a meal, this is largely controlled by the action of the hormone insulin.

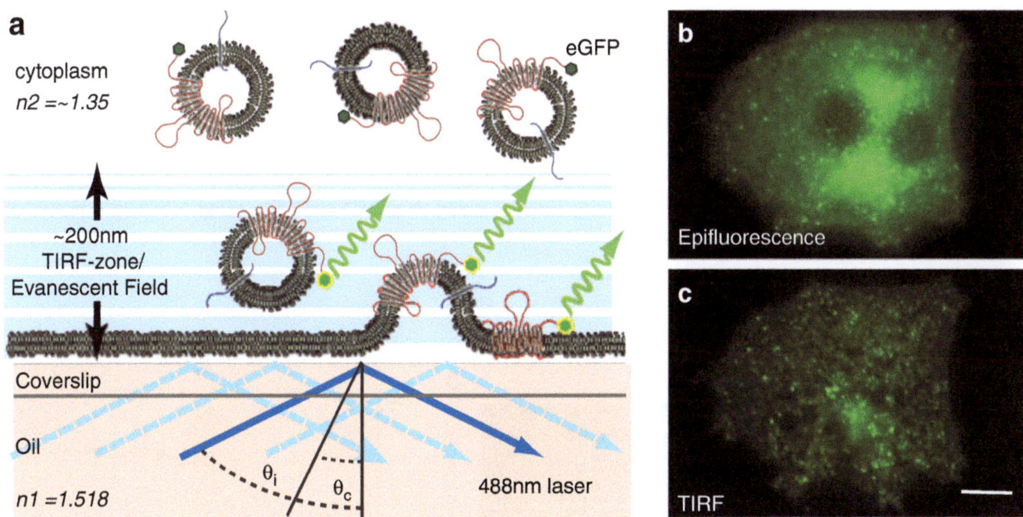

Fig. 1. Total internal reflection fluorescence microscopy (TIRFM). (**a**) Total internal reflection of excitation energy introduced above the critical angle (**c**) leads to the generation of an exponentially decreasing evanescent field capable of exciting fluorescent molecules. Prism or, as depicted here, objective-based TIRF microscope systems are capable of exciting fluorophores within the evanescent field or "TIRF-zone" of between 70 and 250 nm from the interface between the higher (glass) and lower (cell) refractive indices (*n*). Epifluorescence (**b**) vs. TIRF (**c**) images of a 3T3-L1 adipocyte expressing the glucose transporter GLUT4 fused at the carboxyl-terminus to eGFP (GLUT4-eGFP). *Scale bar* represents 10 μM.

Insulin is released from pancreatic β-cells in response to elevated blood glucose. Its major site of action is in muscle and adipose tissue where it triggers a signal transduction pathway that culminates in the exocytosis of the glucose transporter, GLUT4, from intracellular storage vesicles to the plasma membrane (PM). Defects in this process lead to the development of chronic diseases such as type II diabetes, and thus an understanding of the molecular events that regulate these processes and how they become defective during the development of disease is essential.

We have used a number of applications of TIRFM to probe various aspects of insulin action in 3T3-L1 adipocytes (2–4). These include high and low temporal frequency live cell TIRFM and fixed cell TIRFM. The choice of approach is dependent on the desired outcome (Fig. 2). Low frequency TIRFM is very useful for the relatively rapid screening of treatments using larger numbers of cells (Fig. 3). High frequency TIRFM involves working with single cells and is time consuming with a low throughput. Its advantage is in being able to visualise, identify and quantify individual trafficking events in detail and to assess how these change under various conditions (Fig. 4). Fixed cell TIRFM is excellent for rapid screening of conditions/treatments and for looking at effects on endogenous proteins by immunofluorescence (Fig. 5).

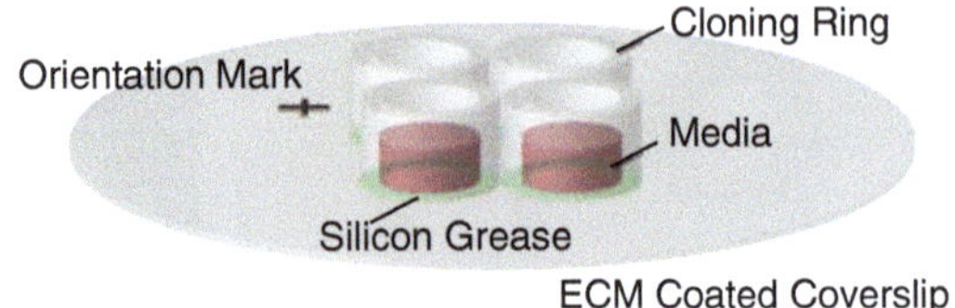

Fig. 2. Cloning ring setup for low resolution and fixed TIRFM. Cloning rings with one end coated with silicon grease are placed in a 2 × 2 configuration in the centre.

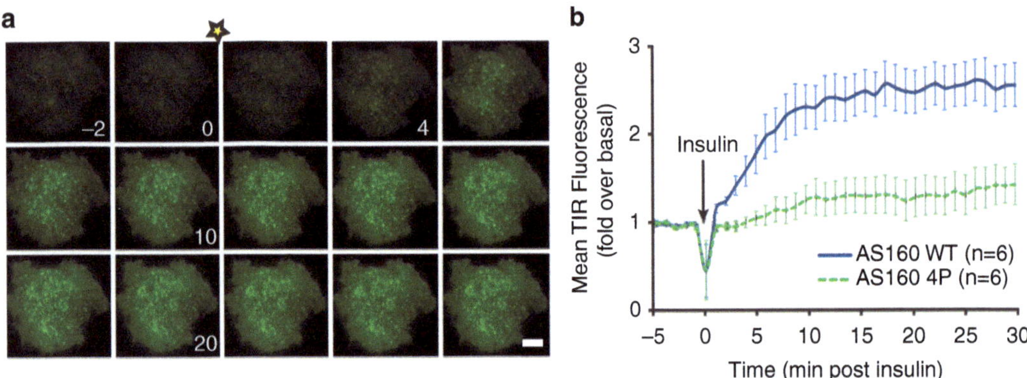

Fig. 3. Low frequency TIRFM data. Example of data that can be generated from a single coverslip of 3T3-L1 adipocytes imaged at low frequency. Adipocytes were co-electroporated with GLUT4-eGFP and either wild-type (WT) or the dominant negative 4P mutant AS160 constructs and seeded into cloning rings on a matrigel coated coverslip as in Fig. 2. Six cells were chosen per condition and imaged at a rate of 1 image/cell/min. (**a**) Representative time course of an adipocyte overexpressing GLUT4-eGFP and WT-AS160. *Star* denotes the addition of insulin at time 0. (**b**) The mean fluorescence profile of cells overexpressing the AS160 constructs. Data is presented as the mean ± SEM. *Scale bar* represents 5 μM.

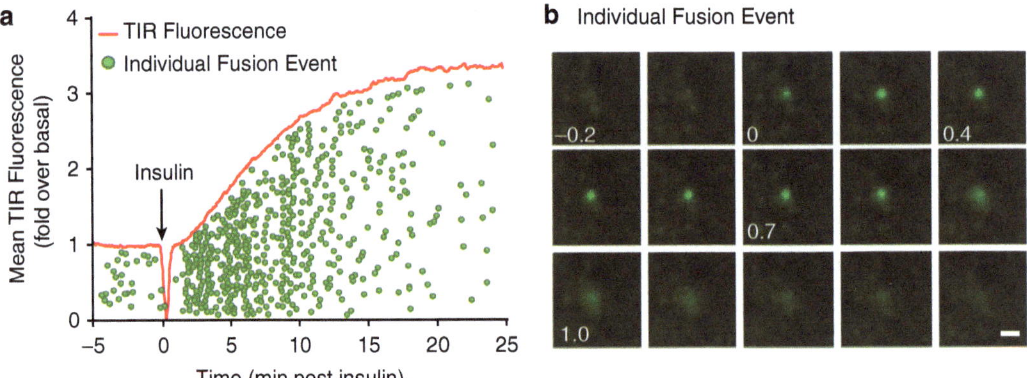

Fig. 4. High frequency TIRFM data. Example of data that can be generated from a single coverslip of 3T3-L1 adipocytes imaged high frequency. (**a**) The fluorescence profile and detected fusion events from a single adipocyte overexpressing IRAP-pHluorin, imaged by TIRFM at a rate of 10 Hz. (**b**) A representative IRAP-pHluorin fusion event. *Scale bar* represents 0.5 μM.

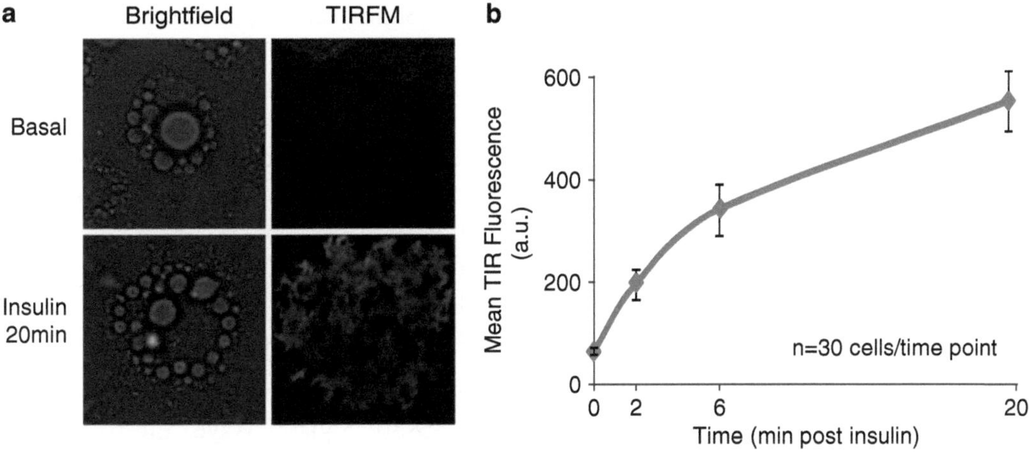

Fig. 5. Fixed cell TIRFM data. Example of data that can be generated from a single coverslip of fixed 3T3-L1 adipocytes. Adipocytes constitutively expressing GLUT4 with a HA-tag in its first exofacial loop were seeded into four cloning rings on a matrigel-coated coverslip. Cells within a cloning ring were stimulated with 100 nM insulin for either 0, 2, 6 or 20 min prior to fixation. Cells were stained with an anti-HA antibody and then an Alexa-488 secondary. (**a**) Representative brightfield and TIRF images for basal and 20 min of insulin stimulation. (**b**) Mean fluorescence from each time point. $n = 30$ cells/time point, data is mean ± SEM.

2. Materials

The products in brackets are used by us. Similar products from alternative manufacturers should also be suitable.

2.1. Required for All Methods

1. Coverslips (LaCon 42×0.17 mm, [LaCon GmbH, Erbach, Germany]) and suitable imaging chamber (POC—[PeCon, GmbH, Erbach, Germany]).
2. TIRF capable imaging system (Zeiss Axiovert 200 M with; TIRF2[Carl Zeiss MicroImaging Gottingen, Germany]; Andor Du888 EMCCD[Andor, Belfast, N. Ireland]) with incubation, motorised stage and motorised focus.
3. Microscope Control and Image Acquisition Software. In this protocol, we use micro-Manager as an example. μManager is open source microscope control software built around the ImageJ platform. It is highly flexible, provides support for most hardware and can be customised for your applications (more information and free download can be found at http://www.micro-manager.org/).

   Image Analysis Software (Image J (5), TIRF explorer (6, 7)).

   Matrigel™ (BD Biosciences, Franklin Lakes, USA) or suitable extracellular matrix protein.

4. 8×8 mm Cloning Pings (Sigma-Aldrich, St. Louis, MO, USA). Silicon Grease (Sigma-Aldrich, St. Louis, MO, USA).
5. PBS.

2.2. Required for Electroporation

1. High quality (260:280 > 1.8), high concentration (>4 mg/mL) stocks of plasmid DNA constructs of interest (e.g., GLUT4-EGFP; IRAP-pHluorin).
2. 5× Trypsin/EDTA (1:2 dilution with PBS of 10× Stock, [Invitrogen, Carlsbad, USA]).
3. Culture Media (DMEM [Invitrogen, Carlsbad, USA]).
4. FBS (Invitrogen, Carlsbad, USA).
5. Electroporator (ECM 830 Square Wave Electroporation System, [BTX Molecular Delivery Systems, Massachusetts, USA]).
6. Electroporation solution (20 mM Hepes, 135 mM KCl, 2 mM $MgCl_2$, 0.5% Ficol 400, 1% DMSO. pH 7.6).
7. 20 mM ATP: 50 mM Glutathione stock (pH ~ 7.4).
8. 0.4 cm Electroporation Cuvettes (BioRad, [Hercules, USA]).
9. BSA > 98% (Sigma-Aldrich, St. Louis, MO, USA) A highly purified grade of BSA should be used. We routinely batch test to ensure undesirable basal activation of insulin signalling due to trace amounts of growth factors.

2.3. Required for Live Cell TIRFM

1. Insulin Stock (70 μM Insulin stock stored in HCl)
2. KRP (120 mM NaCl, 0.6 mM Na_2HPO_4, 0.4 mM NaH_2PO_4, 6 mM KCl, 1.2 mM $MgSO_4$, 12.5 mM HEPES, 1 mM $CaCl_2$, 10 mM Glucose, 0.2% (w/v) BSA, pH 7.4) or other suitable imaging buffer

2.4. Required for Fixed Cell TIRFM

1. 16% paraformaldehyde (Electron Microscopy Supplies, Hatfield, USA)
2. 50 mM Glycine (for fixed Cell TIRFM)
3. Fixed Cell Imaging and Storage Buffer (PBS, 5% Glycerol, 2.5% 1,4-diazabicyclo[2.2.2]octane (DABCO)[Sigma-Aldrich St. Louis, MO, USA])

3. Methods

3.1. Preparation of Plasmid DNA (See Note 1)

1. Prepare plasmid DNA by maxiprep purification (as per manufacturer's instructions).
2. Measure DNA concentration (260:280 ratio) of plasmid DNA using a suitable spectrophotometer and dilute DNAs to 4 mg/mL using sterile milli-Q water.

3.2. Preparation of Matrigel Coated Coverslip (See Note 2)

1. Thaw matrigel overnight on ice at 4°C.
2. The next day, dilute matrigel 1:50 into ice-cold PBS using pre-chilled pipettes (chilled by pipetting ice-cold PBS several times).
3. Aliquot onto coverslips (2 mL per 42 mm coverslip).
4. Leave at room temperature for 2 h.
5. Wash twice with PBS to remove unbound material. Coverslips are ready to be used or can be stored in PBS for up to 1 month at room temperature (*NB:* coated coverslips should be stored in petri dishes sealed with parafilm).
6. Matrigel coated coverslips should be washed twice in culture media prior to use.

3.3. Electroporation Procedure

The following protocol describes the electroporation of 3T3-L1 adipocytes. All steps should be performed in a sterilised tissue culture hood. All solutions should be at room temperature unless otherwise stated. A confluent 10-cm dish of differentiated adipocytes is sufficient for 1–4 electroporations depending on the downstream applications. Cells should ideally be greater than 95% differentiated to obtain optimal results and should be electroporated at 5–7 days post-initiation of differentiation (the efficiency of electroporation and/or cell recovery decreases after this time due to increases in lipid content/cell buoyancy).

1. Trypsinise cells from each dish using 5× Trypsin/EDTA for 5–10 min in a 37°C incubator (check trypsinisation by gently tapping the cells and assessing the effectiveness of dislodgment using a microscope. NB: Excessive trypsinisation will result in a loss of cell viability and a poor electroporation efficiency).
2. Resuspend cells in 10 mL of DMEM culture medium containing 10% FCS and transfer to a 50-mL Falcon tube and make up the volume to 30 mL with DMEM culture medium (dishes can be pooled here).
3. Centrifuge cells at 500 × *g*, room temperature for 2 min.
4. Aspirate medium and resuspend cells in 30 mL of PBS.
5. Centrifuge cells at 500 × *g*, for 2 min.
6. Aspirate medium and resuspend cells in 30 mL of PBS.
7. Centrifuge cells at 500 × *g* for 2 min.
8. During this spin, thaw single use 100× ATP:Glutathione stock and add to Electroporation solution. Filter sterilise using a 0.2-μM filter.
9. Add 10–40 μg of plasmid to each electroporation cuvette.
10. Aspirate medium and resuspend cells in the electroporation buffer (approximately 400 μL per electroporation with a bit extra).

11. Transfer 0.4 mL volumes of resuspended cells to each cuvette containing the DNA; tap gently to mix, close lid and invert briefly. Continue with next step immediately.
12. Electroporate each cuvette with a single 20 ms, 200 V square wave pulse.
13. Immediately add 2 mL of DMEM culture media containing 10% FCS, penicillin/streptomycin/glutamine and mix gently. A white viscous material (representing a mixture of lipid and coagulated protein from dead adipocytes) may be present after the electroporation procedure. This should be removed using a pipette.
14. Seed ~100–200 μL of cells per cloning ring. Whilst cells can be added directly onto coated coverslips, we have found it more practical to seed into cloning rings (see Fig. 2). Sterile cloning rings are coated on one end with silicon grease to form a seal on the coverslip and placed centrally in a suitable configuration. The use of cloning rings reduces the number of cells required per coverslip to achieve a suitable cell density. They also allow multiple treatments to be imaged on the same coverslip. This dramatically improves the throughput of low frequency and fixed cell experiments.

3.4. Preparation for TIRF Microscopy

Cells can be used for microscopy from 12 to 72 h post-electroporation, although generally we leave them for a minimum of 48 h to ensure good recovery, attachment and stable expression levels.

1. Wash the cells three times in serum-free DMEM.
2. Prepare a working stock of insulin (10×) diluted from 70 μM stock on the day of the experiment.
3. Add a suitable amount of serum-free DMEM and incubate for approximately 2–3 h at 37°C/10% CO_2 before imaging or stimulation.
4. Transfer coverslips to individual POC chambers.
5. Add 3 mL of KRP/0.2% (w/v) BSA/10 mM glucose pH 7.4 buffer.
6. Wash the bottom of the coverslip carefully with dilute detergent on a kimwipe.
7. Repeat this with RO water and dry carefully.
8. Coverslips are now ready for imaging.

3.5. Low Frequency TIRFM of Multiple Cells

Low frequency TIRFM is useful for rapid screening of relatively large numbers of cells. It is ideally suited for the simulataneous imaging of multiple treatment regimes simultaneously when using cloning rings (Fig. 2).

1. μmanager and open "Multi-dimensional Acquisition" and select "Edit position List".

2. Using wide-field epifluorescence, select 10–20 healthy cells per condition and record their location using the mark function.
3. Save locations. This will avoid you having to search for cells again if something goes wrong.
4. Using brightfield/transmitted light imaging, discard unhealthy looking or apoptotic cells. Look for membrane blebbing and/or excessive granular appearance (see Note 6).
   (a) Select each position and use "Go To"
   (b) To remove use "Remove"
5. Narrow selection to 5–10 cells/condition (based on appearance). The number of cells chosen will dictate the temporal frequency at which you will be able to acquire data.
6. Ensure cells are in proper focus for TIRFM (use "Update") to save new position information to the list.
7. Set up image acquisition
   (a) The critical factors are now to choose:
      (i) Laser Power Exposure time.
      (ii) Duration of experiment.
      (iii) The combination of these settings will have a dramatic impact on cell viability, photobleaching and ultimately the quality of your data (see Note 5).
   (b) In the "Multi-dimensional Acquisition" panel ensure that the following options are ticked
      (i) Time Points—define the number of time points and the interval. Typically you can hope to achieve a rate of 10 cells/min when using autofocus
      (ii) Multiple positions
      (iii) Autofocus—this will require setup (see Sect. 4)
      (iv) Channels
      (v) Save Images
         - Ensure display is set to single window
         - Specify the directory you wish to save in and a suitable file name (see Sect. 4)
   (c) Begin Acquisition.
   (d) Add insulin (1–100 nM final) at an appropriate time.

3.6. High Frequency TIRFM of Single Cells

1. Select 10–20 suitable cells by wide-field epifluorescence (as above Sect. 3.5, step 1).
2. Using brightfield, discard unhealthy looking cells. Look for membrane blebbing and excessive granular appearance (as above Sect. 3.5, step 2 and see Note 6).

3. Look at cells by TIRFM and remove those that are not well attached.
4. Select your "favourite".
5. Set up image acquisition
   (a) The critical factors are now to choose
      (i) Laser power (see Note 5)
      (ii) Exposure time
   (b) Open the "Burst Acquisition" panel. You will now need to specify
      (i) Time points.
      (ii) Specify the directory you wish to save in and a suitable file name (see Sect. 4).
      (iii) Ensure that the 'Output to disk' and 'display while aquiring' options are checked.
   (c) Begin acquisition.
   (d) Keep an eye on the focus. This can be adjusted manually as required.
   (e) Add insulin (1–100 nM final) at an appropriate time.

3.7. Fixed Cell TIRFM

The strength of this application of TIRFM is the rapid generation of data, the ability to look at endogenous proteins and the relative ease of looking at multiple proteins/markers. This approach is at its most powerful when looking at constitutively expressed proteins (endogenous or via viral transduction and selection). This allows cells to be selected by brightfield, avoiding any user bias towards protein expression levels and localisation.

1. If necessary transfect cells (see Sect. 3.4) and seed onto coverslips.
2. Treat as necessary.
3. Wash twice with ice cold PBS.
4. Fix in 3.8% paraformaldehyde at for 20 min at room temperature.
5. Quench with 50 mM glycine in PBS.
6. Wash twice with room temperature PBS.
7. If imaging an overexpressed fluorescent protein skip to Sect. 3.7, step 14.
8. Block and permeabilise with BB (blocking buffer—PBS with 2% BSA and 0.1% saponin) for 30 min.
9. Incubate coverslips for 1 h in BB and primary antibody(ies) at correct dilution.
10. Wash coverslips with BB 5 × 5 min.

11. Incubate coverslips for 1 h BB and the corresponding fluorochrome-conjugated secondary antibody(ies) at correct dilution. Keep in the dark as much as possible from now on.
12. Wash coverslips with BB 5 × 5 min.
13. Wash coverslips with PBS.
14. The coverslips can be viewed immediately or store at 4°C in the dark. Storage and imaging should be performed in PBS containing 5% glycerol and 2.5% DABCO (1,4-diazabicyclo[2.2.2]octane).

3.8. Analysis

The routines below are described using ImageJ unless otherwise stated. ImageJ is an indispensable tool for image editing and analysis. While there are many commercial analysis software packages available, few, if any have the depth and level of customisation that can be achieved through this freely available tool.

1. Analysis of low frequency TIRFM data (change in fluorescence over time)
   (a) Open image sequence
   (b) Select region of background
   (c) Subtract background (*Plugins> ROI> Background Subtraction from ROI*)
   (d) Draw ROI over cell
   (e) Measure mean intensity over time (*Plugins> Stacks-T-functions> Intensity vs. Time Monitor)*
   (f) Copy and paste results into excel
2. Analysis of High Frequency TIRFM
   (a) Change in fluorescence over time
      (i) Open image sequence
      (ii) Average frames to reduce the data points (*Plugins> Stacks-T-functions> Grouped Stack Projector)*
      (iii) Continue from Sect. 3.7, step 1b
   (b) Analysis of fusion events
   With the right construct (e.g., IRAP-pHluorin, VAMP2-pHluorin), fusion events will be readily visible in high frequency TIRF data. These will appear as a rapid localised increase in fluorescence intensity, followed by spreading of the signal as the fluorophore diffuses laterally into the cell membrane (1). Over the course of an insulin stimulation (30 min) there will be in the order of 1,000 events/cell, making the manual counting of these events tedious to say the least. We have developed an automated approach for the detection and analysis of these events (2, 6, 7)

(c) Analysis of movement

(i) The tracking of individual compartments is readily achievable in high frequency images. There are several competent software packages that can return reasonable tracking data.

4. Notes

1. *DNA quality.* High quality DNA (260:280 > 1.8) is essential. Poor quality DNA will result in poor electroporation efficiency and increased cell death. DNA should also be kept at a high concentration in order to minimise the volume added to the electroporation solution.
2. *Coverslips.* For optimal TIRFM, coverslips should be individually inspected (for imperfections) and cleaned with milli-Q water (to remove dirt and glass fragments that result from manufacture) prior to sterilisation. The coating of coverslips with ECM proteins (such as Matrigel) will dramatically improve both cell attachment and the quality of your TIRF image (see Fig. 6). Matrigel should be stored in single-use aliquots to avoid multiple freeze–thaws.
3. *Evaporation* at 37°C in a non-humidified incubator will be significant, thus care should be taken to ensure this is kept to a minimum. This is most readily addressed by having a transparent lid over your sample at all times.
4. *AutoFocus* is essential for multi-position TIRF experiments, as a small shift in focus causes loss of focus. The implementation in μmanager is fairly effective although it comes at the cost of a significant decrease in speed. Time should be taken to optimise these settings to minimise the time spent acquiring an accurate focus.
5. *Laser Intensity, exposure time and image frequency and experiment duration* determine the amount of electromagnetic radiation each cell is exposed. Consequently, this impacts on photobleaching and cell viability. Laser intensity must be kept to the absolute minimum required to generate a reasonable image (this is crucial for high frequency experiments). We use 100 mW laser at approximately 25% output with a 2.5% ND filter. There is a further approximately 50% reduction in laser power through the couplings and to the objective. In our hands, cells can be exposed continuously to this intensity for up to an hour with no signs of phototoxicity and minimal photobleaching. It is important to pay attention to how cells look pre- and post-imaging to assess any potential damage, and modify your protocol accordingly.

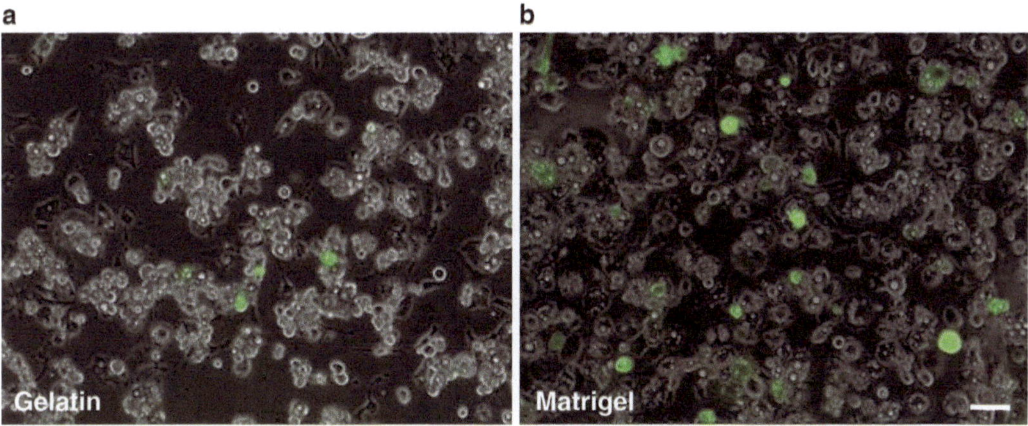

Fig. 6. Comparison of adipocytes seeded onto glass or matrigel-coated coverslips. Overlays of epifluorescence and brightfield images of 3T3-L1 adipocytes electroporated with GLUT4-eGFP and seeded in equal volumes onto glass coverslips coated with (**a**), gelatin or (**b**), ECM (matrigel). *Scale bar* represents 50 μm.

6. *Cell selection* is critical for generating quality TIRFM data. Time and care should be taken when selecting cells to image. Avoid choosing cells highly overexpressing proteins as these are most likely to display aberrant localisation and trafficking. When selecting adipocytes ensure the cells are well differentiated, look healthy and are well attached.

7. *Generating additional data.* It is worth taking brightfield/transmitted light and epifluorescence images of cells prior to imaging for later reference. When imaging at high frequency, it is worthwhile taking a basal and stimulated TIRF image of the other cells you have chosen. This will give you the ability to more readily compare treatment groups and will provide a measure of the healthiness and responsiveness of your cells. When imaging at low frequency, acquire epifluorescence images in conjunction with the TIRF images. This will enable you to normalise your TIRF data to the total cell fluorescence.

References

1. Burchfield JG, Lopez JA, Mele K, Vallotton P, Hughes WE (2010) Exocytotic vesicle behaviour assessed by total internal reflection fluorescence microscopy. Traffic 11:429–439
2. Lopez JA et al (2009) Identification of a distal GLUT4 trafficking event controlled by actin polymerization. Mol Biol Cell 20:3918–3929
3. Ng Y et al (2010) Cluster analysis of insulin action in adipocytes reveals a key role for Akt at the plasma membrane. J Biol Chem 285:2245–2257
4. Zhao P et al (2009) Variations in the requirement for v-SNAREs in GLUT4 trafficking in adipocytes. J Cell Sci 122:3472–3480
5. Rasband WS, ImageJ, U.S. National Institutes of Health. (Bethesda, Maryland, USA, http://imagej.nih.gov/ij/, 1997–2010)
6. Mele K, Burchfield JG, James DE, Vallotton P, Hughes WE (2009) Towards fully automated identification of vesicle-membrane fusion events in TIRF microscopy. Int J Comput Aided Eng Technol 1:502–515
7. Mele K et al (2009) Automatic identification of fusion events in TIRF microscopy image sequences. In: ICCV09 1st workshop on Video-oriented Object and Event Classification (VOEC'09), Kyoto, Japan, 2009

Chapter 6

Live-Cell Quantification of Mitochondrial Functional Parameters

Marco Nooteboom, Marleen Forkink, Peter H.G.M. Willems, and Werner J.H. Koopman

Abstract

Mitochondria are semi-autonomous organelles, which are central to cellular energy production and signal transduction. Given the tight integration between mitochondrial and cellular physiology, experimental strategies are required to study mitochondrial (dys)function in living cells. For this purpose one can use various chemical and protein-based fluorescent reporter molecules (probes), which are introduced into the cell using specific incubation protocols or transfection techniques. These probes include reporters to monitor mitochondrial membrane potential (Δψ), cytosolic and mitochondrial free calcium concentration (Ca^{2+}), reactive oxygen species (ROS), cytosolic and mitochondrial pH, glucose and ATP. However, proper interpretation and quantification of the above readouts is not trivial. Here, we present our protocol for automated temporal analysis of mitochondrial position in living cells and explain how it can be used for computer-assisted quantification of mitochondrial morphology and Δψ. We further discuss how this approach can be applied for simultaneous quantification of multiple mitochondrial and cellular parameters.

Key words: Human skin fibroblasts, Live-cell imaging, Baculovirus transduction, TMRM, AcGFP1, tagRFP

1. Introduction

Mitochondria are organelles that are among the prime suppliers of cellular energy in the form of ATP in most mammalian cells. Mitochondria are also crucially involved in cellular Ca^{2+} homeostasis, redox signaling, generation of reactive oxygen species (ROS), and induction of programmed cell death (1). Individual mitochondria can move through the cell, and mitochondria can mix their content by continuous fusion and fission events. Depending on the

Emilio Badoer (ed.), *Visualization Techniques: From Immunohistochemistry to Magnetic Resonance Imaging*, Neuromethods, vol. 70, DOI 10.1007/978-1-61779-897-9_6, © Springer Science+Business Media, LLC 2012

cell type, metabolic state, cell-cycle phase and differentiation stage, mitochondria display different motility patterns and a variety of morphologies ranging from spherical objects, elongated filaments, and/or interconnected (sub)network (1, 2). At the ultrastructural level, the mitochondrion consists of a matrix compartment that is surrounded by a highly folded mitochondrial inner membrane (MIM) and an outer membrane (MOM). The structure of the MIM and composition of the mitochondrial matrix also depend on metabolic state (1, 2). Embedded within the MIM lies the oxidative phosphorylation system, which consist of the electron transport chain (ETC) and F_0F_1-ATP synthase (3). The ETC consists of four complexes (CI, CII, CIII, and CIV), which extract electrons from NADH (at CI) and $FADH_2$ (at CII). Subsequently, the electrons are transported to CIII and CIV by ubiquinone and cytochrome-*c*, respectively. The energy that is generated by the electron transport is used to pump protons across the MIM, thereby establishing an inward proton-motive force (PMF) consisting of an electrical ($\Delta\psi$) and chemical (ΔpH) component. The energy stored in the proton gradient is used by a fifth complex (CV) to facilitate the synthesis of ATP. In general, proper mitochondrial functioning requires the presence of a sufficiently inside negative $\Delta\psi$ across the MIM (150–180 mV). Therefore, $\Delta\psi$ can be considered as a key estimator of "mitochondrial health."

Mitochondrial dysfunction is involved in many human disorders including multiple sclerosis, Parkinson's disease, cancer, type II diabetes, metabolic syndromes, and inherited respiratory chain disorders (3–8). Moreover, mitochondria are important pharmacological targets (9–11). Given the tight integration between mitochondrial and cellular metabolic and signaling pathways, understanding mitochondrial (dys)function necessitates its analysis in living cells. The latter has several prerequisites: (1) using an appropriate cell type, (2) using a suited fluorescent biosensor, (3) using a microscopy system allowing optimal image acquisition under physiological conditions, and (4) using validated procedures for image processing and analysis. Considering these prerequisites, we first present some background information (Sects. 1.1–1.4) followed by protocols describing our strategy to noninvasively study key mitochondrial parameters (i.e., morphology and $\Delta\psi$; Sects. 2–4). Finally, we discuss how this strategy allows simultaneous quantitative analysis of multiple mitochondrial and cellular parameters.

1.1. Live-Cell Imaging of Mitochondrial Function

Fluorescent biosensors for cellular and mitochondrial function (a.k.a. "probes" or "reporter molecules") can be divided in two groups: (1) chemical fluorescent probes, which are introduced into the cell using a specific "loading" protocol, and (2) protein-based ("proteinaceous") fluorescent sensors, which are translated by the cell after introduction of sensor's DNA by cell transfection.

Chemical probes have two advantages over proteinaceous sensors: (1) they can be introduced into the cell with relative ease, and (2) they stain all cells in a relatively equal intensity. Proteinaceous sensors, however, can be more efficiently targeted to intracellular compartments like mitochondria. In our research, we have mainly focused on primary human skin fibroblasts of healthy individuals and patients with an isolated deficiency in complex I, the first complex of the ETC (1, 12). These cells attach well to glass microscopy coverslips, allowing their use in superfusion experiments. Moreover, the fibroblasts displayed a height ≤3 μm in the axial (z) direction, facilitating image quantification (13). Primary cell lines can be resistant to transfection by "standard" cationic lipid-based methods (e.g., Superfect®, Qiagen, Venlo, The Netherlands or Lipofectamine™, Invitrogen, Breda, The Netherlands). In the case of primary human skin fibroblasts, we have used an insect virus system (baculoviral transfection) (14). For neurons lentiviral transfection can be used (15, 16), but this approach poses safety risks and requires specialized lab equipment.

1.2. Mitochondria-Specific Staining

Ideally, the applied staining should exclusively label mitochondria. In human skin fibroblasts, we have used a monomeric green fluorescent protein (GFP) (AcGFP1) targeted to the mitochondrial matrix by attaching an N-terminal cytochrome-*c* oxidase subunit 8 (cox8) leader sequence (mito-AcGFP1). Using this protein-based probe has the distinct advantage that its staining is independent of $\Delta\psi$, also allowing mitochondrial labeling and proper visualization under conditions that $\Delta\psi$ is (partially) depolarized. We have generated insect viruses (baculoviruses) for expression of mito-AcGFP1 in primary cells, yielding mitochondria-specific fluorescence signals (17). If a mitochondrial label with a red fluorescence is required (see Sect. 5), a mitochondria-targeted variant of a red fluorescent protein (mito-tagRFP) can be used (18). Alternatively, mitochondrial morphology in living cells can be determined using chemical cations, which accumulate in the mitochondrial matrix in a $\Delta\psi$-dependent manner (Sect. 3). These cations include rhodamine 123 (R123; Invitrogen), TMRE (tetramethyl rhodamine ethyl ester; Invitrogen), and TMRM (tetramethyl rhodamine methyl ester; Invitrogen). We used the latter molecule for simultaneous quantification of mitochondrial shape, position, and $\Delta\psi$ for individual mitochondria (17, 19). Importantly, TMRM concentrations need to be as low as possible (<100 nM) to prevent artifacts like cytosolic staining, fluorescence auto-quenching, phototoxic effects, and inhibition of mitochondrial respiration (17, 20).

1.3. Probes for Quantification of Mitochondrial Membrane Potential

Lipophilic cationic molecules, such as rhodamines, rhodamine derivatives and carbocyanines, can freely diffuse across membranes. Due to their positive charge, these molecules sequester in the negatively charged mitochondrial matrix according to the Nernst equation. Therefore, these cations are also called "Nernstian dyes" (19). However, toxicity and specificity of the different probes should be taken into consideration during visualization of mitochondrial function in living cells. R123 is a fluorescent marker for mitochondria, which sequesters in the mitochondrial matrix in a $\Delta\psi$-dependent manner at approximately 4,000:1 ratio relative to the cytoplasm (21–23). R123 undergoes a spectral red-shift of 11 nm and its fluorescence can be quenched at high concentrations in energetically coupled mitochondria (23). The latter phenomenon also can occur for TMRM and means that R123/TMRM fluorescence intensity increases, instead of decreases upon $\Delta\psi$ depolarization. Evidence has been provided that the extent of dequenching can be used as a sensitive measure of $\Delta\psi$ (24). The presence of probe dequenching can be demonstrated by adding (e.g., pipetting) of the protonophore carbonyl cyanide-*p*-trifluoromethoxyphenylhydrazone (FCCP; 1 μM) to R123/TMRM-stained cells. When FCCP induces an increase in mitochondrial R123/TMRM fluorescence, followed by a decrease, the probe is auto-quenching itself. When FCCP only induces a decrease in mitochondrial R123/TMRM fluorescence, the probe is present in non-quenching concentrations. Because the FCCP effect can occur rapidly, time-lapse experiments need to be carried out using a short image acquisition interval (~1 s). In our research, we have exclusively used TMRM in the absence of dequenching to obtain steady-state mitochondrial fluorescence signals that are positively correlated to $\Delta\psi$. By trial and error, we determined that a TMRM concentration of ≤25 nM was suited for this purpose (Sect. 3).

Both R123 and TMRM also redistribute across the plasma membrane according to the Nernst equation. R123 can bind proteins and redistributes slowly, whereas TMRM quickly redistributes across the membranes. This makes TMRM less toxic and ideal to monitor fast depolarization events that would be missed by R123. However, fast redistribution of TMRM is more dependent on the plasma membrane potential (ΔV). Ideally, TMRM recordings of $\Delta\psi$ should therefore be paralleled by ΔV measurements. For this purpose, an elegant method was presented based on the combined use of a plasma membrane potential indicator (PMPI) and TMRM in cerebellar granule neurons (25).

1.4. Microscopy Imaging of Mitochondria

Probe visualization can be carried out using an automated epifluorescence video microscope (Fig. 1). Our system consists of a monochromator for excitation of the fluorescent probes (TILL Photonics, Gräfelfing, Germany), a fully motorized inverted microscope (Carl Zeiss, Jena, Germany), an environmental control system (temperature, O_2, CO_2, medium superfusion) to maintain cell health (Zeiss), a collection of dichroic mirrors (Omega Optical Inc., Brattleboro, VT, USA), a motorized filter wheel (Sutter Instrument Company, Novato, CA, USA), containing multiple

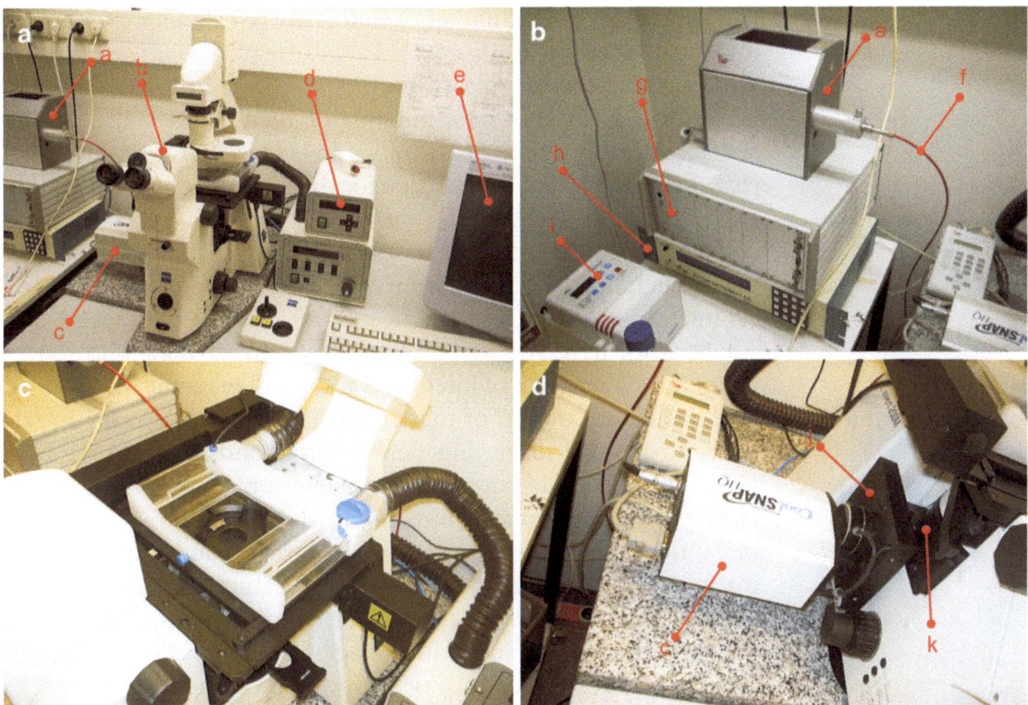

Fig. 1. Automated epi-fluorescence videomicroscope. (**a**) Overview of the system. (**b**) Magnification of the excitation hardware and filter-wheel control hardware. (**c**) Magnification of the environmental control system. (**d**) Magnification of the fluorescence detection hardware. *a*, monochromator; *b*, motorized inverted microscope; *c*, CCD-camera; *d*, environmental control system (*top* = temperature, *bottom* = CO_2); *e*, monitor part of the computer that controls the microscopy hardware; *f*, optic fiber to direct excitation light to the microscope; *g*, monochromator control hardware; *h*, filter-wheel control hardware; *i*, peristaltic pump for cell superfusion; *j*, emission filter-wheel; *k*, motorized dichroic mirror. A color version of this image can be found in the online version.

filters for selection of fluorescence emission wavelengths (Omega Optical Inc.), and a CCD-camera for image acquisition (Roper Scientific, Evry Cedex, France). This hardware is remotely controlled by MetaFluor® 6.1 software (Molecular Devices Corporation, Downingtown, PA, USA).

Our mitochondrial imaging strategy requires information about the subcellular localization of mitochondrial objects within the cell. In this approach mitochondrial position and morphology is assessed from a binary image (BIN), where mitochondria are represented as white objects on a black background. The BIN image is calculated from a microscopy image of cells stained with an appropriate mitochondria-specific probe. In order to obtain BIN images and/or images ready for quantification it is important to record high quality images using proper focusing, magnification, and a maximal signal-to-noise (SNR) ratio. Although settings for acquiring high quality images will have to be optimized for each individual experiment (choice of microscope, light source, and wavelength filters), a number of factors are relevant for experiments involving imaging.

First, the cells or mitochondria have to be in focus. This note of caution seems obvious, but out of focus light will introduce artifacts during image processing (26). It is therefore important that the right type of microscope for the correct cell type is chosen. Large flat cells like human skin fibroblasts can be imaged under conventional wide-field fluorescent microscopes, whereas cells with greater height in the *Z*-direction can best be imaged using a confocal microscope (27).

Second, it is important to use the correct magnification for imaging of the object of interest. A too small magnification will not give proper separation of mitochondria in cells, especially in relatively small cells displaying dense mitochondrial networks (e.g., HEK293 cells). This may impair generation of the mitochondrial BIN image. Additionally, mitochondrial imaging should be carried out using short illumination times (to prevent light-induced phototoxicity) and, if needed, at lower temperature (e.g., 20°C instead of 37°C) to prevent artifacts induced by mitochondrial movement during acquisition of individual images. To further minimize light-induced phototoxicity and/or probe photobleaching, the excitation intensity should be kept as low as possible. It is further recommended that the recorded fluorescent signals are substantially lower than the maximum gray value allowed by the CCD camera used.

Third, the fluorescent probe should have no toxic effect on the organelle and cellular level. Loading of a fluorescent probe or expression of a proteinaceous sensor should therefore be optimized to achieve the lowest possible concentration/expression without affecting/losing the signal of interest.

Fourth, in case of a chemical probe, leakage from the mitochondria/cells should be absent. This can be assessed by first measuring the fluorescence intensity histograms of the probe in a collection of cells imaged at a certain time point (T_0), followed by storing the cells in the incubator for a certain time (e.g., 15 min), and re-measuring the fluorescence intensity histogram (T_{15}). In the absence of probe leakage, the histograms for T_0 and T_{15} will be identical.

Finally, a high SNR is essential for quantitative analysis of mitochondrial function. In the ideal case, this means that mitochondrial staining intensity is: (1) far higher than nonspecific fluorescence signals (e.g., the cytosol) and (2) far higher than the intensity of the background. The noise in the images will increase when the speed of image recording increases. This can be minimized using image averaging (26). Taken together, this means that best quantitative results are obtained using large flat cells (e.g., fibroblasts), imaged at a relatively high magnification (×40, ×60 objective) and displaying a high mitochondria-specific fluorescence signal.

2. Protocol 1: Visualizing Mitochondria with MITO-ACGFP1

2.1. Materials

1. A baculovirus stock solution for cell transfection with cox8 mito-AcGFP1 (mito-AcGFP1).
2. Human skin fibroblasts of a low passage number cultured in medium 199 (Invitrogen, Breda, The Netherlands) + 10% fetal calf serum (Greiner Bio-One, Alphen a/d Rijn, The Netherlands) + 1% penicillin/streptomycin (Invitrogen) on a glass coverslip (ϕ24 mm, Menzel GmbH, Braunschweig, Germany) or disposable incubation chamber (Willco Wells BV, Amsterdam, The Netherlands).
3. A microscope allowing excitation with 470 nm light (monochromator) and containing a 505 nm dichroic mirror (Chroma Technology Corporation, Bellows Falls, VT, USA; XF2031) and 565 long pass emission filter (Chroma; XF3085).
4. HEPES/Tris (HT) buffer containing 10 mM HEPES, 132 mM NaCl, 4.2 mM KCl, 1.0 mM $MgCl_2$, 5.5 mM D-glucose, 1.0 mM $CaCl_2$ (pH adjusted to 7.4).

2.2. Methods

2.2.1. Cell Culture

1. Seed the cells at such densities that the cells are 60–70% confluent at the time of imaging. This allows subtraction of the background signal.
2. Keep the cells in an incubator close to the microscope system prior to imaging.

2.2.2. Baculoviral Transfection (17)

1. Baculoviral expression vectors were constructed using the Gateway® Cloning system (Invitrogen, Breda, The Netherlands). To obtain an AcGFP1-Baculo destination vector, the ScrfI/XbaI restriction fragment of pcDNA5/FRT/TOAcGFP destination vector was cloned into the ScrfI/XbaI site of the mammalian pFastBacDual destination vector (Invitrogen).
2. The mitochondrial targeting sequence of human cytochrome-*c* oxidase subunit 8A (first 93 basepairs of the sequence NM_004074) flanked by Gateway® AttB sites (Invitrogen) was created using PCR. The flanked sequence was cloned into pDONR201 entry vector using the Gateway® BP Clonase II Enzyme Mix (Invitrogen).
3. The mitochondrial targeting sequence was transferred to the AcGFP1-Baculo destination vector using the Gateway® LR Clonase II Enzyme Mix (Invitrogen) to obtain a mitochondria-targeted AcGFP1 (mito-AcGFP1). Baculoviruses were then produced as described in the Bac-to-Bac® Baculovirus Expression System manual (Invitrogen).

4. In order to express mito-AcGFP1 in human skin fibroblasts, seed the cells onto ϕ24 mm coverslips or disposable incubation chambers and cultured for 24 h.
5. Infect the cells with mito-AcGFP1 baculovirus (5–7% v/v) for 48 h prior to imaging.

2.2.3. Live-Cell Microscopy

1. Switch on the system at least half an hour in advance to allow heating up of the lamp. To prevent damage, never repeatedly switch on the excitation source within a short period of time.
2. Wash the cells with HT buffer to remove excess culture medium.
3. Cover the cells by HT buffer and mount the cells on the microscope.

3. Protocol 2: Visualizing Mitochondria with TMRM

3.1. Materials

1. Human skin fibroblasts of a low passage number cultured in medium 199 (Invitrogen) + 10% fetal calf serum (Greiner Bio-One) + 1% penicillin streptomycin (Invitrogen) on a glass coverslip (ϕ24 mm, Menzel GmbH) or disposable incubation chamber (Willco Wells BV, Amsterdam, The Netherlands).
2. A TMRM super stock solution (concentration + solvent). Dissolve 1 mg of TMRM powder (Invitrogen) in 3.99 mL DMSO to reach a concentration of 0.5 mM. Make aliquots of 5 μL in brown Eppendorf tubes. Storage life at −20°C is 1 year. Note! Always keep TMRM powder and solution protected from light.
3. A TMRM working solution. Thaw an aliquot of TMRM superstock solution and add 495 μL of imaging buffer to make the TMRM working solution 5 μM. Vortex and keep on ice during the time of experiment.
4. A video microscope system allowing excitation with 540 nm light (monochromator) and containing a 560 nm dichroic mirror (Chroma; XF2017) and 565 long pass emission filter (Chroma; XF3085). The microscope should be equipped with an environmental control system.
5. HEPES/Tris (HT) buffer containing 10 mM HEPES, 132 mM NaCl, 4.2 mM KCl, 1.0 mM $MgCl_2$, 5.5 mM D-glucose, 1.0 mM $CaCl_2$ (pH adjusted to 7.4).

3.2. Methods

3.2.1. Cell Culture

1. Cells should be seeded 3 days before the measurement in different densities. Ideally, the density of control and patient or treated cells should be the same.

2. Mitochondrial membrane potential measurements are very sensitive to environmental changes (e.g., temperature, pH). Culture the cells in separate dishes to minimize pH differences.
3. Transfer the cells to an incubator close to the microscope system at least 1 h prior to imaging.
4. Load the cells with TMRM in a dish-by-dish manner.

3.2.2. TMRM Staining (in the Dark)

1. Pipette 5 μL (final concentration of 25 nM) of TMRM working solution in an Eppendorf tube. Lower the concentration if the signal is too intense.
2. Take 1 mL of medium from the culture dish, add it to TMRM and vortex for 5–10 s.
3. Replace the rest of the medium from the dish by the medium/TMRM solution.
4. Incubate the cells in a 37°C, 5% CO_2 incubator for 25 min.
5. Wash the coverslip/culture dish three times with 1 mL phosphate-buffered saline (PBS).
6. Replace the PBS by imaging buffer and mount the coverslip/culture dish on the microscope.
7. Start loading of the next coverslip/culture dish immediately, because imaging of TMRM signals can only occur for limited amount of time.

3.2.3. Live-Cell Microscopy

1. Select an area of interest where the mitochondrial network can be seen in focus. Optionally confocal stack images should be acquired in thick cells.
2. Set microscope settings to first record an image for mitochondrial mask recording (such as mito-AcGFP), followed by an image for TMRM. Short illumination times (100 ms) should be applied, because light induction of TMRM can cause cellular photo-damage. TMRM labeling in human skin fibroblasts has previously shown that short acquisition times of TMRM at low- and auto-quenching concentrations (described below) did not alter the fluorescence intensity histogram.
3. After acquiring 10–20 individual cells, collect a background image either by finding an empty field or scraping of cells in the field of view. This is to prevent environmental changes affecting TMRM signal quantification.
4. For time-lapse recordings, carry out additional control experiments for photo-bleaching. In our experience, TMRM showed no leakage in these cells for up to 15 min and extended illumination of TMRM did not result in photo-bleaching during 10 min recordings (17).

3.2.4. Determining Auto-Quenching

1. Prepare a 2 μM FCCP work solution in imaging buffer.
2. After TMRM incubation and washing, add 0.5 mL imaging buffer to the coverslip/culture dish. Mount the coverslip/culture dish on the microscope as previous.
3. Locate an appropriate field of view and set up a time-lapse recording with an interval of 2 s. Record ten iterations and subsequently add 0.5 mL of 2 μM FCCP to the cells (final concentration 1 μM). Optionally perfusion can be used to administer FCCP, although different tubing should be used, since FCCP will bind most tubing materials.
4. Monitor the mean fluorescent intensity of the cell for an increase (auto-quenching present) or decrease (absence of auto-quenching) of TMRM fluorescent intensity.

4. Protocol 3: Quantifying Mitochondrial Position, Shape, and ΔΨ by Image Analysis

4.1. Materials

1. Image analysis software like Image Pro Plus (Media Cybernetics Inc., Bethesda, MD, USA), Metamorph® (Molecular Devices Corporation, Palo Alto, CA, USA), or an open-source alternative like ImageJ/FIJI (http://rsbweb.nih.gov/ij/ or http://pacific.mpi-cbg.de/wiki/index.php/Fiji).

4.2. Methods

1. The first step in image quantification is correction of the image for the background signal. When the background is homogeneous, background subtraction can be carried out by measuring the mean fluorescent intensity selecting an extracellular region of interest (ROI) or by subtraction of an "empty" image without cells. The resulting background corrected (COR) image (Fig. 2a) is then converted to 8-bit (256 gray values).
2. Next, the contrast in the COR image is optimized using a linear contrast stretch (LCS) operation to cover the full range of gray values.
3. To highlight mitochondrial particles from the image, a top-hat spatial filter is applied. It is crucial to choose the proper size of the top-hat filter because its performance depends on the used objective and size of the mitochondrial objects. In case of TMRM-stained fibroblasts, imaged using a ×40 objective, a top-hat filter with a size of 7×7 pixels has to be applied three times to achieve optimal results (28).
4. As an undesirable side effect, top-hat filtering can introduce artifacts into the image due to amplification of nonmitochondrial (noise) pixels. In TMRM-stained fibroblasts these artifacts can be removed by applying a 3×3 median (MED) filter,

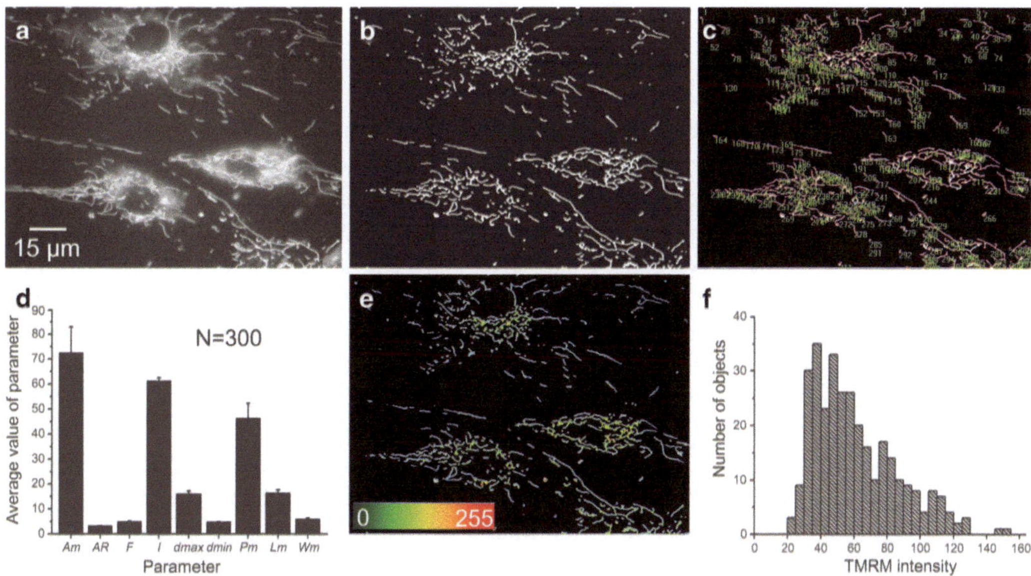

Fig. 2. Automated quantification of mitochondrial morphology and membrane potential. (**a**) Typical example of primary human skin fibroblasts (#5120), stained with the $\Delta\psi$-sensitive cation TMRM. This image (COR) was acquired with the video microscopy system depicted in Fig. 1 and background corrected. (**b**) Binary (BIN) image calculated from the image in (**a**), representing white mitochondria on a *black* background. (**c**) Automated analysis of mitochondrial position, number and morphology of the image in (**b**) using image analysis software. Each identified mitochondrial object is surrounded by a *red line* and marked by a number. In total, 300 objects were present in this image, meaning that each cell contained ~100 mitochondrial objects. (**d**) Average value (+standard error) of key descriptive parameters introduced previously (26) for the image in (**c**). (**e**) Color-coded image obtained by masking the COR image in (**a**) with the BIN image in (**b**). Given the membrane potential sensitivity of TMRM staining, this image can be used to estimate $\Delta\psi$. (**f**) Histogram of TMRM fluorescence intensity for the objects in (**e**). $\Delta\psi$ mitochondrial membrane potential; *Am* area of mitochondrial object (pixels); *AR* aspect ratio of mitochondrial object (arbitrary units); *F* form factor of mitochondrial object (arbitrary units); *I* TMRM fluorescence intensity of mitochondrial object (*gray value*); *dmax* maximal diameter of mitochondrial object (pixels); *dmin* minimal diameter of mitochondrial object (pixels); *Pm* perimeter of mitochondrial object (pixels); *Lm* length of mitochondrial object (pixels); *TMRM* tetramethyl rhodamine methylester; *Wm* width of mitochondrial object (pixels).

followed by application of a threshold operation, without degradation of mitochondrial pixels. The resulting image is of a binary (BIN) nature, in which mitochondria are represented as white objects on a black background (Fig. 2b).

5. This image can be used to directly asses the number and position of mitochondria (Fig. 2c). Moreover, mitochondrial shape descriptors can be quantified at the level of individual organelles, cells, or images (Fig. 2d) as described previously (26).

6. By combining the BIN image with the COR image using Boolean logic ("AND"), a masked (MSK) image is obtained. The MSK image includes only mitochondrial objects, each with its individual TMRM fluorescence intensity, on a black background (Fig. 2e). This allows quantification of TMRM intensity for each mitochondrial object as reflected by the histogram in Fig. 2f.

5. Simultaneous Quantification of Multiple Mitochondrial and Cellular Readouts

Above, we describe our strategy that combines specific mitochondrial staining and subsequent image processing to allow automated quantification of mitochondria-specific fluorescence signals. In this way, information about mitochondrial number, position, shape, and $\Delta\psi$ can be extracted from the same image. However, because mitochondrial staining with cations is $\Delta\psi$-dependent, mitochondrial depolarization can affect image quantification. Although modest $\Delta\psi$ depolarization does not affect our analysis, a more extensive $\Delta\psi$ loss will prevent proper calculation of the BIN and thereby the MSK image. However, mitochondrial labeling with mito-AcGFP1 can be combined with simultaneous TMRM staining allowing calculation of a BIN image from the mito-AcGFP1 image. The latter image can then be used to calculate a MSK-version of the TMRM image (19). This strategy can also be applied with other chemical probes that are spectrally compatible with mito-AcGFP1. Alternatively, mito-tagRFP can be used in case of green-fluorescent chemical probes. Table 1 provides an overview of how mito-AcGFP1 or mito-tagRFP can be simultaneously used with chemical probes for $\Delta\psi$, Ca^{2+}, ROS, pH, NADH, endoplasmatic reticulum localization and lysosome localization.

In addition to the chemical probes, a large variety of protein-based sensors are available. These sensors are often based on the FRET (fluorescence resonance energy transfer) principle, in which a ligand binding protein is sandwiched between two different fluorescent proteins. The first protein (the fluorescence donor) is excited by light of a specific wavelength and emits fluorescence light of a longer wavelength. If the donor protein comes within close vicinity of the second protein (the fluorescence acceptor), and its absorbance overlaps the emission spectrum of the donor, FRET occurs which induces fluorescence of the acceptor protein. In this way, FRET induces a decrease in donor fluorescence emission and an increase in acceptor fluorescence emission. The FRET sensor is engineered in such a way that the distance between the donor and acceptor protein depends on the conformation of the ligand binding protein. Depending on the structure of the FRET-based sensor, ligand binding will either increase or decrease the FRET efficiency. Cyan fluorescent protein (CFP) and yellow fluorescent protein (YFP) are often used as a donor and acceptor, respectively. A circular permuted fluorescent protein (cpFP) can be used as an alternative. The fluorescence signal of cpFPs is very sensitive to conformational changes in the attached ligand binding protein. Examples of protein-based probes are the ratiometric Ca^{2+} sensors Pericam (33) and Cameleon (34), based on a calmodulin-based Ca^{2+} binding protein combined with either cpYFP or the FRET pair CFP/YFP. Other sensors include the ratiometric H_2O_2

Table 1
Simultaneous quantification of mitochondrial and cellular parameters using chemical fluorescent probes and mitochondria targeted mito-AcGFP1 or mito-tagRFP fluorescent proteins

		Wavelengths (in vitro)		Mask generation (Em/Ex)		
Parameter	**Probe**	**Excitation**	**Emission**	**mito-AcGFP1 (475/505)**	**mito-tagRFP (555/584)**	**References**
$\Delta\psi$	TMRM (mitochondria)	540	590	x		Invitrogen (T668), (17)
Calcium (Ca^{2+})	Fura-2 (cytosol)	340 + 380	505		x	Invitrogen (F1221), (29)
	Indo-1 (cytosol)	350	405 + 485		x	Invitrogen (I1223), (30)
	Rhod-2 (mitochondria)	540	590	x		Invitrogen (R1245MP), (31)
ROS	HEt (cytosol)/Mito-HEt (MitoSOX) (mitochondria)	396 + 510 505	580 535	x	x	Invitrogen (D11347/ M36008), (32)
	DCFDA (cytosol)					Invitrogen (C6827), (13)
pH	BCECF (cytosol)	503	528	x	x	Invitrogen (B1150)
	SNARF®-1 (cytosol)	575	640			Invitrogen (C1272)
NADH	Autofluorescence (mitochondria)	360	480	x	x	(31)
Endoplasmatic reticulum	ER-Tracker® Red (ER)	587	615	x	x	Invitrogen (E34250)
	ER-Tracker® Green (ER)	504	511			Invitrogen (E34251)
Lysosomes	LysoTracker® Red (lysosome)	577	592			Invitrogen (L7528)
	LysoTracker® Green (lysosome)	504	511	x	x	Invitrogen (L7526)

One or more probes are given for each parameter with cellular localization between parentheses and the predicted combination of the probe with either mitochondria targeted mito-AcGFP1 or mito-tagRFP for mitochondrial mask generation. The maximum excitation and emission wavelengths of commonly used chemical fluorescent probes are given under in vitro conditions. For optimized spectral separation, intracellular in vivo spectra should be recorded (see Sect. 5)

Table 2
Simultaneous quantification of mitochondrial and cellular parameters using proteinaceous fluorescent sensors and mito-targeted mito-AcGFP1 or mito-tagRFP fluorescent proteins

Parameter	Probe	Wavelengths (in vitro)		Mask generation (Em/Ex)		References
		Excitation	Emission	mito-AcGFP1 (475/505)	mito-tagRFP (555/584)	
Calcium (Ca^{2+})	Ratiometric-pericam (cytosol)	415 + 494	517		x	(33)
	Ratiometric-pericam-mt (mitochondria)				x	
	Cameleon-1 BFP-GFP (cytosol)	381	445 + 510		x	(34)
	Cameleon-2 CFP-YFP (cytosol)	433	472 + 527		x	
ROS	HyPer (cytosol)	420 + 500	516		x	(35)
	MitoHyPer (mitochondria)				x	
pH	SypHer (cytosol)	420 + 490	535/25[a]		x	(36)
	MitoSypHer (mitochondria)				x	
ATP	ATeam (cytosol)	435	475 + 527		x	(37)
	Mito-ATeam (mitochondria)					

For each cellular and mito- parameter, the proteinaceous sensors and its cellular localization is given between parentheses. Following the in vitro maximum excitation and emission wavelengths, these sensors could be combined with mito-tagRFP for mitochondrial mask generation. In case of FRET-based sensors, both channels combined could function as mitochondrial mask image. For optimized illumination, intracellular in vivo spectra should be recorded (see Sect. 5)

[a]The maximum emission wavelength of SypHer is not yet known, although probably similar to HyPer. Instead, the emission filter set used in reference (36) is given

sensor HyPer, based on the bacterial protein OxyR, and its pH sensing synthetic mutant SypHer (35, 36). The free (ATP) can be measured by the ATP sensor ATeam. This sensor is based on the bacterial ε subunit derived from the F_0F_1-ATP synthase of *Bacillus subtilis* and *Bacillus* sp. PS3 (37). In principle, all of these reporters should be able to be combined with mito-tagRFP for mitochondrial localization (Table 2). However, the optimal excitation and emission wavelengths given for both the chemical and protein-based probes are generally determined under in vitro conditions. The latter might not realistically mimic the cellular environment. Therefore it is recommended to record intracellular in vivo spectra of each fluorescent reporter to allow proper selection of excitation and emission filters for microscopy, in order to avoid fluorescent bleed-through of either wavelengths. This can for instance be performed using a spectrophotofluorometer (e.g., Shimadzu RF-5301PC, Shimadzu Scientific Instruments, Tokyo, Japan).

Two aspects have to be considered when one intends to combine the use of two different proteinaceous fluorescent probes. First, both probes have to be introduced into the cell. This can be achieved either by transfection of the cells with two plasmids or by a vector encoding both proteins. Correct protein synthesis and protein folding has to be examined, for example using native gel electrophoresis and in-gel fluorescence analysis (27). Second, certain ratiometric protein-based sensors are pH-sensitive. For instance, in the case of the ATP:ADP ratio sensing reporter Perceval, pH correction was carried out by paralleled measurements of with the pH indicator SNARF-5F (38). In case of ATeam, its YFP/CFP emission ratio was virtually pH-independent under physiological pH ranges (37). The individual wavelengths of ratiometric sensors can be assessed as internal control. Synchronization of both wavelengths (i.e., increase or decrease in fluorescence signal) will indicate an environmental or morphological change, whereas asynchronous behavior of both wavelengths is a strong indicator of ligand sensing. Currently, so-called "dead" sensors that are identical to the normal sensors but unable to bind the ligand are considered the optimal controls for sensor specificity.

Acknowledgments

This work was supported by an equipment grant of NWO (Netherlands Organization for Scientific Research, No: 911-02-008), the Dutch Ministry of Economic Affairs (Innovative Onderzoeks Projecten (IOP) Grant: #IGE05003), and by the CSBR (Centres for Systems Biology Research) initiative from NWO (No: CSBR09/013V).

References

1. Koopman WJ, Nijtmans LG, Dieteren CE, Roestenberg P, Valsecchi F, Smeitink JA, Willems PH (2010) Mammalian mitochondrial complex I: biogenesis, regulation, and reactive oxygen species generation. Antioxid Redox Signal 12:1431–1470
2. Benard G, Rossignol R (2008) Ultrastructure of the mitochondrion and its bearing on function and bioenergetics. Antioxid Redox Signal 10:1313–1342
3. Smeitink J, van den Heuvel L, DiMauro S (2001) The genetics and pathology of oxidative phosphorylation. Nat Rev Genet 2:342–352
4. del Hoyo P, Garcia-Redondo A, de Bustos F, Molina JA, Sayed Y, Alonso-Navarro H, Caballero L, Arenas J, Agundez JA, Jimenez-Jimenez FJ (2010) Oxidative stress in skin fibroblasts cultures from patients with Parkinson's disease. BMC Neurol 10:95
5. Campbell GR, Ziabreva I, Reeve AK, Krishnan KJ, Reynolds R, Howell O, Lassmann H, Turnbull DM, Mahad DJ (2010) Mitochondrial DNA deletions and neurodegeneration in multiple sclerosis. Ann Neurol 69(3):481–92.
6. de Moura MB, dos Santos LS, Van Houten B (2010) Mitochondrial dysfunction in neurodegenerative diseases and cancer. Environ Mol Mutagen 51:391–405
7. Maximo V, Lima J, Soares P, Sobrinho-Simoes M (2009) Mitochondria and cancer. Virchows Arch 454:481–495
8. McFarland R, Taylor RW, Turnbull DM (2010) A neurological perspective on mitochondrial disease. Lancet Neurol 9:829–840
9. Moncada S (2010) Mitochondria as pharmacological targets. Br J Pharmacol 160:217–219
10. Wallace DC, Fan W, Procaccio V (2010) Mitochondrial energetics and therapeutics. Annu Rev Pathol 5:297–348
11. Zhang E, Zhang C, Su Y, Cheng T, Shi C (2010) Newly developed strategies for multifunctional mitochondria-targeted agents in cancer therapy. Drug Discov Today 16:140–146
12. Distelmaier F, Visch H, Smeitink J, Mayatepek E, Koopman W, Willems P (2009) The antioxidant Trolox restores mitochondrial membrane potential and Ca^{2+}-stimulated ATP production in human complex I deficiency. J Mol Med 87:515–522
13. Koopman WJ, Verkaart S, van Emst-de Vries SE, Grefte S, Smeitink JA, Willems PH (2006) Simultaneous quantification of oxidative stress and cell spreading using 5-(and-6)-chloromethyl-2′,7′-dichlorofluorescein. Cytometry A 69:1184–1192
14. Condreay JP, Kost TA (2007) Baculovirus expression vectors for insect and mammalian cells. Curr Drug Targets 8:1126–1131
15. Ahmed BY, Chakravarthy S, Eggers R, Hermens WT, Zhang JY, Niclou SP, Levelt C, Sablitzky F, Anderson PN, Lieberman AR, Verhaagen J (2004) Efficient delivery of Cre-recombinase to neurons in vivo and stable transduction of neurons using adeno-associated and lentiviral vectors. BMC Neurosci 5:4
16. Fleming J, Ginn SL, Weinberger RP, Trahair TN, Smythe JA, Alexander IE (2001) Adeno-associated virus and lentivirus vectors mediate efficient and sustained transduction of cultured mouse and human dorsal root ganglia sensory neurons. Hum Gene Ther 12:77–86
17. Distelmaier F, Koopman WJ, Testa ER, de Jong AS, Swarts HG, Mayatepek E, Smeitink JA, Willems PH (2008) Life cell quantification of mitochondrial membrane potential at the single organelle level. Cytometry A 73:129–138
18. Kuznetsov AV, Hermann M, Saks V, Hengster P, Margreiter R (2009) The cell-type specificity of mitochondrial dynamics. Int J Biochem Cell Biol 41:1928–1939
19. Koopman WJ, Distelmaier F, Esseling JJ, Smeitink JA, Willems PH (2008) Computer-assisted live cell analysis of mitochondrial membrane potential, morphology and calcium handling. Methods 46:304–311
20. Scaduto RC Jr, Grotyohann LW (1999) Measurement of mitochondrial membrane potential using fluorescent rhodamine derivatives. Biophys J 76:469–477
21. Johnson LV, Walsh ML, Chen LB (1980) Localization of mitochondria in living cells with rhodamine 123. Proc Natl Acad Sci USA 77:990–994
22. Johnson LV, Walsh ML, Bockus BJ, Chen LB (1981) Monitoring of relative mitochondrial membrane potential in living cells by fluorescence microscopy. J Cell Biol 88:526–535
23. Emaus RK, Grunwald R, Lemasters JJ (1986) Rhodamine 123 as a probe of transmembrane potential in isolated rat-liver mitochondria: spectral and metabolic properties. Biochim Biophys Acta 850:436–448
24. Voronina SG, Barrow SL, Gerasimenko OV, Petersen OH, Tepikin AV (2004) Effects of secretagogues and bile acids on mitochondrial membrane potential of pancreatic acinar cells: comparison of different modes of evaluating DeltaPsim. J Biol Chem 279:27327–27338
25. Nicholls DG (2006) Simultaneous monitoring of ionophore- and inhibitor-mediated plasma

and mitochondrial membrane potential changes in cultured neurons. J Biol Chem 281:14864–14874

26. Koopman WJ, Visch HJ, Smeitink JA, Willems PH (2005) Simultaneous quantitative measurement and automated analysis of mitochondrial morphology, mass, potential, and motility in living human skin fibroblasts. Cytometry A 69:1–12
27. Dieteren CE, Willems PH, Vogel RO, Swarts HG, Fransen J, Roepman R, Crienen G, Smeitink JA, Nijtmans LG, Koopman WJ (2008) Subunits of mitochondrial complex I exist as part of matrix- and membrane-associated subcomplexes in living cells. J Biol Chem 283:34753–34761
28. Willems PH, Smeitink JA, Koopman WJ (2009) Mitochondrial dynamics in human NADH: ubiquinone oxidoreductase deficiency. Int J Biochem Cell Biol 41:1773–1782
29. Visch HJ, Rutter GA, Koopman WJ, Koenderink JB, Verkaart S, de Groot T, Varadi A, Mitchell KJ, van den Heuvel LP, Smeitink JA, Willems PH (2004) Inhibition of mitochondrial Na^+-Ca^{2+} exchange restores agonist-induced ATP production and Ca^{2+} handling in human complex I deficiency. J Biol Chem 279:40328–40336
30. Paredes RM, Etzler JC, Watts LT, Zheng W, Lechleiter JD (2008) Chemical calcium indicators. Methods 46:143–151
31. Visch HJ, Koopman WJ, Zeegers D, van Emst-de Vries SE, van Kuppeveld FJ, van den Heuvel LW, Smeitink JA, Willems PH (2006) Ca^{2+}-mobilizing agonists increase mitochondrial ATP production to accelerate cytosolic Ca^{2+} removal: aberrations in human complex I deficiency. Am J Physiol 291:C308–C316
32. Forkink M, Smeitink JAM, Brock R, Willems PHGM, Koopman WJH (2010) Detection and manipulation of mitochondrial reactive oxygen species in mammalian cells. Biochim Biophys Acta Bioenergetics 1797:1034–1044
33. Nagai T, Sawano A, Park ES, Miyawaki A (2001) Circularly permuted green fluorescent proteins engineered to sense Ca^{2+}. Proc Natl Acad Sci USA 98:3197–3202
34. Miyawaki A, Llopis J, Heim R, McCaffery JM, Adams JA, Ikura M, Tsien RY (1997) Fluorescent indicators for Ca^{2+} based on green fluorescent proteins and calmodulin. Nature 388:882–887
35. Belousov VV, Fradkov AF, Lukyanov KA, Staroverov DB, Shakhbazov KS, Terskikh AV, Lukyanov S (2006) Genetically encoded fluorescent indicator for intracellular hydrogen peroxide. Nat Methods 3:281–286
36. Poburko D, Santo-Domingo J, Demaurex N (2011) Dynamic regulation of the mitochondrial proton gradient during cytosolic calcium elevations. J Biol Chem 286(13):11672–11684
37. Imamura H, Nhat KP, Togawa H, Saito K, Iino R, Kato-Yamada Y, Nagai T, Noji H (2009) Visualization of ATP levels inside single living cells with fluorescence resonance energy transfer-based genetically encoded indicators. Proc Natl Acad Sci USA 106:15651–15656
38. Berg J, Hung YP, Yellen G (2009) A genetically encoded fluorescent reporter of ATP:ADP ratio. Nat Methods 6:161–166

Chapter 7

Functional Imaging Using Two-Photon Microscopy in Living Tissue

Ivo Vanzetta, Thomas Deneux, Attila Kaszás, Gergely Katona, and Balazs Rozsa

Abstract

Over the last 20 years, neuroscientists have become increasingly interested in two-photon microscopy. One of the reasons for this interest is that two-photon fluorescence excitation allows counterbalancing the deterioration of the optical signals due to light scattering, and this opens the door for high-resolution imaging at considerable depth in living tissue. Due to progress in fluorescent marking techniques, to date, two-photon microscopy allows the functional exploration of neuronal activity at multiple scales, from the subprocesses of a single cell (dendrites, single spines, etc.), through single cells or small networks of a few neurons, up to large neuronal populations in the order of a cortical column. Here, we provide some information on the practical aspects of two-photon microscopy applied to imaging neurons in living tissue. We first discuss the advantages, shortcomings, and possible developments of the technique. We provide some practical considerations on the choice of the microscope itself, as well as on its principal elements. Because of recent progress in tackling high-speed imaging of 3D objects, we devote particular attention to the discussion of *z*-axis scanning techniques. Next, we illustrate some common applications, such as calcium imaging of neuronal activity, in vitro and in vivo. We also briefly illustrate how two-photon microscopy can be used for the imaging of erythrocyte flow in individual capillaries. Some practical considerations on specific protocols are provided in the form of self-consistent text boxes.

Key words: Two-photon, Laser scanning, In vivo, Fluorescence microscopy, Functional imaging, Calcium imaging, 3D scanning, Erythrocyte movement

1. Introduction

Fluorescence microscopy has proven an essential tool for the examination of biological specimens, both fixed or alive (1). One of the reasons for its great success is that fluorescent objects can be selectively visualized at good signal-to-background ratio even at small fluorophore concentrations.

Emilio Badoer (ed.), *Visualization Techniques: From Immunohistochemistry to Magnetic Resonance Imaging*, Neuromethods, vol. 70, DOI 10.1007/978-1-61779-897-9_7, © Springer Science+Business Media, LLC 2012

When applied to thick (over ~100 μm) samples, conventional fluorescent microscopy suffers under the shortcoming that the fluorophore is excited in a comparatively large volume, giving rise to a large out-of-focus component in the fluorescence reaching the detector. Confocal microscopy (2, 3) deals with this problem by using a pinhole to reject the photons originating from all locations other than from one point in the focal plane, providing an optical section of the sample. The focal plane is then scanned in two dimensions (*x*, *y*). The same procedure is repeated, serially, for each desired depth (*z*-dimension) until the relevant part of the biological specimen is covered. Subsequently, the imaged volume is visualized using three-dimensional reconstruction. It is important to realize that, although in confocal microscopy the out-of-focus photons do not contribute to the image, the corresponding molecules are still excited, leading to a high imaging cost in terms of photodamage and photobleaching. A partial solution to this problem consists in recording through many pinholes in parallel (e.g., using Nipkow disk-based approaches). In addition to collecting, for a given excitation volume, photons emitted from more than one location, the same sample locations are visited multiple times, leading to a higher signal rate. This allows the lowering of laser powers used and finally leads to less photodamage and bleaching.

Another shortcoming, however, remains: detection through the pinhole(s) blocks not only the fluorescence originating from out-of-focus excitation, but also the photons that were indeed emitted from the in-focus-point, yet underwent scattering on their way out of the tissue. This loss of "good" photons further reduces the yield of the instrument. All these problems are exacerbated by the strong light scattering of biological tissue, which makes high-resolution depth-imaging impossible with conventional and confocal microscopy.

The application of two-photon excitation to biological microscopy can thus be considered as a true breakthrough in fluorescent microscopy depth-imaging. Several excellent reviews of two-photon functional imaging, and calcium imaging in particular, can be found in the literature. They either focus on its technical aspects (4–8), on its applications (9–12), or on its place in the wider context of the various recent technological developments, which provide tools for a "reverse engineering" on the brain (13). In this introduction, we therefore sketch only briefly the basic principles and advantages of two-photon microscopy over more conventional approaches.

Two-photon excitation occurs when two low-energy photons together excite a fluorescent molecule, which then decays back to its fundamental state by emission of a photon of somewhat lower energy than the sum of the two exciting ones. The theoretical grounds of two-photon absorption were identified as early as 1931 by Maria Göppert-Mayer (14). However, experimental confirmation had to wait for progress in laser technology; in the 1960s that progress made ultrashort pulsed lasers available, which finally allowed the design of two-photon microscopes (15).

Why has two-photon microscopy become so popular over the last 10 years or so?

First, living tissue scatters light less in the infrared than at visible wavelengths, at which most common fluorophores are typically excited in single-photon mode. It thus became possible to optically examine fluorescent objects within living specimen up to depths that were previously inaccessible (5, 16). These reach from typical depths of several hundreds of microns up to 1 mm (17), although in the latter case the power of the excitation pulses has to be boosted via regenerative amplification (18).

Second, two photons must be absorbed simultaneously for excitation to occur. Therefore, the probability that a fluorescent molecule is excited depends quadratically on light intensity, rather than linearly as in the case of single-photon fluorescence. In simple terms, significant excitation occurs only where the local concentration of photons is very high, which happens essentially only at the focal point of the microscope, even in the presence of scattering. Photobleaching and photodamage to the tissue are thus greatly reduced. Most important, no fluorescence is emitted from out-of-focus locations resulting in "automatic" optical sectioning, without any need for out-of-focus rejection strategies like that implemented in confocal microscopy by the usage of one or more pinholes.

Finally, in two-photon microscopy all emitted fluorescence photons are useful—even the ones that were scattered on their way out of the tissue. Collecting them allows a dramatic increase in signal-to-noise ratio (SNR) with respect to situations where the scattered photons are rejected, as in confocal microscopy. Note that multiple-scattered photons are emitted from a seemingly large field of view (an area of ~1.5 times the recording depth, see: (19)) requiring a collection optics with a sufficiently large acceptance angle.

For the above reasons, two-photon excitation can selectively excite (or photochemically activate or inactivate) microscopic parts of a biological sample, at depths up to 1 mm and only at the in-focus plane, avoiding the photodamage and photobleaching that would result from a comparable excitation with confocal microscopy.

Overall, two-photon microscopy is thus a unique tool for imaging biological samples—in vivo in particular—in depth, as well as for local photochemistry. However, its performance in imaging transparent or very thin specimen is worse than that of confocal or even wide field fluorescence microscopy, because the longer wavelengths used for imaging reduce the maximal achievable spatial resolution. As often in biological research, the choice of the optimal imaging technique shall thus depend on the particular scientific question that needs to be addressed.

2. Materials

2.1. Microscopes: General

(a) Build or buy?
For high-end techniques such as multiphoton fluorescence, the instrument is obviously of fundamental importance. Two-photon microscopes have become commercially available from most major producers, as well as from a few smaller companies, providing the user with turn-key solutions implemented in ready-to-go setups. "In-house" solutions sometimes constitute a valuable alternative to purchasing an all-finished instrument, both because of economical considerations and because this allows for more flexibility in the mechanical design, which is particularly important for in vivo applications. If it is possible to convert confocal microscopes into two-photon ones (15), it is somewhat cumbersome. Therefore, it is often preferable to build a two-photon microscope essentially from scratch. Detailed descriptions of the needed components and how to proceed to assemble them can be found in excellent publications (e.g., (6, 7, 20, 21)). Importantly, software has also become available that allows to reliably control laser scanning (22, 23). Summarizing, building one's own instrument has the important advantage that it allows to customize the microscope to one's specific needs. However, a good degree of experience and technical know-how should be available in-house and/or experienced, qualified assistance is highly commendable (but see also (24) for a user-friendly adaptation of a low-cost commercial microscope to two-photon imaging). Figure 1 shows a typical setup for a two-photon microscope.

(b) Lasers and objectives
Excitation light must be provided by a femtosecond pulsed laser source, which nowadays are made commercially available by several producers. Critical parameters are wavelength, pulse width and power, which depend to some extent on the specific application. Available since the 1980s, Ti:sapphire lasers appear to comply with the needs posed by most applications, warranting a more than reasonable compromise between the different requirements. Commercial Ti:Sapphire lasers are wavelength-tunable over a large range (~690–1,040 nm), allowing to match the excitation wavelength of a wide range of fluorescent molecules. However, for wavelengths above 1,000 nm, as needed for the excitation of red fluorescent proteins, Ti:sapphire lasers perform poorly, making other light sources preferable, such as Ti:Sapphire laser pumped optical parametric oscillators (OPOs, ~1,000–1,600 nm), mode-locked Ytterbium-doped lasers (25, 26) (1,030–1,060 nm, nontunable) or Cr:Forsterite lasers (27) (tunable, 1,200–1,300 nm).

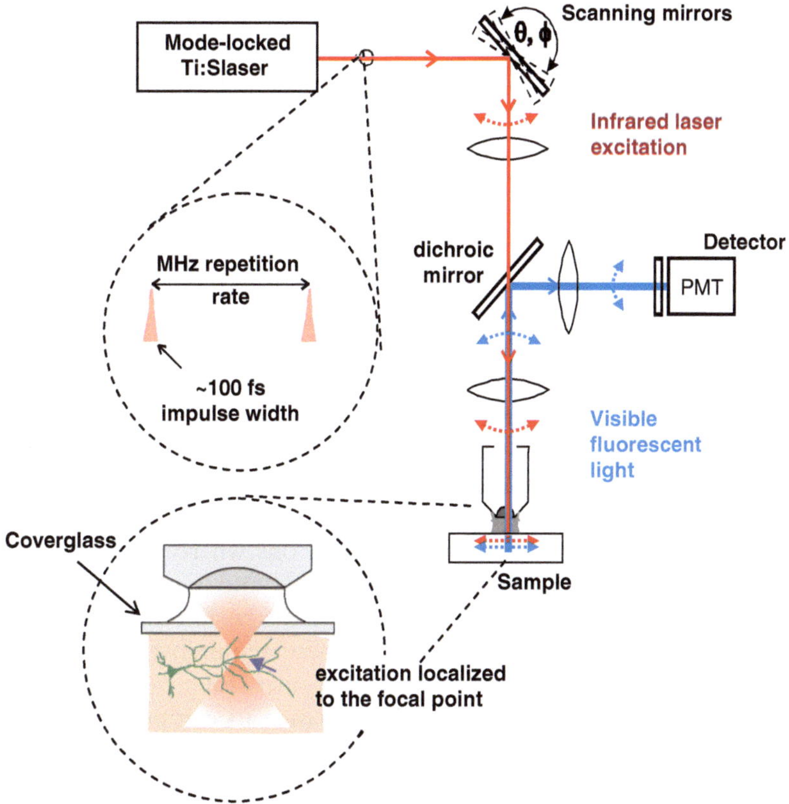

Fig. 1. A typical setup for a two-photon microscope. A MHz femtosecond infrared pulsed laser beam is sent onto a set of galvanometric scanning mirrors, providing lateral deflection of the excitation beam, which is thus focused onto different *xy* locations in the sample preparation by a high numerical aperture objective, often of the water-immersion type. Fluorescence, in the visible domain, is then collected by the same objective and sent toward the detector, usually a PMT through a dichroic mirror, which eliminates backscattered excitation light. The *z*-position (depth of focus) has to be adjusted independently, either manually, or using piezoelectric actuators (see text). Recently, it has been shown that scanning in all three (*x*, *y* and *z*) dimensions can be performed without mechanical movement, using opto-acoustic deflectors (44).

Ti:Sapphire lasers are pumped with a laser beam from an argon or frequency-doubled Nd:YVO4 laser and yield outputs of 2 W average power or more in the central part of their wavelength range. This power is usually enough, except for ultradeep imaging (500–1,000 μm), where the laser pulses have to be powered-up by regenerative amplification, at the expenses of pulse frequency (typically by a factor of 1,000 (17, 18)). Note however, that, ultimately, a depth limit is imposed on imaging by the fluorescence generated at the surface of the sample by out-of-focus light, with a consequently worse localized excitation and a deterioration of contrast. In gray matter, this limit appears to be in the order of 1,000 μm (17).

Pulse rates have to be such as to balance the fluorophore's excitation efficiency and onset of saturation. This criterion is met by the typical Ti:Sapphire pulse rates (~80 MHz) for most common molecules, which is fortunate because in mode-locked lasers pulse rate is difficult to change. Stable excitation rate being an important consideration, it is essential that a constant number of pulses arrive on a pixel. Assuming a typical pulse rate of 80 MHz and a typical dwell time of 1 μs, more than 80 pulses arrive on a pixel. The number of pulses per pixel will thus be stable at the 1.25% level.

The objective generates the laser focus required for localization of excitation and is therefore a critical element of the microscope. A large choice of objectives suitable for two-photon microscopy can be found on the market. Important criteria for a choice include: (1) the numerical aperture, which has to be large because it determines resolution and the solid angle within which fluorescence is collected; (2) the magnification, which has to be chosen according to the desired field of view; (3) the working distance, which has to be relatively large, especially in in vivo applications where several media have to be stacked between objective and sample (coverglass, agarose, etc.); and (4) a good transmission efficiency in the near-IR. Note that some of these requirements are to some extent conflicting (for instance, increasing the working distance of an objective without increasing its exit pupil leads to a decreased numerical aperture), so that in most cases the optimal choice consists in making the best compromise between the different requirements. Given the considerable price of two-photon microscope objectives, budgetary issues are often part of this compromise. A good discussion of lasers and objectives can be found in (5, 7, 11).

(c) Pulse length and power

Optimal pulse length is more difficult to define. The optical components in the excitation path cause dispersion, pulses reaching the sample are thus broadened ("chirped") with respect to their initial duration. Indeed, in an optical medium the propagation speed of electromagnetic radiation depends on wavelength, and this effect is stronger for short pulses (typically around or below 100 fs) than for long ones, because of their different envelop in Fourier space. Being constant and known for a given microscope's excitation path, such dispersive broadening can be precompensated by "negatively dispersing" the pulses. These "prechirped" pulses are thus restored to short ones when exiting the objective, which maximizes two-photon absorption at the sample level (5). Although this procedure is relatively easy to implement (28, 29), the added optical components unavoidably result in some power loss of which one has to take into account.

More problematic is the limitation on peak power and thus on pulse duration posed by photobleaching and photodamage, which appear to depend more than quadratically on excitation intensity (8, 30, 31). The bottom line is that the pulse length warranting the optimal trade-off between fluorescence yield and photodamage has to be determined somewhat empirically for each given application.

(d) Scanning methods (*x–y* plane)

Once the desired pulse duration has been set and power of the laser beam into the excitation path has been adjusted—usually with the help of a Pockels cell (but a much simpler way is to use just a polarizer and a quarter-wave plate (5)), the beam has to be scanned through the sample.

The most common solution to do this is to use mirrors moved by galvanometers, which allow rapid, arbitrary positioning of the focus in the focal plane (*x–y*). When used in raster scanning application, galvanometric mirrors not only have excellent optical properties, but also allow zooming and image rotation. However, such applications are limited to the study of relatively slow phenomena, because of their limited speed (at best ~1 ms per line, resulting in typical frame durations of ~1 s). Speed can be increased at the cost of the resolution and the area covered, but still a large portion of the scanning is wasted on measuring the background part, i.e., the noninteresting regions of the image.

There are raster scanning methods for in vivo imaging that are intrinsically faster. For example, the scanning speed along the critical of the two in-plane dimensions can be increased using rotating polygonal mirror (32) or resonant galvanometer (33, 34) technology. In rapid raster scanning the whole *x–y* plane is scanned, yielding images of excellent quality. Yet, neither scanning speed adjustment nor image zoom or rotation is possible. For a discussion of scanning methods, including multifocal scanning, see (11).

A classical alternative to scanning the whole *x–y* plane is the so-called "line scan." There, the laser beam is scanned by the galvanometer-based mirrors along a line placed by the user on top of the structure of interest, rather than across the whole 2D plane. High speeds can be reached in this way; for instance, multiple nearby dendritic spines can be visited in less than 5 ms (35). However, the price to pay is that much of the spatial information is lost. Also, selection and adjustment of the scanning trajectories should be made efficiently within the time of the experiment, which can be quite cumbersome because in most systems the structure of interest has to be a straight line (e.g., a dendritic spine, an "adequate" dendritic segment, etc.).

Moreover, usually, measurement speed cannot be controlled. Therefore, precious measurement time is wasted, because extracellular space of no interest between regions of interest (ROIs) is swept over at the same speed as the ROIs themselves.

X–y scanning and line scanning become increasingly wasteful when the structures of interest are sparsely distributed. Indeed, galvanometer-based approaches are still mechanical systems, where the speed of scanning is limited by the acceleration/deceleration rates of the motors. Most important, all the structures to be scanned have to lie in the focal plane (lines can be scanned only at constant *z* value), with the exception of some more advanced approaches (see Sect. 3.2.2). As a final consideration, line scans—and in general all ROI scanning methods—need extensive software support, which is not always provided by microscope manufacturers.

Random-access scanning follows a different concept: the focused laser beam is directed selectively to predetermined targets (points). Therefore, this approach is of interest when the ROIs are sparsely distributed within the sample. There, scanning the whole *x–y* plane as done in raster scanning is not only inefficient in terms of time, but also counterproductive because of useless fluorophore excitation outside the ROI and consequent bleaching and photodamage. Although this is possible using galvanometric mirror technology, speeds reach at best ~0.5 ms/point because of mechanical inertia. A valid alternative consists in using acousto-optic devices to direct the laser beam.

An acousto-optic deflector is a transparent crystal, within which an optical grating is created by sound waves. The frequency of the sound wave determines the spacing of the grating, and thus the angle of diffraction of the laser beam on this grating. In this way the laser beam can be diverted without mechanical movements, allowing positioning speeds that are orders of magnitude faster than those obtainable with galvanometric mirrors. However, the involved optical materials are highly dispersive, leading to substantial temporal and angular dispersion, which, for typical pulse durations (<1 ps) deteriorates the excitation efficiency and point spread function (36) and thus image quality. Compensations are possible using gratings, prisms, and/or additional acousto-optic devices (37–40). Importantly, acousto-optic deflectors have also been proposed (41, 42) for tackling scanning along the *z*-axis, which so far has been only marginally explored (but see (43–45)), (see Sect. 2) for more details.

(e) Objective filling

Before impinging onto the sample, the beam is expanded to “fill” the back aperture of the microscope objective, which finally focuses the light into the sample. Since the radial intensity profile of a laser beam is Gaussian, the question arises of what an optimal filling is. A laser beam is said to "overfill" the back aperture of the objective when the latter accepts the central part of the beam up to the radius R at which its intensity decays to 13.7% ($1/e^2$). The "fill factor" for such a ratio between the laser beam's size and the objective back aperture is assigned the value of 1. In this case, 84% of the beam's power is transmitted and lateral and axial resolution (defined as HWHH, $1/e$ or $1/e^2$ excitation light intensity radius around the point of focus (6)) are nearly as good as with a uniform excitation beam (92 and 96%, respectively). Underfilling the objective using a narrower beam increases the amount of power transmitted by the objective, but results in a reduced effective numerical aperture (NA) and thus in a loss of resolution (for more details including calculations, see (5, 6)). However, it has to be considered that, with increasing depth the effective NA is reduced: large angle photons travel considerable longer distances in the tissue than photons near to the optical axis and are thus more likely to be scattered away or to be absorbed (e.g., by surface blood vessels). Depending on the application, underfilling the objective might thus be the way to go, in particular, if the objects to be observed are comparatively large and their detectability does not suffer under the slightly reduced accuracy. Indeed, as the size of the focal volume grows with decreasing effective NA, the peak intensity therein decreases, thus decreasing photobleaching and photodamage.

(f) Detection

In two-photon microscopy, optical sectioning is achieved by spatially ultraselective excitation. Therefore, all photons emitted by the excited fluorescent molecules are useful and should thus be collected, including scattered ones. With this respect, it should be kept in mind that fluorescence has shorter wavelengths than excitation light, therefore it is more heavily scattered and beyond depths of a few 100 μm, the amount of ballistic photons emerging from the sample is negligible. Therefore, large numerical aperture and low-magnification objectives (e.g., 20× at a NA of 0.9) are preferable to capture as many photons as possible and large sensitive area photomultipliers (PMTs—a performance comparison between different types can be found in (6)) are disposed in an arrangement that allows to detect as much of them as possible, either scattered in the forward direction ("trans"-collection geometry) or backwards ("epi"-collection geometry). For in vivo imaging, samples can in most cases be considered semi-infinite thus impeding trans-collection, which makes an efficient epi-detection particularly important (19). Of most common use is the so-called "whole-area" configuration, where the back of the objective is projected onto the PMT's sensitive area and thus allows all photons collected by the objective to end up on a detector preferably placed as close as possible to the objective (7). Importantly, new approaches emerge for in vivo applications, trying to collect even the light that does not enter the objective (46).

2.2. The z-Axis: Imaging in 3D

Ideally, one would like to investigate physiological functions at high spatial and temporal resolution in many arbitrary 3D positions, simultaneously. For example, simultaneous (i.e., millisecond-near) measurement of hundreds of neurons (somata and/or cellular processes) are required in order to understand the functioning of even just a small patch of cortex, hippocampus, or any other brain structure. Even at the single neuron level, in order to detect and image individual local dendritic spikes at high temporal and spatial resolution, hundreds of pixels must be acquired every few milliseconds from the same dendritic segment, over at least several tens of micrometers *z*-scanning range.

(a) Moving Objectives, linear regime
One of the main limitations of "classical" laser scanning microscopy is the sequentiality of scanning in 3D, which is essentially limited to so-called "3D stacks." There, the *xy* plane is scanned for a given *z* position. The *z* position is then changed and the *xy* plain is scanned again, and this is repeated for all *z* positions of interest. Whereas this procedure allows obtaining quite complete 3D information for anatomical (structural) imaging, the time needed for such a 3D stack is very long: seconds up to tens of minutes, depending on the *x*, *y* and, especially, *z* range and resolution. Such long times are unworkable for functional imaging and, although they can be reduced if only a subset of the *xy* plane is scanned for each *z* position, a strong limitation results on the objects of interest that can be studied: they have to lie within one and the same *xy* plane.

Recently, two papers have reported the development of methods for 3D functional imaging that overcome this limitation—at least to some extent. In the first one (43), galvanometric mirror scanning in the *xy* plane is combined with axial scanning relying on a piezoelectric actuator. Light amplitude drop at increasing depth is easily compensated for by increasing the excitation beam at the source level, via an adjustment of the Pockels cell limiting the laser power. What is important here is that the objective physically moves up and down the *z*-axis. Mechanical inertia is thus a major constraint on the possible spatiotemporal trajectories of the focal point within the sample. In a nutshell, one assumes the $z(t)$-movement of the objective (a sinusoidal movement (43)) as given by the interplay between the piezoelectric actuator's characteristics and the mechanical inertia of the system, and the *xy* trajectories are chosen, such as to cover the desired sample points in 3D in the least possible time. Although performance is ideally optimized by defining such 3D trajectories case-by-case, depending on the particular spatial distribution of points of interest, in practice, stereotypical movements sampling the whole 3D volume are often preferable (e.g., spirals, square spirals or Lissajous patterns), essentially due to the conspicuous user input required for the selection of all points of interest in 3D. For instance, when imaging neuronal populations (43), showed that 90% of the neurons in volumes up to cubes of 250 μm of side could be imaged with acquisition rates on the order of 10 Hz, using either a spiral-like or a custom, user-defined, trajectory.

An intrinsic limitation of the above approach is the dependence upon galvanometer-based and piezoelectric actuator scanning technologies, in particular the latter ones. In fact, at least in their linear range, even the fastest piezoelectric actuators cannot move faster than 100 Hz and the inertia of the microscope objective easily reduces this speed by a factor of 10. At the higher speeds, the vibration from physical movement of the objective lens can easily disturb image quality and disrupt biological samples (but for *z*-ranges in the order of 10–30 μm see the "Roller Coaster" approach described below).

(b) Moving Objectives, nonlinear regime ("Roller Coaster" Scanning)
Despite considerable technological advances over the last years, functional imaging of multiple dendritic spines (or more than 16 points in 3D) "simultaneously," that is, in the time scale of the millisecond, remains problematic (43, 44, 47, 48). To address these shortcomings, a new imaging technique has been developed by combining multiple-line scanning (35) with a rapid, nonlinear, piezo-based *z*-scanning strategy that is relatively simple (49). The general strategy consists in restricting the measurement only to points located on a 3D trajectory that crosses locations of interest (e.g., dendrites and spines) in a given volume. This results in an increase in the product of the signal collection efficiency (signal/pixel) and measuring speed (repetition rate of the measurement: $v_{\text{measurement}}$) that is proportional to the ratio of the smallest quadratic volume (3D sample) containing the 3D trajectory divided by the volume illuminated during 3D trajectory imaging:

$$\frac{\text{signal}}{\text{pixel}} \times v_{\text{measurement}} \propto \left[\frac{\iiint\limits_{\text{3Dsample}} \mathrm{d}V}{\iint \text{PSF}\,|_{yz}\, \mathrm{d}y\mathrm{d}z} \right] \Big/ \int\limits_{\text{3Dtrajectory}} g(s)\mathrm{d}s$$

Here, $PSF|_{yz}$ is the *yz* projection of the normalized PSF (point spread function). The formula is calculated for volume scanning when raster scanning with fast *x* scanning was used. Galvanometric scanners are able to follow the *x* and *y* projection coordinates of the 3D trajectory up to scanning speeds of a few milliseconds (~kHz range scanning). However, scanning in the axial direction is more challenging. Fast *z*-drive by piezo-positioners has been proposed previously (see above: (43)), where the piezo-positioner was driven by a sinusoidal signal and the phase shift and amplitude drop at higher repetition rates was compensated for. As a result, a 10 Hz *z*-scanning speed could be achieved over a range of 250 μm. The question is: what about the nonlinear response of the piezo crystals?

Speeding up the driving frequency of the piezo-positioner leads naturally to a drastic drop in amplitude of the—ideally sinusoidal—movement (49), as a consequence of the increasing presence of nonlinear components. The *z*-movement therefore differs increasingly from a sinusoid and compensating for the deviation becomes increasingly complicated. However, when exploring the nonlinear regime of a high force, cylindrically symmetric, piezo actuator, one encounters resonances (e.g., at 150 Hz and at 690 Hz in (49)) in the objective movement without servo feedback. If, at the same time the position signal is measured as a function of time, one realizes that after 50–100 cycles "warm-up" period the position of the objective reaches a steady-state nonlinear response function $z(t)$ at a mean position. At this point, one can take the measured shape of the $z(t)$ function and calculate, for each t, the *xy* drive signal to be sent to the galvanometric mirrors, such that the $xy(t)$ position together with $z(t)$ yield the desired 3D trajectory. This procedure is exemplified in Fig. 2.

Suppose the desired trajectory has to sample the point (x_1, y_1, z_1). First, we start by measuring the $z(t)$ response function (after warm-up). Then, a time t_1 is determined from $z(t)$, at which the focal plane of the objective will be at $z(t_1) = z_1$ height (Fig. 2a, b), then at this time point t_1 the *xy* drive signal for the galvanometric mirrors will be set to (x_1, y_1) (Fig. 2c). The same procedure is repeated for the each "critical" point (i.e., at which the trajectory was user-defined in the user interface) and the total curve will be interpolated at the end for the necessary resolution and smoothness. Note that, due to the way the trajectory has been constructed, the scanning speed is not the same along the whole cycle. Inspired by the typical dynamics-dictated shape of its trajectories, this procedure has been termed "Roller Coaster Scanning."

The stability of the 3D trajectories was verified by bleaching 3D curves in a homogeneous fluorescent plastic sample (Fig. 3a). A *z*-stack of the fluorescence was taken afterwards,

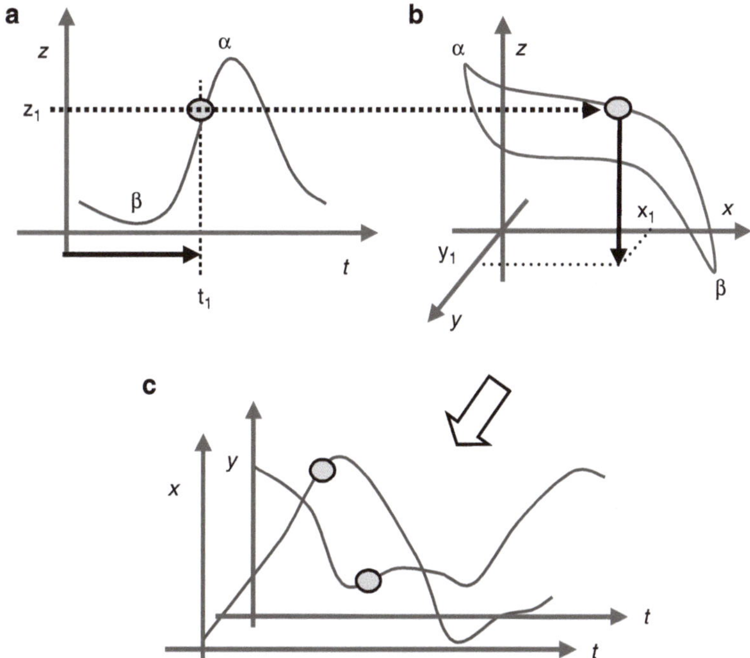

Fig. 2. Roller coaster scanning method. (**a**) After 50–100 cycles of the nonlinear drive (warm up period) the nonlinear position response of the objective reached a steady-state periodic response function $z(t)$, the amplitude of which was set to be larger than the required *z*-scanning range. (**b**) A part of this response function (typically one full period) was projected to the selected 3D trajectory. (**c**) The $x(t)$ and $y(t)$ input signals of the digital servo of the scanner motors were determined as the *x* and *y* coordinate projections of the 3D trajectory (x_1, y_1). The descending and ascending phase of the oscillation could cover different parts of the selected 3D trajectory as illustrated in (**b**). The selected 3D trajectory was the result of an interpolation algorithm which was based on user-selected points on the image *z*-stack of the chosen dendritic region. Due to the typical shape of the trajectories, the technique goes under the name of "roller coaster scanning." Figure modified with permission from (49).

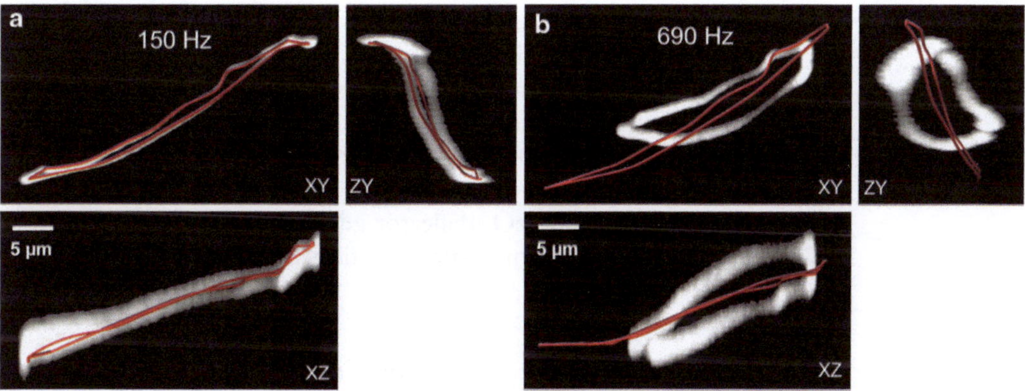

Fig. 3. Stability of the 3D trajectory during a roller coaster measurement. (**a**) Measurement performed at 150 Hz. Maximum intensity image stack projections of an originally homogeneous fluorescent sample after bleaching it along the user-selected 3D trajectory also used later in (**a**, **b**). Measured intensities were inverted. User-selected 3D trajectory is overlaid in *red* showing the reliability of the 3D measurement. Projection planes are noted in image corners (XY, ZY, XZ). (**b**) Measurement performed at ~700 Hz. Note that the agreement between the user-selected and the actual trajectories is lost, yet, trajectories remain highly stable in time. Figure modified with permission from (49).

and the image was grayscale-inverted, representing the time-averaged laser irradiation in the 3D space. As can be seen, replaying the same drive values (x, y and z) many times with the same speed resulted in the same trajectory, at high precision in all three dimensions. This trajectory was used for imaging neuronal dendrites curving in 3D. The spatial and temporal resolution and S/N warranted by Roller Coaster Scanning at 150 Hz are illustrated by the data shown in Figs. 7 and 8 (in the Sects. 3 and 3.2).

When the z-speed is increased even further, another resonance peak is found at 690 Hz. As opposed to the 150 Hz movement, in this case, the bleached trajectory deviates significantly from the planned one. Yet, the trajectory is stable (Fig. 3b), suggesting that optimized mechanical design and more sophisticated software corrections should allow better control of this second resonance peak, therefore allowing to extend Roller Coaster Scanning with automatic algorithms up to high-frequency regimes.

In conclusion, the Roller Coaster Scanning Method is ideal for rapid scanning extensive and continuous dendritic segments with high spatial resolution. Even the fine details of the spatial and temporal structure of small individual physiological events and their different compound responses could be measured simultaneously. Its shortcoming is the relatively limited scanning range, i.e., in the order of only a few tens of microns (25 and 15 μm, at 150 and 690 Hz, respectively).

(c) Static Objectives, acousto-optic deflectors for random access scanning

To overcome the limitations of piezo-driven z-scanning, such as the constraints imposed by only few available mechanical resonances and the small z-scanning ranges obtainable at high scanning speeds, it is advantageous to scan without mechanically moving parts. In this perspective, acousto- and electro-optical scanning open new outlooks for rapid scanning along the z-axis ((41): axial at 400 kHz; (42): tens of kHz (44): several kHz). In case of acousto-optical scanning an acoustic wave propagates thorough the TeO_2 deflector generating an optical grid, which results in a deterministic deflection angle Θ:

$$\Theta = \lambda f / v$$

Here, λ is the wavelength of incident laser light, and v and f are the velocities and frequencies of the propagating acoustic waves. Generation of chirped acoustic waves (41) by linearly changing f as a function of time results in an optical grid with cylindrical lens focusing effect. In order to eliminate lateral movement of the focal point two deflectors with counter-propagating chirped acousto-optical are required. Three-dimensional scanning can be achieved by changing incident laser light divergence using four deflectors generating two cylindrical lenses (44). Three-dimensional scanning of 16 regions in a $200 \times 200 \times 50$ μm^3 volume with 3 kHz acquisition rate has been demonstrated, although quality of measurements is currently limited by technical problems such as the angular dispersion of the optical grids generated.

Random access scanning is a promising, novel approach. However, scanning ranges have so far been limited, e.g., $50 \times 200 \times 200$ μm^3 (44). Technical improvements are likely to soon increase the accessible volumes to theoretical predictions of large volumes up to $350 \times 350 \times 200$ μm^3 (42), or even more (1 mm axial at low magnification, 10× objective (44)).

The second report on fast RD imaging (44) makes use of acousto-optical deflector (AOD)-based scanning. Due to the absence of physical movements (apart from sound waves propagating within a crystal), AOD-based scanning techniques do not share the shortcomings of galvanometric/piezoelectric scanning described above. When applied to functional two-photon microscopy in neurons, Duemani-Reddy and co-workers (44) easily reach acquisition rates in the kHz range, without any of the artifacts that would result from physical vibrations of the sample or the image. In Duemani-Reddy and coworker's report, the scanning range is limited to 50 μm range. This is considerably smaller than the axial range of many piezo-actuators, at least in their linear regime (up to a few hundreds of microns). However, using low magnification objective lenses, and/or wide angular aperture AODs allows extending the *z*-range achievable with AOD scanning by at least a factor of 10 (50). This makes AOD-based scanning approaches particularly promising for the in vivo imaging of large neuronal populations in volumes comparable to the size of the fundamental information-processing modules (e.g., cortical columns).

(d) Random access vs. Roller Coaster Scanning

How do the two 3D imaging strategies Roller Coaster Scanning and random access acousto-optical scanning (44, 47, 48) perform, compared to each other?

A general problem with the AOD-based 3D scanning is that, when good spatial resolution needs to be maintained, the measurement is limited to a small number of points (~40 points/ms). In this case, in fact, deflectors with large apertures are needed to maintain a large number of grating lines in the aperture during operation. Switching time on the other hand is determined by the time the (new) acoustic waves propagate through the entire crystal, which can be as long as ~25 μs. Suppose, e.g., that a dendritic segment has to be scanned "continuously" (i.e., at high spatial sampling rate, say one point every 0.1 μm). With the above switching times, the dendritic segment scannable per ms is then limited to a few μm only (in our case ~4 μm/ms). An additional shortcoming of the long switching times typical of random-access scanning is the reduction of SNR, because a large percentage of the total measurement time is wasted during switching. Indeed, typical switching times (10–35 μs) are considerably longer than the minimum pixel dwell time used in two-photon microscopes with moving mirrors (~1–4 μs).

In contrast, Roller Coaster Scanning at high resolution along long neuronal processes provides a continuous measurement of dendritic length and does not go on the expenses of pixel dwell time. In practice, currently available galvanometric scanners limit this value to a few 100 μm/ms, but this still provides about two orders of magnitude higher resolution for the measurement of contiguous structures than acousto-optical scanners, because switching time is equal to zero and pixel dwell time is technically not limited (although typically above 0.1 μs).

Current AOD-based techniques also place a limitation on either the spatial resolution or on the field of view (AOD: ~200 × 200 points resolution and ~200 × 200 μm field of view as opposed to Roller Coaster Scanning: 2,200 × 2,200 points resolution and 650 × 650 μm field of view).

Finally, the optical pathway of the Roller Coaster microscope is simple and does not contain any material beyond the objective that introduces angular, linear dispersion or laser intensity loss, therefore: (1) experiments requiring high-energy pulses like two-photon uncaging, bleaching, ablation or in vivo imaging of deep tissue (5) are achievable, (2) the excellent spatial resolution characteristics of two-photon microscopy are preserved, and (3) the combination of Roller Coaster Scanning with new optical methods such as high-resolution imaging techniques is possible (51, 52).

Yet, several laboratories are working on solutions to overcome the drawbacks of AOD technology, which, being a new technology, is expected to have many unexplored resource reserves. On the contrary, the mechanical limitations of Roller Coaster Scanning are difficult to improve. At some point, random access scanning is thus likely to equal or even outperform piezo-actuator-based Roller Coaster Scanning. Most important, the expected *z*-scanning ranges that can be tackled by acousto-optic scanning are considerably larger than with Roller-Coaster technology.

(e) Holographic microscopy

Holographic microscopes can produce spots of variable sizes and numbers, and in shaped patterns to match user-selected biological structures, using a liquid-crystal spatial light modulator to generate a holographic pattern of illumination. Use of a specific pattern (can be generated with Fresnel lenses) generates *z*-shift of the focal point making this method capable of three-dimensional pattern generation and also 3D point scanning. However, its use in imaging is limited by the low refresh rate of the available high-resolution liquid-crystal spatial light modulators, typically being in the refresh rate range of LCD displays (~60 Hz). Such low refresh rates rule out single-point illumination scanning needed for two-photon detection.

Holographic illumination can be used in combination with imaging detectors (such as CCD cameras) to generate various forms of structured light illuminations giving the ability to discriminate depth information, but these approaches perform poorly deep in scattering tissues. Holographic illumination can be effectively used in the generation of multiple spots or complex patterns for two-photon uncaging or other light stimulation methods, where refresh rate is not that critical. Development of novel high-resolution—yet high refresh rate—liquid-crystal spatial light modulators fueled by the advancements in image projection technologies on the market is likely to greatly increase the currently available usage ranges.

Other *z* scanning strategies include: variable-focus lenses (53) and adaptive optics.

2.3. Fluorophores

For the most common in vivo application, that is, calcium imaging, several well-experimented fluorescent dyes are available, which can be easily found on the market (e.g., Fura-2, Fluo-4 AM, Oregon Green BAPTA). Voltage-sensitive dyes (VSD) suitable for two-photon imaging have also become available, e.g., the Stark-effect-based ANNINE-6plus (54). The classical dye for glial staining is Sulforhodamine (55).

In general, however, it has to be kept in mind that a tight link exists between the choice of the fluorescent probe and the particular application it is used for, which poses constraints on the necessary sensitivity, possible wavelengths, bleaching dynamics, etc. Unfortunately, so far only few two-photon excitation spectra have been published, as opposed to single-photon excitation spectra, which have been published for most common fluorophores. Therefore, in most cases when one uses a new fluorescent indicator, its two-photon excitation spectra must be determined experimentally. Indeed, the physical processes underlying one and two-photon excitation of a molecule are not identical, two-photon excitation spectra cannot be simply calculated from their corresponding single-photon ones. A method to measure those is illustrated in (7).

3. Methods: Two-Photon Fluorescence Imaging in Living Tissue

In this section, we present applications of two-photon microscopy to the study of functional activity in living tissues, and in particular for in vivo studies in the mammalian neocortex. Indeed, among all physiological processes investigated using two-photon imaging, the functional activity of neurons is probably the one that is most widely studied. Moreover, one of the major advantages of two-photon microscopy over conventional microscopy and/or other optical techniques is its ability to penetrate deep into scattering tissues, with a depth-scale that approximately matches the thickness of the cortical sheet—or at least of several of its layers. The same considerations apply to some other preparations, such as the hippocampus and the spinal cord. Therefore, here, we focus particularly on two-photon imaging of neuronal activity, as well as on the specific method of bolus loading of the marker dye (56), which allows uniform staining of large groups of neurons with calcium markers in a selected volume of interest (up to nearly 1 mm^3).

An important consideration here is that, for in vivo imaging, the structures of interest must be optically accessible, such as the cerebral cortex. Other structures, which do not lie on the surface of the brain, can be reached by ablating the overlying tissue, eventually using small diameter, microlens or grin-lens-based objectives, to minimize the ablation area. Another option is "in toto" imaging, where the structure of interest is extracted in its entirety from the preparation. As opposed to imaging in slices, here, all connectivity internal to the structure of interest is left intact, which puts the in toto preparation somewhat in an intermediate position between slices and in vivo studies. For instance, the in toto preparation is usefully applied to the hippocampus. We succinctly provide information about this preparation in the form of a text box (Box 2).

We also point out that the images shown here for rapid 3D imaging (Sect. 3.2.2: Figs. 7 and 8) have been obtained in hippocampal slices. However, from the methodological point of view, exactly the same considerations developed there also apply to in toto and in vivo imaging.

The optimal choice of a particular fluorescent probe for a given purpose (e.g., which to use for calcium imaging: Fura-2, Fluo-4, Oregon Green, etc.?) will of course depend on the wavelengths available for excitation. Every particular application will also set its specific constraints onto the choice of the dye to be used, for instance, due to its bleaching dynamics and sensitivity, solubility, average stained volume, etc. Together, all these parameters are often difficult to quantify and thus to properly take into account. In most cases, the best way to find the optimal solution is through trial and error.

3.1. Some Practical Guidelines on How to Improve Signal Generation and Its Detection

- Use a proper cage: Sometimes scientists do not want to waste time for proper shading of their microscopes or the stage of the shading box made in the lab is just not stable enough. The price for bad shading is high background noise that will decrease image quality.
- Water-cooled detectors are superior in some very low light applications.
- Use a local amplifier instead of sending weak signals through long cables: high quality amplifiers with preferentially local AD converters provide good measurement quality.
- Novel GaAsP PMTs provide high QE and low noise.
- Check the input aperture of the detectors: it should be large enough for imaging scattered photons.
- Order optimal filter sets matching the requirement of your measurements. But beware: delivery of custom-made filters takes time and prices might be high.
- Detectors should be as close as possible to the objective (and condenser, if applies).
- Use the same PMT voltage for a given measurement type but remember that changing from one measurement type to the other might be useful. For example, high voltages are required for line scans but lower voltages are preferred for *z*-stacks.
- PMT values in a given measurement type could be optimized for SNR in a measurement series by using background noise SD and the peak amplitude of signal to calculate SNR as function of voltage. Some well repeatable signal such as backpropagating action potentials (bAPs) is ideal for these measurements. Once the voltage is determined no further changes are required.
- Use lenses with high NA for optimal photon collecting but large enough working distances to be practical.
- Thermal drift and normal aging of materials can affect detection efficiency. Track and record detection efficiency at least on a monthly basis, using standard samples (fluorescent beads).
- Lasers are also subject to “aging.” Check the laser power at objective level at least monthly.
- Keep the working environment clean, thermally stable and—above all—dust-free. Filtered air and thermally stabilized room temperature are highly desirable.

3.2. Application of the Technique

3.2.1. Calcium Imaging

Calcium imaging is probably the most common application of functional two-photon microscopy (Fig. 4). Calcium imaging is primarily used to study the spiking activity of a neuron or a small network of neurons. Indeed, it consists of loading neurons with

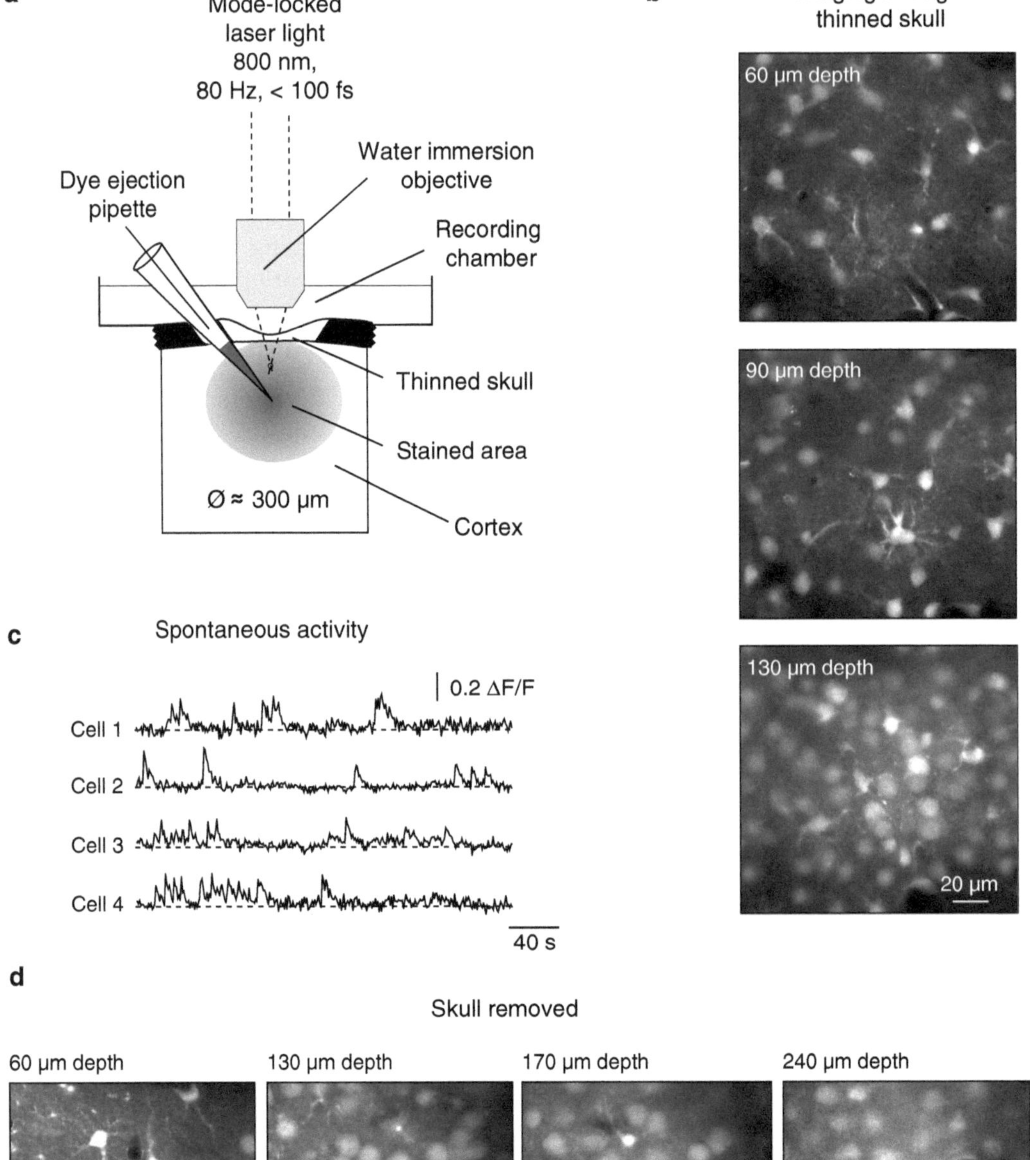

Fig. 4. In vivo calcium imaging of neuronal populations. (**a**) Schematic drawing of the experimental arrangement. (**b**) Images taken through a thinned skull of a postnatal day 13 mouse at increasing depth. (**c**) Spontaneous Ca^{2+} transients recorded in a different experiment through a thinned skull in individual neurons (P5 mouse) located 70 µm below the cortical surface, from a region similar to that shown in (**b**). (**d**) Images obtained as in (**b**) in an experiment (P13 mouse), in which the skull was removed before imaging. Reproduced with permission from (56). Copyright (2003), National Academy of Sciences, U.S.A.

specific markers that bind to calcium, the marker acquiring its fluorescent properties only upon binding. Since ionic calcium channels open when the cell emits an action potential, the intracellular calcium concentration temporarily increases and spiking activity of the cell results in increased fluorescence signals. The typical calcium response to a single action potential is a transient with a rise time of ~20 ms, and a slow decay with a time constant as long as ~1/4 s (Fig. 5), due to the slow evacuation of calcium ions back toward the extracellular medium.

Using simultaneous electrophysiology recordings, it has been shown that single action potentials can be detected in vivo (57). Such resolution is possible only in cells with low enough spiking rate. Otherwise, the smoothing effect of the slow calcium decays prevents detection of single action potentials. Nevertheless, sub-second resolution can be obtained using specific deconvolution techniques ((58), Supplemental Materials of (59)), and even in the case of cells with high spiking activity, the monotonous (possibly linear) relation between calcium levels and number of spikes allows a precise quantification of the output activity of the cell. Note that functional calcium signals have also been observed in astrocytes, which probably play a role in the metabolic-vascular coupling to neural activity (60, 61). With this respect, it is advisable to stain the preparation with sulforhodamine, a specific glial marker, such as to differentiate between those cells and neurons.

An advantage of two-photon calcium imaging as compared to microelectrode recordings is that it allows to acquire signals simultaneously from many cells: imaging 50–100 cells can be achieved without excessive difficulty and imaging of up to as much as 300 cells has been demonstrated using specific three-dimensional scanning techniques (4).

On the other hand, the use of these optic probes to monitor electrical activity in the cortex has limitations. Simply getting them into the cortex can be damaging, e.g., a mechanical stress occurs when the dye is injected into cells or into the extracellular space (tissue). Moreover, a number of side effects limit dye usage, including photobleaching, toxicity to the cell, and photodynamic damaging. An even more intrinsic limitation is that these probes act as an artificial calcium buffer, because their very principle is to bind to intracellular Ca^{2+} ions. Therefore, calcium probes can alter intracellular signaling pathways and ultimately induce an abnormal behavior of the cell.

For the above reasons, considerable research efforts are devoted to improving the calcium markers. Commonly used, commercially available, calcium markers include: Oregon Green (perhaps the most used one), Fluo-4, Fura-2, and others. For a review on calcium imaging that enters the details of different markers, their chemical machinery and their particularities see (62). A guide on fluorescent proteins can be found in (63).

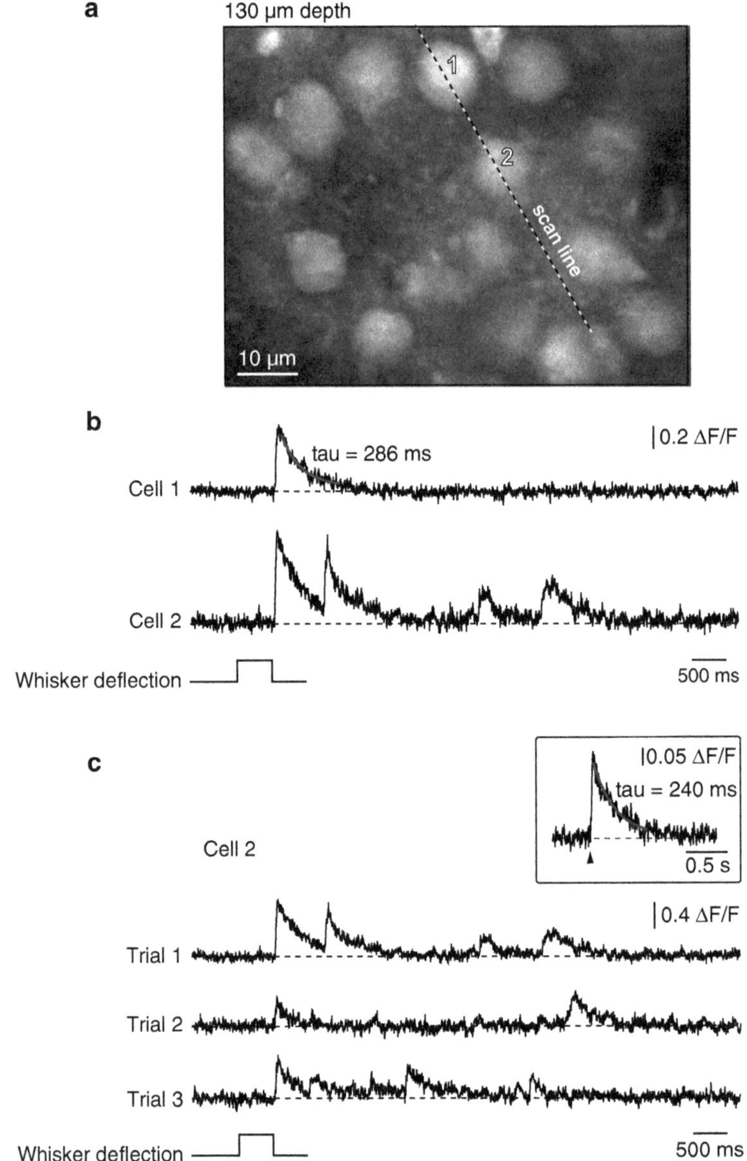

Fig. 5. In vivo recordings of Ca^{2+} transients evoked by whisker deflection. (**a**) A high-magnification image of layer 2–3 neurons in vivo in the barrel cortex of a P13 mouse. (**b**) Line-scan recordings of Ca^{2+} transients evoked in two neurons by a deflection of the majority of whiskers on the contralateral side of the mouse's snout. The position of the *scanned line* and the cells analyzed are indicated in (**a**). Note that the Ca^{2+} transients occurred 17–22 ms after the termination of the stimulus and therefore probably represent stimulus-offset responses. Here and in (**c**), the *solid line* represents a mono-exponential fit of the decay phase of the transient. (**c**) Ca^{2+} transients evoked in cell 2 during three consecutive trials. The top trace is from the trial illustrated in (**b**). (Inset) A Ca^{2+} transient in a P14 layer 2–3 neuron evoked in vivo by single-shock electrical stimulation (70 V, 0.2 ms) average of five consecutive trials. Reproduced with permission from (56). Copyright (2003), National Academy of Sciences, U.S.A.

To minimize the buffering capacity of a dye, its concentration in the cell has to be kept low, which goes at the expense of imaging quality. Whatever the aimed compromise, it has to be considered that the upper limit of imaging quality is attained at a dye concentration at which its calcium buffering capacity equals the intrinsic calcium buffering capacity of the cell. See (4) for details, and in particular on some relevant properties of the probe.

The problems resulting from invasive delivery of the dyes into the cortex can be avoided using genetically encoded markers, which, in addition, in certain cases even allow to selectively target specific cell types. Large research efforts are therefore being devoted to develop new ones. We will describe this promising approach in more detail below.

Still, in many cases, neurons have to be stained with calcium markers and there are multiple ways to do so, which we will detail below. Particular care will be taken in detailing the popular method of multicell bolus loading (MCBL) (56) used for studying small networks of neurons.

(a) Filling of single cells

A common calcium imaging method is to stain a single neuron by allowing the diffusion of the calcium marker directly into its cytoplasm through a glass pipette. Since the dye will diffuse into the whole arborescence of the neuron, this allows acquiring signals from different components of the same neurons. As the method requires the user to punch a small opening on a patch of the cell membrane, the technique is naturally applied when using combined whole cell patch clamp and calcium imaging, yielding correlated local calcium data and either local dendritic or global, somatic electrophysiological recordings. Examples are back-bAPs observed in the dendrites of a cell, as shown on Fig. 7.

More sophisticated techniques of electroporation allow delivering the marker in the extracellular medium in the cell's proximity, while an electric field is created that temporarily breaks the cell membrane and allows the neuron to take up the dye with minimal mechanical damage (64).

(b) Bolus loading in the extracellular medium

The so-called technique of bolus loading (56, 65) consists of delivering the calcium marker into the extracellular medium (tissue injection).

In such case, an acetoxymethyl (AM) ester is added to the fluorescent molecule, which allows it to enter neighboring cells through ordinary uptake mechanisms (62). Importantly, the calcium-binding site of the molecule is masked; only inside the cell this part is cleaved by enzymes and the molecule's calcium binding ability is activated.

A solution containing the dye is pressure-injected directly inside the tissues through a glass micropipette. The complex mechanisms briefly described above result in the staining of all cells and cell processes inside a sphere of diameter ~300 μm, including axons and dendrites of cells whose soma is even more distant.

Since both neurons and astrocytes are stained using this technique, it is useful to dissolve a second dye in the pipette solution, sulforhodamine, which selectively stains astrocytes (55). Using the appropriate set of optical filters to select the appropriate fluorescence emission wavelength, sulforhodamine fluorescence signals at 605 nm can be observed in a separate channel from calcium fluorescence, which is usually around 520 nm.

The success of the staining depends strongly on a large set of physiological parameters that are usually collectively termed as the "quality of the preparation." A detailed protocol can be found in (65), and a video publication at the Journal of Visual Experiments allows visualizing a calcium experiment in detail (66). This preparation consists in three steps. First, the surgical procedures: animal anesthesia, craniotomy (and in general duraectomy), and building a chamber fixed on the skull such as to stabilize the brain with agarose pressed by a cover glass. Second, the injection of the fluorescent calcium marker, after which, it takes about 1 h for the dye to stain the cortical region. Third, two-photon calcium fluorescence signals can be acquired. In Box 1, we highlight a number of points that critically affect the quality of the two-photon calcium signal upon bolus-loading the sample. Box 2 shows an application of bolus loading the hippocampal in toto preparation.

Box 1
Critical Steps for Two-Photon Imaging Using Bolus-Loaded Ca^{2+} Probes

The quality of the acquisitions primarily relies on the quality of the staining, and this requires the correct preparation of the dye solution. Then, to allow normal dye uptake, a healthy and stable physiological condition of the cells is of primary importance. These and additional factors that contribute to the quality of imaging are listed below.

Factors contributing to the quality of the staining:

- *Dye preparation*: as the dye is hydrophobic, it is necessary to first dissolve it in a solution of DMSO/Pluronic (Invitrogen, or can be prepared by dissolving

(continued)

Box 1
(continued)

20% Pluronic acid in DMSO); its mixing can be facilitated by a combination of vortexing, centrifuging and sonicating; then it has to be diluted into a pipette solution, and mixed again.

- *Quality of the surgery*: mechanical or thermal shock to the cortex during craniotomy has to be avoided. Use cold ACSF to cool upon drilling the craniotomy/thinning the skull; do not touch the cortical surface; bleeding from the brain can severely affect the quality of two-photon imaging; fast preparation is preferable, in particular the dura mater removal (when needed), during which the brain surface is exposed.
- *Careful injection*: since the dye molecules and the DMSO/Pluronic solution can be toxic to cells, it is preferable to minimize the quantity that is injected; on the other hand, this decreases the quantity of fluorescence signal; a good injection protocol is given in (65).
- *Physiological conditions of the animal*: monitor temperature, ECG, respiration, etc. to ensure that the animal is in optimal physiological condition.

Factors contributing to the quality of imaging:

- *Shot noise*: the system should be optimized, in particular for light collection efficiency; in general shot noise can then be sufficiently reduced by increasing the laser power (provided imaging is not too deep); do not reduce shot noise below the level of other noise (such as pulsation) to avoid unnecessary bleaching and phototoxicity due to excessive (and unnecessary) dye excitation.
- *Cardiac pulsation*: open a small cranial window, maintain the brain stable with low melting temperature agarose that is applied in liquid phase and covered with a coverslip before its solidification; different groups have developed different solutions for building a chamber, in some cases a gap is left free of coverglass, to allow insertion of an electrode for injection through the agarose or for simultaneous electrophysiological recording.
- *Spatial resolution* (and in particular, the z resolution) is highly important to minimize the contamination of the neuronal signals by the neuropil signals: best resolutions are obtained with a large numerical aperture objective and

(continued)

Box 1
(continued)

complete filling of its back aperture by the laser beam (see the optics part in this chapter for more details); to completely benefit from high numerical aperture, stain and image in a region devoid of blood vessels as much as possible and do not image on the side of the craniotomy: consider that for a numerical aperture of 0.95, a laser beam focused on a point situated at depth z below the surface, is penetrating the surface through a disk of radius equal to z—which thus should be as devoid of blood vessel as possible.

- *Compromise between temporal resolution, spatial resolution, extent of the region imaged, and shot noise*; theoretically, the key factor here is the dwell time, i.e., the time spent in each pixel. The more it can be reduced, the better. Indeed, short dwell times allow acquiring more pixels per second, thus allowing increasing either the rate at which images are acquired, or the total number of pixels in the image, or both. However, a lower limit on dwell results from the need to collect enough light to overcome shot noise limitations. In practice, limitations on temporal resolution arise also from scanning requirements (speed of scan, possibility to access random locations and thus gather signal only from relevant points, e.g., neurons), and from the low temporal resolution of calcium transients.

Box 2
Multicell Bolus Loading Applied to In Toto Hippocampal Preparations

Staining of neuronal populations is usually applied in vivo, but there are no practical limitations to using the technique for in vitro loading of cells (67). Here, we describe a method for staining a large neuronal population in the whole hippocampal preparation. The dye preparation and loading has been detailed above and is well described in (10, 65). Figure 6 shows an example of this preparation, as through a two-photon fluorescence microscope.

(continued)

Box 2 (continued)

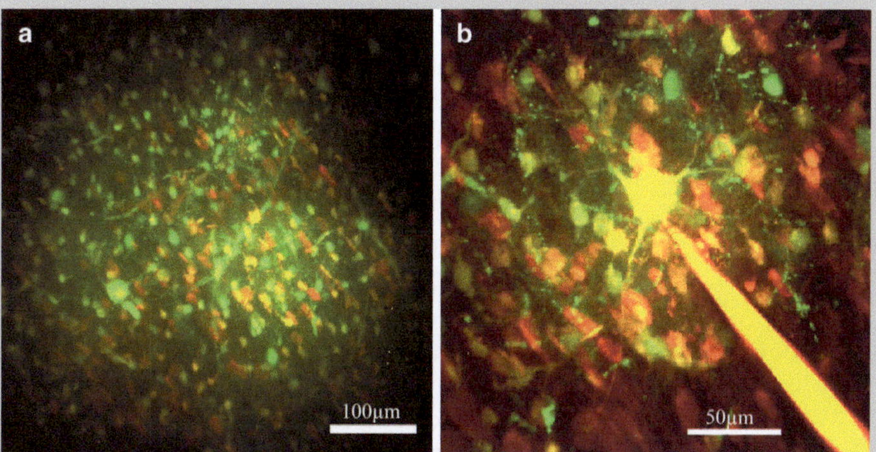

Fig. 6. In toto hippocampal preparation with MCBL and patch clamping. An example loading is shown below on (**a**), where neurons are green (2 mM *Oregon green* BAMTA-AM) and glial cells are *red* (180 μM Supforhodamine-101). (**b**) Shows a whole cell patch clamped neuron filled with 200 μM Fluo5 and 50 μM Alexa-594.

Extraction of the whole hippocampus:

- Mice are briefly anesthetized with isoflurane and swiftly decapitated.
- The brain is gently but rapidly removed from the skull and placed into a Petri-dish containing carbogen-bubbled, ice-cold ACSF (for composition, see (49)).
- The hemispheres are separated and cut sagittally with a surgical scalpel.
- With the help of two plastic spatulae, the hippocampus of one hemisphere is gently exposed by blunt dissection starting from the third ventricle.
- While gently keeping the brain in position, the hippocampus is disconnected from the rest of the brain at the septal and dorsal areas (the septum can be kept by using gentle dissection around it).
- Using the long edge, one of the spatulae is placed under the longitudinal axis of the hippocampus and used to gently roll out the whole structure towards the cortex.
- The long edge is pushed down to cut loose the preparation (along with the entorhinal cortex).
- The whole hippocampus is sucked up using a wide mouth Pasteur pipette and placed in a submerged tissue holding chamber filled with ACSF at room temperature, bubbled with carbogen.

(continued)

Box 2
(continued)

Advantages of using the in toto hippocampal preparation:

1. It can be placed in any conventional in vitro slice chamber and the dye loading can be performed. The process does not require any modification to the in vitro microscope setups and the calcium imaging and patch clamping can commence as with brain slices.
2. As the preparation is quite translucent, both trans- and epifluorescent photons can be collected, greatly increasing the SNR.
3. A major advantage and a major difference to brain slices is that the in toto hippocampal preparation contains most of the morphological connections that the hippocampus had in vivo. Most of the axonal projections toward the septum and the entorhinal cortex are intact or much less damaged than in brain slices.

(c) Calcium-dependent autofluorescence obtained through genetic modifications

Genetically encoded calcium indicators were developed recently (68), and allow raising strains of animals with specific cell types expressing Ca^{2+}-sensitive fluorescent proteins. These calcium probes are thus delivered in a minimally invasive manner, thus allowing to image calcium dynamics in the intact brain by two-photon imaging through a thinned bone. A short but comprehensive review on genetically encoded sensors of neural activity can be found in (69), whereas a more detailed review focusing on calcium sensors and detailing their performances and optimization can be found in (70).

At the current stage of their development, the performance of genetic indicators remains inferior to those of synthetic dyes directly delivered to the cells with respect to several criteria, including SNR, kinetics, linearity, photostability, and ion selectivity. However, it has been shown that genetic indicators can be used in vivo (in the mouse cortex) to study Ca^{2+} dynamics at single-cell and even subcellular resolution. Moreover, they appear to allow detecting suprathreshold depolarizations consisting of as few as two to three action potentials (71, 72).

Additional advantages of genetic indicators are the possibility of selectively targeting specific types of cells, or even combining the genetic modifications expressing Ca^{2+}-sensitive fluorescence with modifications of other genes, e.g., involved in the plasticity of the cell, or in neurodegenerative diseases.

Another way of using genetically engineered calcium indicators is to encode them into the genome of a virus, and then use this virus to contaminate cells with the indicator. The use of viruses that transfect from a cell to the next by crossing synapses is already quite well established for tracing synaptic connections and network connectivity in the cortex (73). Using genetically encoded calcium sensors in conjunction with them allows reporting of the activity of connected neurons (74).

3.2.2. 3D Imaging of Dendritic Elements Using Roller Coaster Scanning

The spatial and temporal resolution and S/N characteristic of Roller Coaster Scanning are illustrated by the data shown in Fig. 7. A spiny dendritic segment of a CA1 pyramidal neuron was followed by a 3D trajectory crossing 18 regions in 12 spines and the neighboring mother dendrite at 150 Hz. Back-bAP induced Ca^{2+} transients in all of the total 18 regions had a SNR comparable to that of the transients measured in one region by a single line scan using the same conditions as a classical two-photon microscope.

Variability in the local dendritic geometry and misplacement of the user-selected 3D trajectory results in inhomogeneities in the basal fluorescence of the 3D measurement (Fig. 8a, b).

Therefore, during measurement of long dendritic segments a real-time normalization for the first 500 ms (or whatever is available before the relevant event under study) of the raw 3D Ca^{2+} traces are required (Fig. 8c). For better visualization of the 3D Ca^{2+} response, the trajectories are projected into two dimensions (space along the dendrite and time). Ca^{2+} transients (Fig. 8d) were

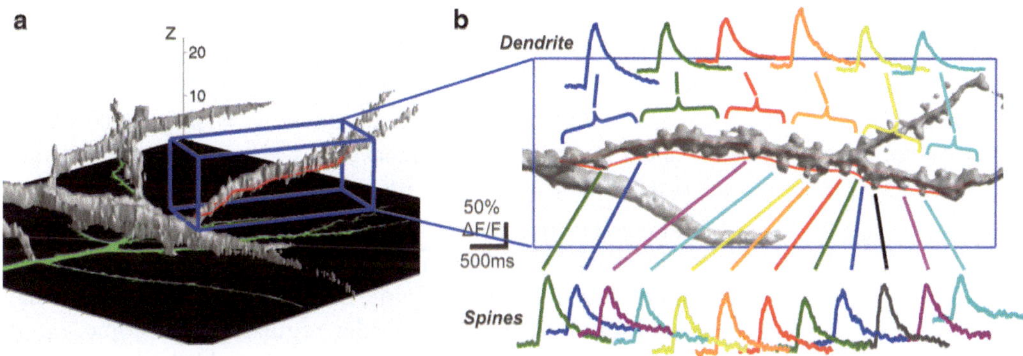

Fig. 7. 3D two-photon dendritic imaging of a CA1 pyramidal cell at 150 Hz temporal resolution. (**a**) Dendritic segment of a CA1 pyramidal cell. The *z*-stack was deconvolved and thresholded for surface fitting. The *blue curve* (*light* and *dark blue*) shows the repeatedly scanned 3D trajectory. The size of the *blue box* is 15 × 15 × 37 μm³. (**b**) Enlarged view of the *blue box* in (**a**). A total of 18 regions including 12 spines located on the 3D scanning trajectory were measured in one sweep, which repeated at 150 Hz. The 12 spines were measured with one part (*dark blue curve*) of the trajectory while the whole parent dendritic segment was measured with the second, approximately parallel running part of the 3D trajectory (*light blue curve*). Ca^{2+} transients were induced by back-propagating action potentials (bAPs) elicited by somatic current injection steps (5 APs, 35 Hz, average of 5 traces). Figure modified with permission from (49).

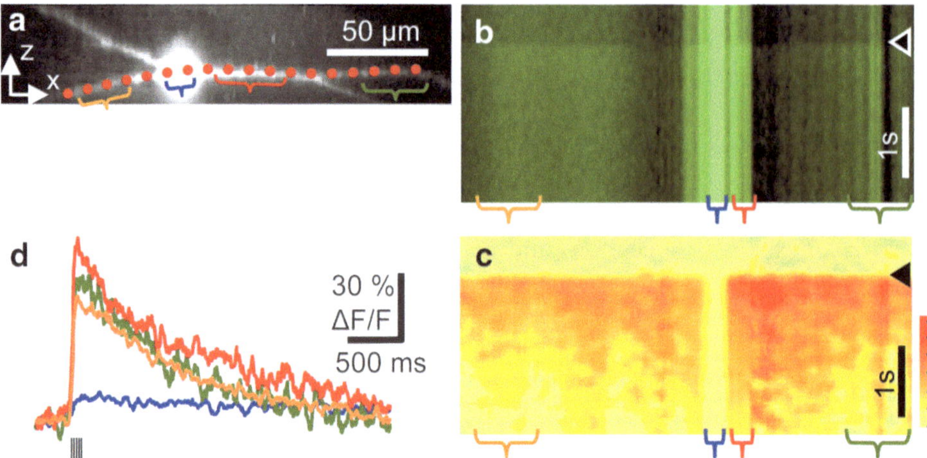

Fig. 8. 3D two-photon dendritic imaging of a >200 μm long interneuron dendrite at 150 Hz temporal resolution. (**a**) Maximum intensity side projection of a long interneuron dendrite (CA1, stratum radiatum). The pipette has been removed. (**b**) Raw fluorescent data measured along the dendrites in (**b**). The *white open triangle* indicates the time when five back-bAPs (35 Hz) were induced by somatic current injections. Time is displayed along the vertical axis. The horizontal axis is the spatial dimension along the measurement line (note that the scanning speed along the line is variable due to the nature of the roller coaster scanning). (**c**) Spatially normalized (and 2D projected) relative fluorescence image (3D Ca^{2+} response) calculated from the linearized 3D trajectory scan in (**b**). Color bar: 0–63%, $\Delta F/F$. (**d**) Individual Ca^{2+} transients measured in the color coded regions in (**b**) (*black bars* indicate the somatic current injections). Figure modified with permission from (49).

derived as a spatial integral of the 3D Ca^{2+} responses calculated for a given dendritic subsegment.

Using detectors with high sensitivity measurements are possible at relatively low laser intensities (3–10 mW), and transients have a high S/N even without averaging and did not show any accelerated time-dependent decrease in amplitude or decay time when compared with "classical" 2D imaging (75). Namely, measurement of more than 500 traces (>20 min total measurement time) is possible from the same dendritic region.

3.2.3. Staining of Other Function-Related Elements

Although two-photon functional imaging is largely performed using calcium markers, other markers are also used to address functions other than calcium activity.

(a) Voltage-sensitive dyes

Performing two-photon microscopy using VSD apparently sounds promising for refining the cortical structures that have been unveiled using VSD under regular, macroscopic, optical imaging. Such a study has been recently published (54), but an important technical limitation applies here, due to low SNRs. Indeed, signal variations ($\Delta F/F$) are typically of the order of 1% when using VSD in two-photon microscopy, which is small as compared to the shot noise level, which is large (typically 2% or more) due to the small amount of collected fluorescence. This implies that extensive averaging is needed: e.g., the responses shown in (54) were obtained after averaging over 200 trials.

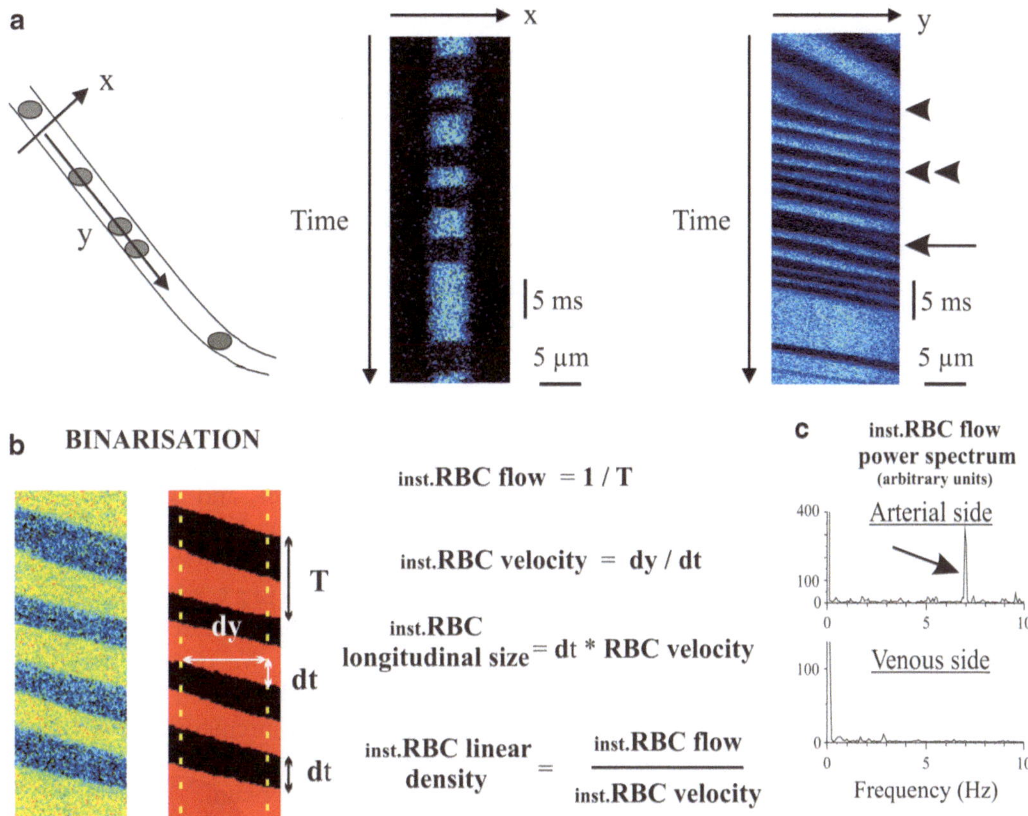

Fig. 9. Technical approaches to measure the parameters of RBC flow in single capillaries. (**a**) Transversal line scans (X direction) through capillaries were used to ensure the absence of lateral movements of capillaries as well as to observe RBC flow. Each RBC is seen as a shadow flowing through the fluorescent plasma. Longitudinal line scans (Y direction) were used to determine all RBC flow parameters. Note that the slope of oblique shadows and thus RBC velocity could vary (*single* and *double arrowheads*) and that occasionally two RBCs were stuck together (*arrow*). (**b**) Raw data were first binarized, and values of instantaneous flow ($1/T$), velocity (dy/dt), and longitudinal size were determined for each RBC. Instantaneous RBC linear density was calculated as inst.RBC flow/inst.RBC velocity. Mean values of RBC flow, velocity, or linear density were determined as the time average of the instantaneous values calculated over 20- to 30-s periods of acquisition. (**c**) Power spectrum analysis of inst.RBC flow reveals that heartbeat fluctuations were essentially observed in capillaries located near arterioles. Reproduced with permission from (77). Copyright (2003), National Academy of Sciences, U.S.A.

(b) Blood-plasma tracers, red blood cell (RBC) flow

Two-photon microscopy also allows studying vascular dynamics as well as neurovascular coupling at so-far unaccessible detail. The cortical vasculature can be imaged by staining the blood plasma with an injection of plasma tracers in the vascular system (e.g., in the tail of the animal) (Fig. 9) (76–78). Several vascular changes related to the neural activity can then be investigated. In these applications, care has to be taken to use large enough tracer molecules as to avoid them leaking out of the vasculature, especially of the capillaries. This can be accomplished by binding the usually small fluorescent molecules to larger molecules, such as dextran (76, 79).

It is then possible to visualize the motion of RBCs in different vessels (venules, arterioles, capillaries), and thus to determine their speed, which is related to blood flow and perfusion (Fig. 9). It is also possible to measure changes in the diameter of the vessels, thus monitoring changes in blood volume. By combining visualization (and quantification) of RBC motion, vessel diameter and at the same time staining the neurons in the studied region with a calcium marker, it is possible, by the adequate use of random scans, to monitor simultaneously neural activity and vascular changes. This methodology offers an interesting platform for the study and modeling of neurovascular coupling.

(c) Using two-photon microscopy for photostimulation
Although two-photon microscopy is mainly used for imaging, it can also be used for other purposes, and in particular for photostimulation of photoactive molecules: the same way that two-photon fluorescence provides a very focused excitation pattern, it can activate such molecules in a very focused region. This way it is possible to simulate glutamate release emerging from a single synaptic transmission. We do not detail this research area, but refer to a review (13) which summarizes all the recent advancements in the measurement and manipulation of neural activity, paving the way to a "reverse-engineering" process of comprehension of the brain.

3.2.4. Analysis of the Data

Adequate data analysis of two-photon imaging data is of critical importance, in particular when imaging calcium in thin processes. First, an "anatomical" image is needed, typically a high-resolution 3D stack, with which the functional data can then be co-registered (80). Next, it is necessary to average the functional signals, because the noise level in individually captured pixels is very high, mostly originating from the quantal properties of light ("shot noise"). In functional calcium imaging of thin processes only about 10–1,000 photons are captured within one image pixel, even though whole field detection captures a significant amount of the light emitted from the sample. To obtain the living cell under the focal spot to emit more photons causes bleaching of the dye and is toxic to the cell. Therefore, careful analysis of the data is essential such as to be able to work with low dye concentration and/or laser intensity, thus reducing the side effects of the imaging, such as photodynamic damage, bleaching, etc.

Analyzing high-speed line scan or random access data is done in three steps. First, in the raw fluorescent images, the pixels whose data originate from the same biological objects should be identified. This is done according to the reference images taken when selecting the trajectory. In line scans, these are usually nearby pixels captured while the focal spot crossed the inspected object. The signal of these pixels is then averaged, resulting in a fluorescence time course for each selected ROI, at better SNR than its individual pixels.

The second step is the calculation, for a given *i*th ROI, of the relative fluorescence time course *fi(t)*, according to the following equation:

$$f_i(t) = \frac{F_i(t) - B_i}{F_{0i} - B_i} - 1$$

Here, F_{0i} is the average baseline fluorescence (before any stimulus is applied), and B_i is the background fluorescence in the same region. Main sources of the background signal are: (1) offset value of the AD converter, (2) dark current of the PMTs, (3) autofluorescence of unstained sample, and (4) out-of-focus fluorescence of highly fluorescent nearby objects (such as the cell body or the patch pipette filled with fluorescent dye). An accurate determination of the background level is critical for the correct evaluation of calcium transient amplitudes but its estimation is not at all straightforward. A good way to do so is to estimate it so-to-say "automatically," by scanning unstained regions of the sample near each of the ROIs, before each measurement (even though this somewhat complicates the measurement protocol). Note that in the above calculation, F_i can be calculated for each pixel *i* of a raw line-scan image, yielding a relative fluorescence image that reveals dendritic compartments (see Fig. 8). Relative fluorescence measurements also allow estimation of calcium concentration levels within small compartments, such as dendritic spines (35, 81, 82). Unfortunately, there are few programs on the market specialized for acquisition and analysis of such physiological measurements.

The third step is to sort, align, pool, and average data coming from different experiments or different regions. This is usually done in specialized programs, such as Matlab, Origin, or IgorPro.

Acknowledgments

We are grateful to C. Bernard for alerting us on the importance of the hippocampal in toto preparation and for teaching us the related protocol. TD is grateful to A. Grinvald for the first encounter with a two-photon microscope and to W. Denk for helpful advice and discussions. IV is grateful to F. Helmchen, W. Göbel, and B. Kampa for helpful discussions. Part of this work has been financed by the Hungarian-French NIH-ANR grant "MULTISCALEFUNIM" to BR and IV.

References

1. Yuste R (2005) Fluorescence microscopy today. Nat Methods 2:902–904
2. Minsky M (1961) US Patent. Number 3013467
3. Conchello J-A, Lichtman JW (2005) Optical sectioning microscopy. Nat Methods 2:920–931
4. Göbel W, Helmchen F (2007) In vivo calcium imaging of neural network function. Physiology 22:358–365
5. Helmchen F, Denk W (2005) Deep tissue two-photon microscopy. Nat Methods 2:932–940
6. Zipfel WR, Williams RM, Webb WW (2003) Nonlinear magic: multiphoton microscopy in the biosciences. Nat Biotechnol 21:1369–1377
7. Tsai PS, Nishimura N, Yoder EJ, Dolnick EM, White GA, Kleinfeld D (2002) Principles, design, and construction of a two-photon laser scanning microscope for in vitro and in vivo brain imaging. In: Frostig R (ed) In vivo optical imaging of brain function. CRC Press, New York
8. Koester HJ, Baur D, Uhl R, Hell SW (1999) Ca2+ fluorescence imaging with pico- and femtosecond two-photon excitation: signal and photodamage. Biophys J 77:2226–2236
9. Kerr JND, Denk W (2008) Imaging in vivo: watching the brain in action. Nat Rev Neurosci 9:195–205
10. Garaschuk O, Milos R-I, Grienberger C, Marandi N, Adelsberger H, Konnerth A (2006) Optical monitoring of brain function in vivo: from neurons to networks. Pflugers Arch Eur J Physiol 453:385–396
11. Svoboda K, Yasuda R (2006) Principles of two-photon excitation microscopy and its applications to neuroscience. Neuron 50:823–839
12. Hübener M, Bonhoeffer T (2005) Visual cortex: two-photon excitement. Curr Biol 15:R205–R208
13. O'Connor DH, Huber D, Svoboda K (2009) Reverse engineering the mouse brain. Nature 461:923–929
14. Göppert-Mayer M (1931) Über Elementarakte mit zwei Quantensprüngen—Göppert-Mayer—2006—Annalen der Physik—Wiley Online Library. Ann Phys 401:273–294
15. Denk W, Strickler JH, Webb WW (1990) Two-photon laser scanning fluorescence microscopy. Science 248:73–76
16. Flusberg BA, Cocker ED, Piyawattanametha W, Jung JC, Cheung ELM, Schnitzer MJ (2005) Fiber-optic fluorescence imaging. Nat Methods 2:941–950
17. Theer P, Hasan MT, Denk W (2003) Two-photon imaging to a depth of 1000 microm in living brains by use of a Ti:Al_2O_3 regenerative amplifier. Opt Lett 28:1022–1024
18. Beaurepaire E, Oheim M, Mertz J (2001) Ultra-deep two-photon fluorescence excitation in turbid media. Opt commun 188:25–29
19. Beaurepaire E, Mertz J (2002) Epifluorescence collection in two-photon microscopy. Appl Opt 41:5376–5382
20. Mainen ZF, Maletic-Savatic M, Shi SH, Hayashi Y, Malinow R, Svoboda K (1999) Two-photon imaging in living brain slices. Methods 18:231–239; 181
21. Majewska A, Yiu G, Yuste R (2000) A custom-made two-photon microscope and deconvolution system. Pflugers Arch Eur J Physiol 441:398–408
22. Tsai PS, Friedman B, Ifarraguerri AI, Thompson BD, Lev-Ram V, Schaffer CB, Xiong Q, Tsien RY, Squier JA, Kleinfeld D (2003) All-optical histology using ultrashort laser pulses. Neuron 39:27–41
23. Pologruto TA, Sabatini BL, Svoboda K (2003) ScanImage: flexible software for operating laser scanning microscopes. Biomed Eng Online 2:13
24. Mancuso JJ, Larson AM, Wensel TG, Saggau P (2009) Multiphoton adaptation of a commercial low-cost confocal microscope for live tissue imaging. J Biomed Opt 14:034048
25. Deguil N, Mottay E, Salin F, Legros P, Choquet D (2004) Novel diode-pumped infrared tunable laser system for multi-photon microscopy. Microsc Res Tech 63:23–26
26. Hönninger C, Morier-Genoud F, Moser M, Keller U, Brovelli LR, Harder C (1998) Efficient and tunable diode-pumped femtosecond Yb:glass lasers. Opt Lett 23:126–128
27. Chen I-H, Chu S-W, Sun C-K, Cheng P-C, Lin B-L (2002) Wavelength dependent damage in biological multi- photon confocal microscopy: a micro-spectroscopic comparison between femtosecond Ti:sapphire and Cr:forsterite laser sources. Opt Quantum Electro 34:1251–1266
28. Diels J-CM, Fontaine JJ, McMichael IC, Simoni F (1985) Control and measurement of ultrashort pulse shapes (in amplitude and phase) with femtosecond accuracy. Appl Opt 24:1270–1282
29. Treacy E (1969) Optical pulse compression with diffraction gratings. IEEE J Quantum Electro 5:454–458
30. Petersen CC, Sakmann B (2000) The excitatory neuronal network of rat layer 4 barrel cortex. J Neurosci 20:7579–7586

31. Hopt A, Neher E (2001) Highly nonlinear photodamage in two-photon fluorescence microscopy. Biophys J 80:2029–2036
32. Kim KH, Buehler C, So PTC (1999) High-speed, two-photon scanning microscope. Appl Opt 38:6004–6009
33. Fan GY, Fujisaki H, Miyawaki A, Tsay RK, Tsien RY, Ellisman MH (1999) Video-rate scanning two-photon excitation fluorescence microscopy and ratio imaging with cameleons. Biophys J 76:2412–2420
34. Nguyen Q (2001) Construction of a two-photon microscope for video-rate Ca2+ imaging. Cell Calcium 30:383–393
35. Lörincz A, Rózsa B, Katona G, Vizi ES, Tamás G (2007) Differential distribution of NCX1 contributes to spine-dendrite compartmentalization in CA1 pyramidal cells. Proc Natl Acad Sci USA 104:1033–1038
36. Iyer V (2005) Fast functional imaging of single neurons using random-access multiphoton (RAMP) microscopy. J Neurophysiol 95:535–545
37. Iyer V, Losavio BE, Saggau P (2003) Compensation of spatial and temporal dispersion for acousto-optic multiphoton laser-scanning microscopy. J Biomed Opt 8:460
38. Lechleiter JD, Lin D-T, Sieneart I (2002) Multi-photon laser scanning microscopy using an acoustic optical deflector. Biophys J 83:2292–2299
39. Roorda RD, Hohl TM, Toledo-Crow R, Miesenböck G (2004) Video-rate nonlinear microscopy of neuronal membrane dynamics with genetically encoded probes. J Neurophysiol 92:609–621
40. Ngoi BK, Venkatakrishnan K, Lim LE, Tan B (2001) Angular dispersion compensation for acousto-optic devices used for ultrashort-pulsed laser micromachining. Opt Express 9:200–206
41. Kaplan A, Friedman N, Davidson N (2001) Acousto-optic lens with very fast focus scanning. Opt Lett 26:1078–1080
42. Reddy GD, Saggau P (2005) Fast three-dimensional laser scanning scheme using acousto-optic deflectors. J Biomed Opt 10:064038
43. Göbel W, Kampa BM, Helmchen F (2006) Imaging cellular network dynamics in three dimensions using fast 3D laser scanning. Nat Methods 4:73–79
44. Duemani Reddy G, Kelleher K, Fink R, Saggau P (2008) Three-dimensional random access multiphoton microscopy for functional imaging of neuronal activity. Nat Neurosci 11:713–720
45. Kirkby PA, Naga Srinivas NKM, Silver RA (2010) A compact acousto-optic lens for 2D and 3D femtosecond based 2-photon microscopy. Opt Express 18:13721–13745
46. Combs CA, Smirnov A, Chess D, McGavern DB, Schroeder JL, Riley J, Kang SS, Lugar-Hammer M, Gandjbakhche A, Knutson JR, Balaban RS (2011) Optimizing multiphoton fluorescence microscopy light collection from living tissue by noncontact total emission detection (epiTED). J Microsc 241:153–161
47. Rózsa B, Katona G, Vizi ES, Várallyay Z, Sághy A, Valenta L, Maák P, Fekete J, Bányász A, Szipocs R (2007) Random access three-dimensional two-photon microscopy. Appl Opt 46:1860–1865
48. Vucini D, Sejnowski TJ (2007) A compact multiphoton 3D imaging system for recording fast neuronal activity. PLoS One 2:e699
49. Katona G, Kaszás A, Turi GF, Hájos N, Tamás G, Vizi ES, Rózsa B (2011) Roller Coaster Scanning reveals spontaneous triggering of dendritic spikes in CA1 interneurons. Proc Natl Acad Sci USA 108:2148–2153
50. Katona G, Szalay G, Maák P, Kaszás A, Veress M, Hillier D, Chiovini B, Vizi ES, Roska B, Rózsa B (2012) Fast two-photon in vivo imaging with three-dimensional random-access scanning in large tissue volumes. Nat Methods 9:201–208
51. Willig KI, Rizzoli SO, Westphal V, Jahn R, Hell SW (2006) STED microscopy reveals that synaptotagmin remains clustered after synaptic vesicle exocytosis. Nature 440:935–939
52. Ding JB, Takasaki KT, Sabatini BL (2009) Supraresolution imaging in brain slices using stimulated-emission depletion two-photon laser scanning microscopy. Neuron 63:429–437
53. Oku H, Hashimoto K, Ishikawa M (2004) Variable-focus lens with 1-kHz bandwidth. Opt Express 12:2138–2149
54. Kuhn B, Denk W, Bruno RM (2008) In vivo two-photon voltage-sensitive dye imaging reveals top-down control of cortical layers 1 and 2 during wakefulness. Proc Natl Acad Sci USA 105:7588–7593
55. Nimmerjahn A, Kirchhoff F, Kerr JND, Helmchen F (2004) Sulforhodamine 101 as a specific marker of astroglia in the neocortex in vivo. Nat Methods 1:31–37
56. Stosiek C, Garaschuk O, Holthoff K, Konnerth A (2003) In vivo two-photon calcium imaging of neuronal networks. Proc Natl Acad Sci USA 100:7319–7324
57. Kerr JND, Greenberg D, Helmchen F (2005) Imaging input and output of neocortical networks in vivo. Proc Natl Acad Sci USA 102:14063–14068

58. Vogelstein JT, Watson BO, Packer AM, Yuste R, Jedynak B, Paninski L (2009) Spike inference from calcium imaging using sequential Monte Carlo methods. Biophys J 97:636–655
59. Greenberg DS, Houweling AR, Kerr JND (2008) Population imaging of ongoing neuronal activity in the visual cortex of awake rats. Nat Neurosci 11:749–751
60. Winship IR, Plaa N, Murphy TH (2007) Rapid astrocyte calcium signals correlate with neuronal activity and onset of the hemodynamic response in vivo. J Neurosci 27:6268–6272
61. Hirase H (2005) A multi-photon window onto neuronal–glial–vascular communication. Trends Neurosci 28:217–219
62. Tsien R (1999) Monitoring cell calcium. In: Carafoli E, Klee C (eds) Calcium as a cellular regulator. Oxford University Press, New York, pp 29–54
63. Shaner NC, Steinbach PA, Tsien RY (2005) A guide to choosing fluorescent proteins. Nat Methods 2:905–909
64. Nevian T, Helmchen F (2007) Calcium indicator loading of neurons using single-cell electroporation. Pflugers Arch Eur J Physiol 454:675–688
65. Garaschuk O, Milos R-I, Konnerth A (2006) Targeted bulk-loading of fluorescent indicators for two-photon brain imaging in vivo. Nat Protoc 1:380–386
66. Golshani P, Portera-Cailliau C (2008) In vivo 2-photon calcium imaging in layer 2/3 of mice. J Vis Exp. 13:681. doi: 10.3791/681
67. Verkhratsky A, Petersen OH (2010) Calcium imaging in neurobiology. Humana Press/ Springer Group, New York
68. Miyawaki A, Llopis J, Heim R, McCaffery JM, Adams JA, Ikura M, Tsien RY (1997) Fluorescent indicators for Ca2+ based on green fluorescent proteins and calmodulin. Nature 388:882–887
69. Barth A (2007) Visualizing circuits and systems using transgenic reporters of neural activity. Curr Opin Neurobiol 17:567–571
70. Hires SA, Tian L, Looger LL (2008) Reporting neural activity with genetically encoded calcium indicators. Brain Cell Biol 36:69–86
71. Hasan MT, Friedrich RW, Euler T, Larkum ME, Giese G, Both M, Duebel J, Waters J, Bujard H, Griesbeck O, Tsien RY, Nagai T, Miyawaki A, Denk W (2004) Functional fluorescent Ca2+ indicator proteins in transgenic mice under TET control. PLoS Biol 2:e163
72. Heim N, Garaschuk O, Friedrich MW, Mank M, Milos RI, Kovalchuk Y, Konnerth A, Griesbeck O (2007) Improved calcium imaging in transgenic mice expressing a troponin C–based biosensor. Nat Methods 4:127–129
73. Song C, Enquist L, Bartness T (2005) New developments in tracing neural circuits with herpesviruses. Virus Res 111:235–249
74. Granstedt AE, Szpara ML, Kuhn B, Wang SSH, Enquist LW (2009) Fluorescence-based monitoring of in vivo neural activity using a circuit-tracing pseudorabies virus. PLoS One 4:e6923
75. Ji N, Magee JC, Betzig E (2008) High-speed, low-photodamage nonlinear imaging using passive pulse splitters. Nat Methods 5:197–202
76. Kleinfeld D, Mitra PP, Helmchen F, Denk W (1998) Fluctuations and stimulus-induced changes in blood flow observed in individual capillaries in layers 2 through 4 of rat neocortex. Proc Natl Acad Sci USA 95:15741–15746
77. Chaigneau E, Oheim M, Audinat E, Charpak S (2003) Two-photon imaging of capillary blood flow in olfactory bulb glomeruli. Proc Natl Acad Sci USA 100:13081–13086
78. Kleinfeld D, Denk W (2005) Two-photon imaging of cortical microcirculation. In: Yuste R, Konnerth A (eds) Imaging in neuroscience and development: a laboratory manual. Cold Spring Harbor Laboratory Press, Cold Spring Harbor, pp 701–705
79. Vanzetta I (2005) Compartment-resolved imaging of activity-dependent dynamics of cortical blood volume and oximetry. J Neurosci 25:2233–2244
80. Losavio BE, Liang Y, Santamaría-Pang A, Kakadiaris IA, Colbert CM, Saggau P (2008) Live neuron morphology automatically reconstructed from multiphoton and confocal imaging data. J Neurophysiol 100: 2422–2429
81. Denk W, Yuste R, Svoboda K, Tank DW (1996) Imaging calcium dynamics in dendritic spines. Curr Opin Neurobiol 6:372–378
82. Maravall M, Mainen ZF, Sabatini BL, Svoboda K (2000) Estimating intracellular calcium concentrations and buffering without wavelength ratioing. Biophys J 78:2655–2667

Chapter 8

Calcium Imaging Techniques In Vitro to Explore the Role of Dendrites in Signaling Physiological Action Potential Patterns

Audrey Bonnan*, Benjamin Grewe*, and Andreas Frick

Abstract

Neurons generate cell-type-specific action potential (AP) patterns as a result of the integration of synaptic inputs received from many other neurons. In neocortical pyramidal neurons, this AP output is not only transmitted to many other postsynaptic neurons, but also back-propagates into the dendritic arbor, thus fulfilling a number of important functions. Back-propagating APs provide sufficient depolarization to activate voltage-gated Ca^{2+} channels at least in the proximal dendrites, thereby generating Ca^{2+} influx into the dendrites and spines of these dendrites. This Ca^{2+} influx, in turn, triggers a variety of signaling cascades that are involved in the regulation of neuronal signaling and plasticity. To better understand the role of Ca^{2+} influx in the regulation of these different neuronal functions, it is important to quantify the dendritic Ca^{2+} dynamics with a high spatial and temporal resolution. Here, we describe techniques that have been optimized to measure $[Ca^{2+}]$ dynamics in neocortical dendrites in response to physiological patterns of APs using two-photon imaging. These physiological AP patterns were previously recorded in vivo in response to a sensory stimulus and then replayed in the same neuron type in vitro using properly timed current injections.

Key words: Dendrites, Two-photon microscopy, Calcium imaging, Action potentials, Back-propagation, Voltage-gated ion channels, Temporal code, Neocortex, Pyramidal neurons

1. Introduction

During information processing in the brain, dendrites of neocortical neurons integrate thousands of incoming synaptic inputs to give rise to cell-type-specific action potential (AP) patterns. These physiological AP patterns are not only transmitted to many other postsynaptic neurons, but they also directly back-propagate into

*The authors Audrey Bonnan and Benjamin Grewe contributed equally for this chapter.

Emilio Badoer (ed.), *Visualization Techniques: From Immunohistochemistry to Magnetic Resonance Imaging*, Neuromethods, vol. 70, DOI 10.1007/978-1-61779-897-9_8, © Springer Science+Business Media, LLC 2012

their dendritic arbors (e.g., (1–4)). The efficacy of AP back-propagation, however, varies in a cell-type- and dendrite-type-specific manner (e.g., (5)), and relies on the morphology and both the passive and active properties of the dendrites. In neocortical pyramidal neurons, for instance, APs back-propagate actively into their dendrites, but their amplitude decreases with distance from soma (for review see (6)).

Studying back-propagating APs (bAPs) is of major interest as bAPs have been implicated in a variety of important functions, such as regulating synaptic efficacy across different time scales, changing the excitability of dendrites, providing synaptic feedback or stabilizing nascent synapses (for reviews see (6–8)). Due to the causal relationship between bAPs and information processing and memory formation in neurons, it is essential to understand the mechanisms of dendritic signaling during physiological AP patterns. Physiological AP patterns can be measured for specific cell types using electrophysiological recordings in vivo, for instance in response to a sensory stimulus, and can then be used to investigate dendritic properties of the same neuron type in an in vitro preparation. Here, we describe such an approach to study the role of dendritic signaling during physiological AP patterns, and provide illustrative examples that were performed on neocortical layer 5 pyramidal neurons in acute slices of the whisker-related barrel cortex (5).

When of sufficient amplitude and/or duration, bAPs are able to activate different types of voltage-gated Ca^{2+} channels in dendrites and spines. The resulting Ca^{2+} influx may trigger diverse signaling pathways, which underlie many neuronal processes, such as the regulation of cellular forms of short-term and long-term plasticity, the properties of Ca^{2+} channels, and dendritic growth or spine formation, amongst others (9–11). The recent development of optical tools for quantifying Ca^{2+} dynamics at unprecedented spatial and temporal resolution enables one to decipher the specificity in Ca^{2+} signaling (for review see Grewe et al. (12)). Since voltage-gated Ca^{2+} channels are good sensors of the amplitude/width of bAPs, imaging changes in $[Ca^{2+}]$ is a valuable way of probing the properties of different dendritic compartments.

In this chapter, we describe an optimized method for calcium imaging in dendrites in acute brain slices using two-photon laser scanning microscopy (TPLSM), as well as illustrative examples from experiments examining the role of dendrites in processing physiological AP patterns (5).

2. Materials

The substances/equipment used are listed below. In some cases alternative products are suggested, but comparable products from other suppliers should also be effective.

2.1. Chemicals and Reagents

Alexa 594	Invitrogen	(#A10438)
Ascorbic acid	Sigma Aldrich	(#95209)
Biocytin	Sigma Aldrich	(#B4261)
$CaCl_2$	Sigma Aldrich	(#C3881)
Fluo-5F	Invitrogen	(#F14221)
Glucose	Sigma Aldrich	(#G7021)
HEPES	Sigma Aldrich	(#54459)
KCl	Sigma Aldrich	(#P9333)
K-gluconate	Sigma Aldrich	(#G4500)
KOH	Sigma Aldrich	(#P5958)
Isoflurane (Belamont)[a]	Nicholas Piramal Limited, UK	

[a]*Controlled substance, see institutional veterinarian*

Mg-ATP	Sigma Aldrich	(#A9187)
$MgCl_2$	Sigma Aldrich	(#M0250)
Myo-inositol	Sigma Aldrich	(#57570)
NaCl	Sigma Aldrich	(#S7653)
Na-GTP	Sigma Aldrich	(#G8877)
$NaHCO_3$	Sigma Aldrich	(#S5761)
NaH_2PO_4	Sigma Aldrich	(#71505)
Na_2-phosphocreatine	Sigma Aldrich	(#P7936)
Na-pyruvate	Sigma Aldrich	(#P2256)
Oregon Green BAPTA-1 (OGB-1)	Invitrogen	(#O6806)
Sucrose	Sigma Aldrich	(#S9378)

2.2. Cutting Solution

$NaHCO_3$: 25 mM
KCl: 2.5 mM
NaH_2PO_4: 1.25 mM
$MgCl_2$: 7 mM
$CaCl_2$: 1 mM
Store at 4°C for up to 1 week.
Add fresh at the day of use:
Sucrose: 240 mM

Saturate with carbogen (95% O_2 and 5% CO_2) for at least 10 min and freeze until ~1/3 of the solution is frozen. Use a hand-held blender to obtain a finely ground slurry and bubble the resulting mixture with carbogen before use.

Note: This solution is used only for the preparation of the slices. For the incubation, slices will be transferred to the "incubation solution."

2.3. Incubation Solution

Glucose: 25 mM

KCl: 2.5 mM

NaCl: 125 mM

$NaHCO_3$: 25 mM

NaH_2PO_4: 1.25 mM

For 1 L of this solution, weigh the aforementioned substances and dissolve in ~800 mL bi-distilled water.

While stirring, add:

$CaCl_2$: 2 mM

$MgCl_2$: 5 mM

Note: To avoid precipitation, saturate solution with carbogen and then slowly add $CaCl_2$ and $MgCl_2$ from a 1 M stock solution.

Adjust pH with KOH to 7.3–7.4. Adjust volume to 1 L and the osmolarity to 300–310 mOsm. Store at 4°C for 2–3 days.

On the day of use, just before preparing the slice, add to this solution ascorbic acid (antioxidant) and Na-pyruvate (nutrient):

Ascorbic acid: 1 mM

Na-pyruvate: 4 mM

2.4. Artificial Cerebrospinal Fluid

Glucose: 25 mM

KCl: 2.5 mM

NaCl: 125 mM

$NaHCO_3$: 25 mM

NaH_2PO_4: 1.25 mM

For 1 L of this solution, weigh the aforementioned substances and dissolve in ~800 mL bi-distilled water.

While stirring, add:

$CaCl_2$: 2 mM

$MgCl_2$: 1 mM

Note: To avoid precipitation, saturate solution with carbogen and then slowly add $CaCl_2$ and $MgCl_2$ from a 1 M stock solution.

Adjust pH with KOH to 7.3–7.4. Adjust volume to 1 L and the osmolarity to 300–310 mOsm. Store at 4°C for 2–3 days.

2.5. Intracellular Pipette Solution

HEPES: 10 mM

K-gluconate: 135 mM

KCl: 4 mM

Mg-ATP: 4 mM

Na-GTP: 0.3 mM

Na_2-Phosphocreatine: 10 mM

Note: Prepare the solution on ice to avoid ATP and GTP degradation.

Prepare 100 mL. Adjust pH with KOH to pH 7.2–7.3, and the osmolarity to ~290 mOsm. Store in 0.5 mL aliquots at -20°C.

Dilute dye stock solution to desired final concentration in ~0.5 mL volumes, add biocytin (1.5–2.5 mg/mL) and use vortex until dissolved. Filter this solution using 0.22-μm pore centrifugal filters (Millipore, #UFC30GV00). Prepare aliquots appropriate for 1 day of experiments (approximately 20 μL) and store them at -20°C. Use 1–2 μL of solution per pipette.

Note: The choice and concentration of the dyes depends on the experimental question. For more detailed discussion see note 2 "Choice of fluorescent dye."

2.6. Dye Stock Solutions

Prepare stock solution of dyes (e.g., 4 mM Alexa 594, 20 mM Fluo-5F, or 20 mM OGB-1) in bi-distilled water and store in small aliquots (1–2 μL) at -20°C for up to 1 year. Add to intracellular pipette solution when preparing aliquots of the intracellular solution.

2.7. Slice Preparation

1. Wistar rats (Charles River)
2. Isoflurane
3. Cutting solution
4. Incubation solution
5. Carbogen (95% O_2 and 5% CO_2)
6. Dissection kit: scissors, forceps, tweezers, blade, scalpel, brush
7. Tissue glue (Use cyanacrylate, e.g., Roti coll 1)
8. Vibrating tissue slicer, e.g., Vibratome 3000 (Vibratome, USA) or Leica VT1200 (Leica, Germany)
9. Circulating water bath for incubation of slices set to 36°C

2.8. Calcium Imaging and Electrophysiology Setup

2.8.1. Two-Photon Laser Scanning Microscope for Calcium Imaging

We recommend the use of a two-photon laser-scanning microscope (TPLSM) because of its superior imaging resolution and depth penetration as compared to a CCD camera-based microscope. Furthermore, TPLSM combines the possibility of imaging thin dendrites or even spines in deeper regions of the slice with a very good signal-to-noise ratio (SNR).

Note: A number of suitable two-photon laser-scanning microscopes exist, ranging from completely custom-made versions (e.g., (13)) *to models based on modified commercial microscopes* (14) *to turnkey options* (e.g., *Leica, Olympus, Nikon*).

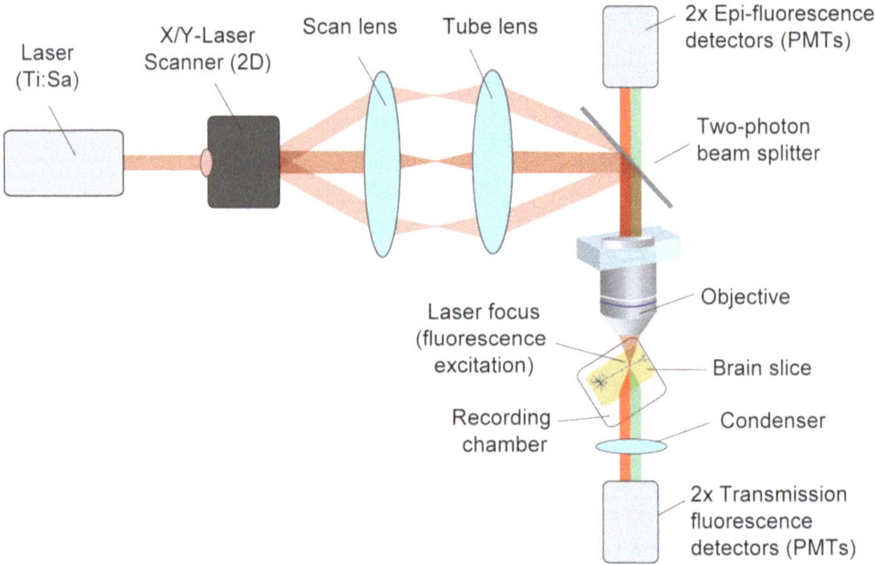

Fig. 1. Schematic of a two-photon laser scanning microscope (TPLSM) for in vitro measurements. The fluorescence is excited with a Ti:sapphire laser running at 870–890 nm or 810 nm. Epi- and transfluorescence for two separate channels (*green* and *red*) are collected through the objective and condenser, respectively. Fluorescence photons for each color are detected by photomultiplier tubes (PMTs).

We employ a Leica TPLSM (schematized in Fig. 1) for high-resolution two-photon Ca^{2+}-imaging as previously described by Koester et al. (15). Briefly, we use a femtosecond Ti:sapphire laser oscillator (Mira 900F, 900 mW average output power; Coherent, Santa Clara, CA) at 870–890 nm (to excite OGB-1 and Alexa 594) or 810 nm (for Fluo-5F combined with Alexa 594). The laser is coupled to a galvanometric scanning unit (TCSNT; Leica Microsystems, Mannheim, Germany). The scanning unit is mounted on an upright microscope (Leica DMLF) equipped with a 40× objective (HCX APO W40x/0.8 NA). Fluorescence signals are acquired using two external nondescanned photomultiplier tubes (R6357, Hamamatsu Photonics, Germany) before the objective (epi-mode) and behind the microscope condenser (transmission-mode) for each channel (green and red). Fluorescence signals from the cell somata and dendrites are integrated and averaged offline using the Leica DMLF Imaging Software in combination with Igor 4 (Wavemetrics, USA).

2.8.2. Electrophysiology Setup for Whole-Cell Recordings

1. Multimeter with resistance measurement function (general electronic equipment) to check the electrical connections and the proper grounding of the recording setup.
2. Microelectrode amplifier, e.g., Axoclamp 2B or Multiclamp 700B (Molecular Devices, USA), BVC-700A (DAGAN corporation, USA), SEC-10LX (npi electronics, Germany).
3. Silver wires for patch-clamp recordings and as bath ground electrode (Ag/AgCl pellets) (Science Products, Germany).

Make sure that the silver wires are properly chlorinated. Use, for instance, an automatic chlorider (ACI-01, npi electronic, Germany) and 2 M KCl solution to chloride silver wires.

4. Motorized *x-/y-/z*-manipulator for precise patch pipette positioning, e.g., Junior 3 Axes or Unit Mini 3 Axes (Luigs & Neumann, Germany), MP-225 (Sutter Instrument, USA), PatchStar Micromanipulator (Scientifica, UK).
5. Patch pipettes made of borosilicate glass tubing (outer diameter 1.5 or 2 mm; e.g., Science Products, Germany).
6. Pipette puller for patch pipettes, e.g., DMZ-Universal-Puller (Zeitz, Germany) or P-97 Flaming/Brown micropipette puller (Sutter Instruments, USA) for making pipettes for somatic whole-cell recordings (4–6 MΩ access resistance).

   Note: Pipettes should be made fresh on the day of the experiments and kept in a dust-free state in a covered container (e.g., *Micropipette storage jar, World Precision Instruments*).
7. Intracellular pipette solution. Add dye stock solution to achieve a final concentration of OGB-1 or Fluo-5F of 200 μM and of Alexa 594 of 20 μM. Add biocytin (1.5–2.5 mg/mL; Sigma, Germany) for *post hoc* analysis of dendritic morphology.
8. ACSF.
9. Perfusion pump (Peri-Star Pro, World Precision Instruments) for continuous bath chamber and slice perfusion.
10. Temperature controller for the perfusion solution (HCC-100A, Science Products, Germany).
11. Slice grid to prevent slice from drifting during perfusion. Use a customized platinum ring (or half ring) with single fibers from dental floss or nylon that are glued onto the ring.

    Note: The size of the grid should be slightly larger than the brain slice.
12. Electrophysiology software and Data Acquisition (DAQ) board. We use custom-written software in Igor 4 in combination with an InstruTECH ITC-18 board (HEKA, Germany). Alternatively, AxoGraph X (http://axographx.com/) or pclamp10 (Axon Instruments, Sunnyvale, USA) can be used for as software for data acquisition/analysis.
13. Windows or Macintosh computer for synchronized DAQ (i.e., DC-7800, Hewlett Packard, USA).

2.9. Data Analysis Software

To analyze the electrophysiological traces, we used custom-written data analysis software that was developed within the Igor 4 programming environment. The software allowed us to extract, display and to analyze relevant electrophysiological measurements of the membrane potential and to overlay these recordings with the accompanying fluorescence measurements. To analyze fluorescence imaging data, we recommend programming a short automated algorithm (as demonstrated in Grewe et al. (5)). This requires, however, some basic skills in programming as well as in data handling and in the processing of extended image sequences. For a semiautomated extraction of dendritic calcium traces from the image sequence, we employed simple macros for the ImageJ Software. To perform subsequent data analysis of calcium data traces including threshold event finding and onset and decay fitting of calcium transients we developed an algorithm within the Igor 5 programming environment, which allows simple data handling and includes all fitting and analysis functions. To run the algorithms for the calcium imaging data, the following software and hardware are required:

1. Image processing software, ImageJ (freeware, www.imageJ.org).
2. Windows or Macintosh PC.
3. Programming environment for subsequent data analysis, Igor 4 or Igor 5 (Wavemetrics, USA) or MatLAB (The Mathworks Inc., USA).

3. Procedures

3.1. Important Methodological Steps

3.1.1. Slice Preparation

The procedure describes preparing coronal slices of the whisker-related barrel cortex of rats in such a way that the apical dendrites of layer 5 pyramidal neurons run close to parallel to the surface of the slice. This ensures stable imaging conditions and depth along the entire apical dendritic axis including most of the dendritic tuft region.

1. *Anesthesia.* Deeply anesthetize the rat with isoflurane; check for absence of the foot reflex and for slowing of the breathing rhythm.
2. *Decapitation.* Decapitate the animal.

   Note: Please ensure that the experimental protocols are in compliance with national and international guidelines for the use of experimental laboratory animals.
3. *Brain.* Open the skull quickly using fine bone scissors and carefully remove the brain. Immediately place the brain in ice-cold and carbogen-saturated cutting solution (bubble for at least 10 min).

   Note: Avoid touching the neocortex or tearing while taking off the skull (see Note 1).
4. *Cutting angle.* Slightly lift the frontal part of the neocortex by placing it onto a 10° ramp. In our hands, this results in coronal slices where the dendrites of L5 pyramidal neurons run close to parallel to the slice surface. Make a vertical (coronal) cut through the caudal part of the neocortex to discard the caudal

part of the neocortex and the cerebellum, and to provide a suitable mounting surface.

Note: This step is essential to allow imaging of distal dendrites.

5. *Mounting*. Mount the brain (both hemispheres together, frontal part of the neocortex upwards) with this surface onto the platform of the vibratome using tissue glue. Ensure a dry surface of the platform before adding the glue. After positioning of the brain, fill up the mounting chamber with ice-cold cutting solution.
6. *Slices*. Cut 300–350 μm thick coronal slices in the barrel cortex region at a blade angle of 14–17°. Slices should be cut at low forward speed and relatively high lateral vibration amplitude of the blade. Gently transfer the slices (e.g., using a glass Pasteur pipette or a plastic Pasteur pipette with the tip cut-off) onto the submerged gauze net of a beaker containing incubation solution.
7. *Incubation*. Incubate the slices at 36°C for 30 min in the oxygenated incubating solution. Use water bath to warm the beaker containing the incubation solution. After incubation, maintain at room temperature until recording.

   Note: Incubation helps the slices to recover from the slicing procedure and to get rid of dead tissue at the slice surface, thereby improving the health of the slices and the depth visibility. Slices from young animals (<3 weeks postnatal) may not require incubation, but slices from older animals will.

3.1.2. Calcium Imaging and Electrophysiology

AP-evoked calcium transients are measured in the dendrites from neocortical pyramidal neurons in brain slices using TPLSM. Whole-cell patch-clamp recordings at the soma are made to inject properly timed current pulses triggering the physiological AP patterns. These AP patterns have previously been measured in the same neuron type in vivo in response to whisker stimulation. The somatic pipette is also used to load the dendrites with a Ca^{2+}-sensitive dye (200 μM OGB-1 or 200 μM Fluo-5F) to measure the AP-associated calcium transients and a red Ca^{2+}-insensitive dye (20–50 μM Alexa 594) to visualize the morphology of the dendritic trees. TPLSM in line-scan mode detects changes in [Ca^{2+}] in the different dendritic compartments, i.e., basal dendrites, main apical dendrites and apical tuft dendrites. For each stimulus trial, a background region next to the dendritic region of interest (ROI) will be measured to correct for the absolute fluorescence at resting [Ca^{2+}], and the changes in fluorescence associated with changes in [Ca^{2+}] induced by the AP patterns. We routinely include biocytin in the patch pipette for *post hoc* morphological reconstruction of the neurons. This analysis serves to confirm the identity of the recorded cell type and can also be used for morphological analysis and computer simulations.

1. *Artificial cerebrospinal fluid (ACSF) perfusion of bath chamber at physiological temperature.* Saturate the ACSF solution for 30 min with Carbogen (95% O_2 and 5% CO_2). Start the bath chamber perfusion with the ACSF solution and adjust for the right perfusion speed (1–2 mL/min) and temperature (32–36°C).

   Note: Near-physiological temperature is important to study back-propagation of AP patterns, because the back-propagation efficacy is strongly dependent on the temperature (16) (see Note 4). *For instance, at room temperature the decay in AP amplitude with distance is strongly reduced! This is largely due to the temperature dependence of the kinetics of voltage-gated ion channels.*

2. *Slice position.* Transfer the slice from the beaker containing incubation solution to the recording chamber of the microscope containing ACSF and position it accordingly to allow imaging within the area of interest. Keep the slice in place with a slice grid.

3. *Brain region of interest.* Use a low magnification objective (5×) to localize your cortical layer of interest, e.g., layer 5A of the whisker-related barrel field of the somatosensory cortex.

4. *Neurons of interest.* Switch to a high magnification objective (40×; NA 0.8) and use infrared differential interference contrast (IR-DIC) or infrared Dodt gradient contrast (17) to visualize the target neurons you wish to record from (e.g., layer 5 pyramidal neuron). Check the orientation of the main dendrites. Dendrites should run close to parallel to the slice surface to ensure that distal dendrites are not cut or too deep below the slice surface for the calcium imaging (see Note 1b).

5. *Whole-cell recording in current-clamp mode.* Use a patch pipette containing intracellular solution and the desired Ca^{2+}-sensitive and Ca^{2+}-insensitive dyes to patch neuronal somata 30–100 μm below the slice surface. Move the pipette with the recording solution under slight positive pressure (~20–30 mbar) into the brain slice and toward the cell body to "clean" its membrane.

   Note: To minimize background fluorescence due to the dye spilling out of the pipette, use low pressure and approach the cell quickly. In addition, the very tip of the pipette can be filled with a dye-free internal solution.

   Once the pipette touches the membrane and a moving dimple is observed (caused by the pipette pressure), release the pressure and if necessary apply a small negative pressure to obtain a high resistance seal between the pipette and the cell membrane ("giga-ohm seal," measured resistance >1 GΩ). The pipette resistance can be measured in the voltage-clamp mode by applying brief and small voltage pulses to the pipette while measuring the resulting current. After a successful "giga-ohm seal" formation give brief suction pulses to break the membrane patch at the pipette tip to obtain the whole-cell configuration. Switch from voltage-clamp mode to the current-clamp mode that allows the injection of current while measuring the membrane potential. Short current pulses are applied to allow compensation of the cell capacitance and measurement of the access resistance between the pipette interior and the cell interior ("bridge").

Note: It is important to keep the access resistance below 20 MΩ to enable proper dye filling of the dendritic arbor! To avoid high access resistance due to blocking of the pipette tip make sure the internal solution is properly filtered. Also use pipettes with a low open tip resistance (*4–6 MΩ*) (see Note 3).

6. *Characterization of type/viability of neuron.* Measure the current–voltage (I/V) relationship by injecting 500 ms long hyperpolarizing and depolarizing current pulses that change in a stepwise fashion (e.g., start at −200 pA, steps: 10–20 pA). This procedure helps to characterize the neuronal cell type (in this case pyramidal neuron) and the neuronal AP firing pattern, and to check the viability of the cell of interest (including membrane potential, AP threshold, AP firing pattern, input resistance).

7. *Dye equilibration.* Wait an appropriate amount of time for the dye to equilibrate within the dendritic compartment of interest before starting the calcium imaging (see Note 3)!

   Note: As a rough guide, one should wait >15 min for very proximal dendrites, >30 min for more distal ones, and >45 min for calcium measurements in the apical dendritic tuft.

   As stated above, a low access resistance is crucial for a good dye loading of the dendritic trees (see Note 5). Monitor the dye loading of the dendrites by measuring the fluorescence intensity.

8. *Dendritic region of interest* (*ROI*). Use the *x/y* motorized microscope stage to move to the dendritic ROI (proximal or distal dendrite). Keep track of the distance from soma of the dendritic imaging site by monitoring the positions of the *x/y*-motorized stage.

9. *Laser intensity.* For the fluorescence measurements, the laser intensity should be kept as low as possible to avoid bleaching of the dye and photo-damage.

   Note: Expose the dye-filled neuron to as little laser light as possible.

   Bleaching can be measured by plotting the fluorescence trace of the Ca^{2+}-insensitive dye (Alexa-594). Photo-damage will lead to an increase in the baseline fluorescence of the Ca^{2+}-sensitive dye (OGB-1, Fluo-5F), the absence of a change in fluorescence associated with APs, or even to swelling of the dendrites (see Note 4).

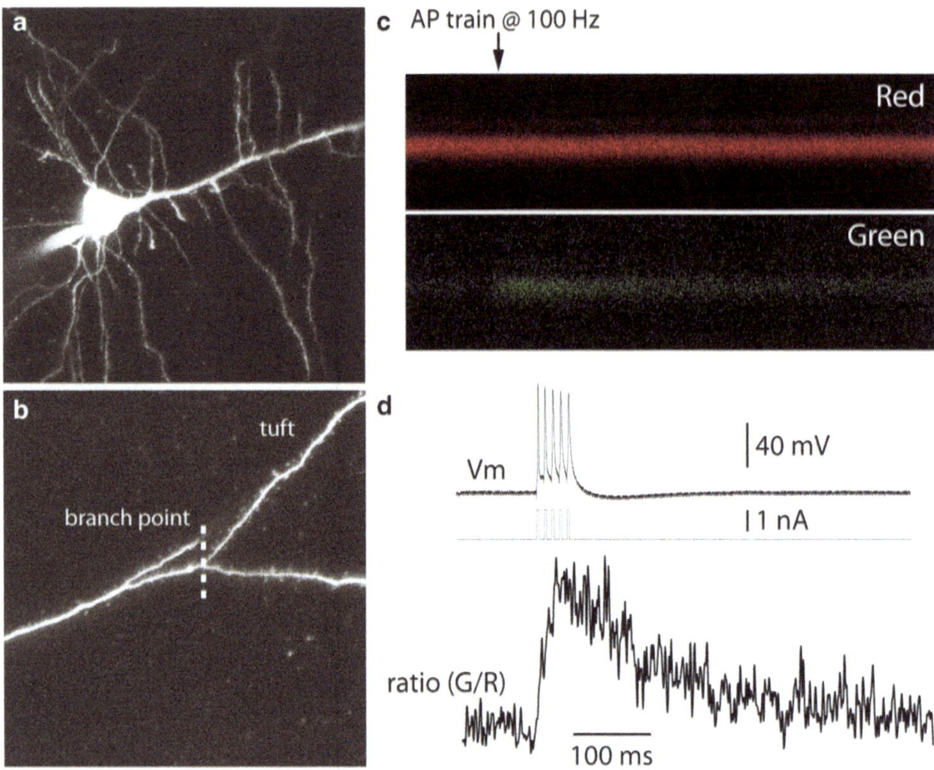

Fig. 2. Two-photon imaging of dendritic calcium transients. Changes in [Ca^{2+}] evoked by a short high-frequent burst of back-propagating action potentials. (**a**) Two-photon microscopy image of a L5B pyramidal neuron. (**b**) Dendritic region showing the major branch point and the tuft region. Neurons were loaded with a *green* Ca^{2+} indicator (200 μM Fluo-5F) and a *red* Ca^{2+}-insensitive dye (20 μM Alexa-594). (**c**) *Red* and *green* fluorescence during a fast line scan (*dashed line* in (**b**)); *horizontal axis* indicates time. (**d**) Time courses of membrane potential (*top*) and calcium transient. The ratio G/R is obtained by dividing the *green* (G) over the *red* (R) fluorescence.

10. *Line-scan measurements of dendritic calcium dynamics.* When the ROI has been positioned, start performing line-scans across the dendrite. For each stimulus trial scan a background region (a region without any fluorescence labeling) next to the dendrite of interest to correct for tissue autofluorescence. A trial consists of a short baseline period (usually 50–100 ms before stimulation) to calculate resting [Ca^{2+}] fluorescence of the dendrite, and of a period containing the calcium transients evoked by the physiological AP firing patterns. APs are triggered by short (1–3 ms) current injections. For each stimulus paradigm we suggest three to six imaging repetitions of a single line scan (256 points, line-scan time 2 ms, zoom factor 8–16) to obtain a better SNR. Intervals between individual sweeps should be at least 30 s to prevent bleaching and photo-damage. For measurement of peak fluorescence see Fig. 2.

11. *Synchronization of electrophysiology and imaging*. Use the TTL trigger output of the Leica Imaging Software to start electric current injections within the electrophysiology software. In our system, the line-scan images are stored using the Leica Microscope Software and the electrophysiology data are stored using the custom-written electrophysiology software in the Igor 4 environment.

3.1.3. Semiautomated Analysis of Fluorescent Calcium Traces Data

After each imaging experiment, the first step is to convert the acquired line-scan image data into numerical fluorescence data traces of the region of interest (here part of the dendrite).

This can be done by using implemented functions of the microscope image processing software such as that available for Leica, Zeiss or Olympus multi-photon systems. Otherwise, we recommend using the ImageJ software (freeware, http://rsbweb.nih.gov/ij/) to draw analysis regions into the image around subjects of interest (here the line-scan position over time, Fig. 2c) and to store fluorescence values as numerical data traces. When using ImageJ, these procedures can be easily recorded and transferred via small macros to automate all processes (Fig. 3a). For semiautomatic data detection and fitting of calcium fluorescence traces (Fig. 3b), we developed an automated calcium event detection algorithm in the Igor programming environment (Igor 4, Wavemetrics, USA). Nevertheless, this data analysis algorithm can be easily implemented in Matlab or other programming languages. The event detection and transient fitting algorithm is based on the following steps:

1. Find calcium events by using a threshold trigger. For more efficient calcium event detection, we recommend a "Schmitt"-trigger approach, where two thresholds (+2 S.D. and −1 S.D.) that have to be delayed by a minimum of 50–100 ms have to be detected for a successful event detection.
2. Fit calcium transient onset with exponential function. Set a window with the width of 50–100 ms around the first detected threshold (+2 S.D.) and fit the onset trace with a rising exponential function. Onsets of detected events were fitted within a window of ±50 ms (±60 ms for <200 Hz cell sampling rate) around the initial guess t_0 from the event detection routine with the following function:

$$
\begin{aligned}
f_{\text{onset}}(t) &= A(1 - \exp(-(t - t_0)/\tau_{\text{on}})\exp(-(t - t_0)/\tau_1) \quad (t > t_0) \\
f_{\text{onset}}(t) &= 0 \quad (t <= t_0) \qquad (1)
\end{aligned}
$$

The free fitting parameters were t_0, t_{on}, and A. t_0 were constrained to the fit window and t_{on} was constrained to 40 ms to avoid artificially steep or shallow onsets. The major goal of onset fitting was to obtain a good estimate of t_0.

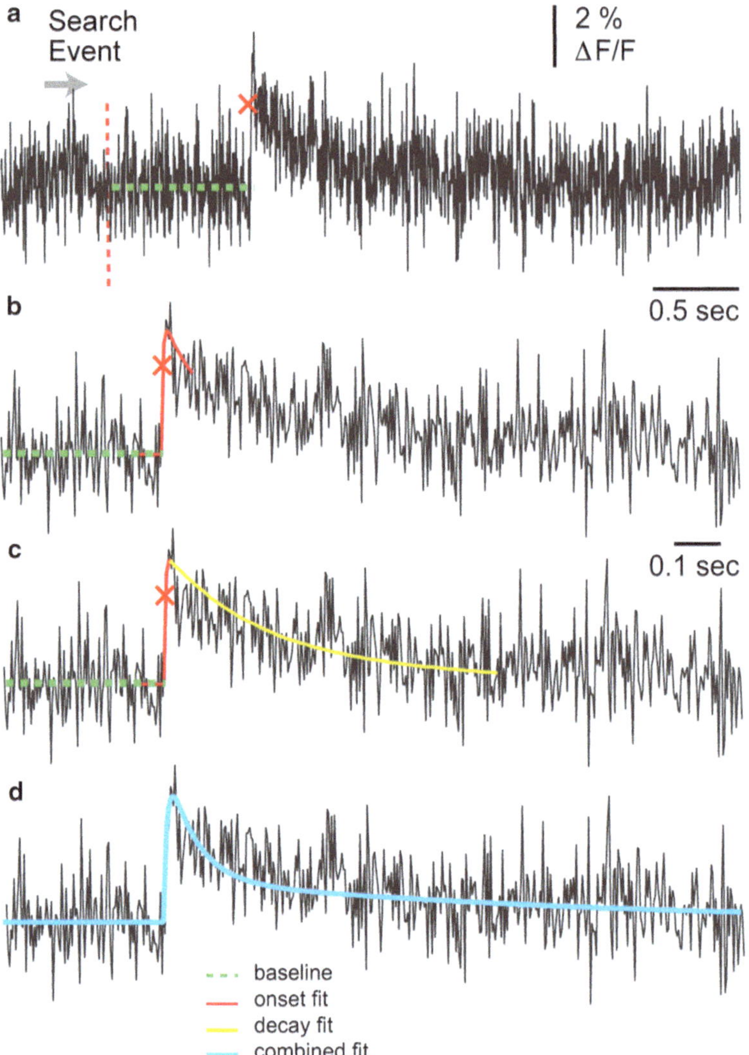

Fig. 3. Automated spike detection algorithm. (**a**) Search event using a threshold-trigger. (**b**) Transient onset-fit. (**c**) Transient decay-fit. (**d**) Combined fits.

3. Fit the calcium decay whether using single or double exponential functions. Find the maximum of the calcium transient and fit an exponential decay from the max value to the second threshold level (–1 S.D.). If necessary, search transient peak amplitude on smoothed (point averaged) calcium traces.
4. Combine the two fits to a single fitting function to extract characteristic parameters such as transient peak amplitude, decay time constants, and onset start times.

$$f_{Ca}(t) = (1 - e^{-(t-t_0)/\tau_{on}})(A_1 e^{-(t-t_0)/\tau_1} + A_2 e^{-(t-t_0)/\tau_2}) \ (t > t_0)$$
$$f_{Ca}(t) = 0 \ (t <= t_0) \quad (2)$$

A_1, t_1, A_2, and t_2 were fitting variables while t_0 and t_{on} were constrained to the values determined by the onset fitting routine. The fit range was from t_0 to 2.5 s thereafter. For the separation of overlapping Ca^{2+} transients, for instance when associated with high-frequency AP patterns, we recommend a "spike-peeling" approach similar to the one described for in vivo application (18).

Another way of analyzing calcium transients, especially if the baseline fluorescence of the Ca^{2+}-indicator is very low (e.g., in the case of Fluo-5F), is to measure the ratio of the changes in green fluorescence (Ca^{2+}-sensitive dye) over red fluorescence (Ca^{2+}-insensitive dye). With this method, the amplitude of the recorded signal is not affected by different surface to volume ratios within the dendrite or by the diffusion of the indicators. In addition, since the red dye, Alexa 594, produces a strong signal even at low concentrations (20 μM), Ca^{2+} signals expressed as ΔG/R tend to be less noisy. Furthermore, the bright Alexa 594 fluorescence signal allows the visualization of fine morphological details of the dendritic tree. Some caution has to be taken with this approach, though, since the diffusion kinetics of the red and the green dye are not identical.

Briefly, for this analysis:

1. Calculate the green fluorescence baseline as the average of the green dye (Fluo-5F or OGB-1) intensity values during the first milliseconds of the recording, before the stimulation.
2. Subtract this baseline from all intensity values of the green dye.
3. Divide the resulting values by the average intensity value of the red dye, Alexa 594.
4. Normalize the G/R_{max} value to 100%. Calculate the G/R_{max} value as follow. Fill a pipette with the regular intracellular solution plus 2 mM Ca^{2+} and perform a bi-directional line scan through this pipette. Calculate an average of the green intensity values (should now be maximal) divided by the red intensity values.

3.2. The Application of Electrophysiology and Imaging Techniques: Imaging of Physiological Relevant AP Patterns (Examples)

We used the described approach, of combining whole-cell recordings and TPLSM for Ca^{2+} imaging in vitro, to explore the active properties of dendrites of layer 5A pyramidal neurons and to investigate how these dendrites process physiological AP patterns (5). These AP patterns were previously recorded in vivo in the same neuron type of the whisker-related "barrel cortex" in response to whisker stimulation (19). Specifically, our approach allowed us to compare the active properties of the different dendritic compartments of L5 pyramidal neurons, i.e., basal dendrites, main apical dendrites, and apical tuft dendrites. We also asked how these properties change during postnatal development and which types of

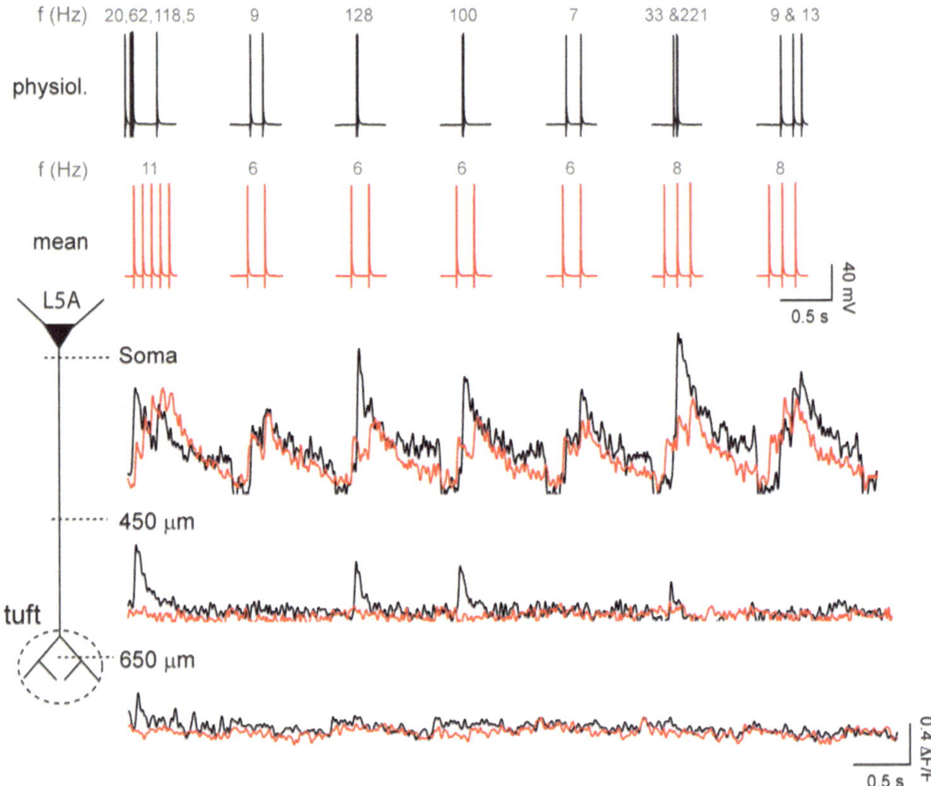

Fig. 4. Physiological firing patterns propagate more efficiently into the distal apical dendrites than mean firing patterns. *Upper two panel*, physiological firing patterns (evoked) and corresponding mean firing patterns (rate). *Lower three panels*, calcium traces of these AP patterns along the apical dendrite at various distances from soma. Neuron was filled with 200 µm OGB-1, and traces are plotted as $\Delta F/F$.

voltage-gated ion channels shape the AP-induced calcium signals. Most importantly, it allowed us to ask whether the AP back-propagation efficacy in the different dendritic compartments relies on the temporal structure of the AP firing pattern. The results from this and other studies emphasize the importance of studying neuronal properties in a manner that reflects the activity of the specific cell type under investigation. In the following paragraphs, we give a few examples of these experiments.

To trigger physiological AP firing patterns, the absolute time of each AP during these patterns was taken and a replica trace consisting of brief current pulses at the respective time points was produced. These physiological firing patterns were compared to those evoked by the same number of action potentials presented at the mean frequency (mean firing pattern). Figure 4 illustrates one such experiment. Several sequences of physiological firing patterns and the corresponding mean firing patterns are shown in Fig. 4a, and the resulting Ca^{2+} signals are depicted at different locations along the apical dendrite in Fig. 4b. From these experiments, it became clear that AP firing patterns containing high-frequency components

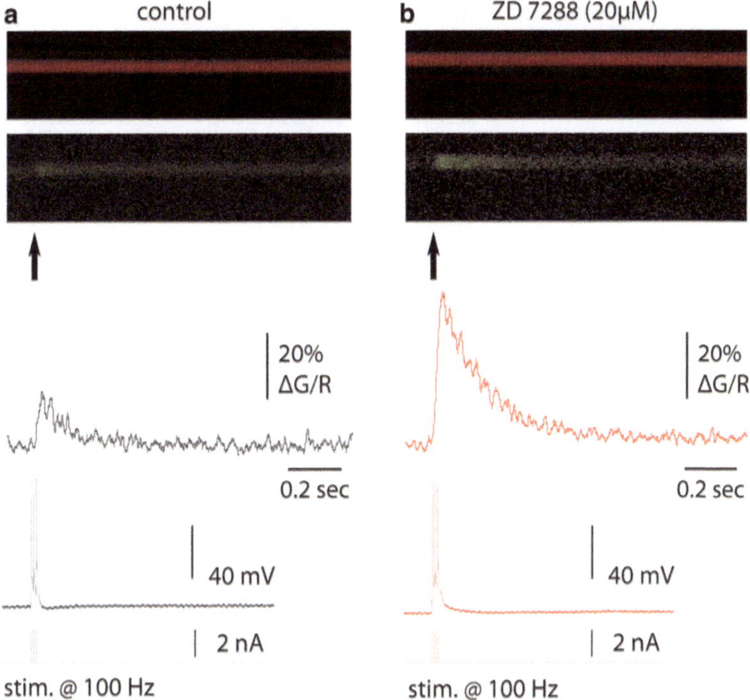

Fig. 5. H channels contribute to dendritic calcium transients. Changes in $[Ca^{2+}]$ evoked by a short high-frequent burst of back-propagating action potentials at the major branch point of the apical dendrite before (**a**) and after the application of the specific h channel blocker ZD7288 (20 μM) as shown in (**b**). Neurons were loaded with a *green* Ca^{2+} indicator (200 μM Fluo-5F) and a *red* Ca^{2+}-insensitive dye (20 μM Alexa-594). The ratio G/R is obtained by dividing G over *red* fluorescence (average of three traces).

propagate much more efficiently into the distal dendrites when compared to those without high-frequency components (including mean firing patterns). These experiments also revealed that basal dendrites and apical tuft dendrites have very different active properties (5).

Figure 5 illustrates the effect of the specific h channel blocker ZD7288 (20 μM) on changes in $[Ca^{2+}]$ associated with short high-frequent (100 Hz) trains of three bAPs. Calcium transients were measured in line-scan mode close to the major branch point ~400 μm from soma in the apical dendrite of a L5B pyramidal neuron.

3.3. Outlook

The method described can be used to address many questions regarding calcium signaling in dendrites, spines, and axons in vitro with excellent spatial and temporal resolution. Super-resolution microscopy, such as STED or PALM microscopy, enables the measurement of Ca^{2+} dynamics at even higher resolutions. In addition, TPLSM can be used in vivo to image calcium transients in dendrites in response to sensory stimuli (e.g., (20, 21)). Alternatively, dendritic calcium signals in vivo can also be measured using a fiberscope approach (22)—even in moving animals. These approaches can be aided by methods to genetically express Ca^{2+}-sensitive dyes (21, 23). Combined with viral transfection methods, opto-genetics and trans-synaptic tracing approaches, researchers have now a large toolset at hand to measure calcium dynamics in specific neuron populations both in vitro and in vivo (12).

4. Notes

Although the described experiments were performed with tissue from rats, they work equally well with mouse brain slices or other tissue.

1. Slices
   - (a) Slice quality

     Take great care to avoid any damage to the neocortex during the surgery and slicing procedure. A good slice quality is crucial for the experiment. Since it deteriorates with time, use slices soon after the incubation period. If the slice looks unhealthy, discard it and take a new one. Signs of unhealthy slices include an abundance of dead or swollen neurons, very prominent nuclei, neuronal surfaces with a high contrast, etc. It is advisable to cut fresh slices rather than trying to obtain measurements from unhealthy ones.
   - (b) Orientation of dendrites

     To image calcium transients in apical dendrites hundreds of μm from the soma, their orientation has to be parallel or close to parallel to the surface of the slice. A steep dendrite angle means that the distal dendrites are either cut or very deep within the slice reducing the imaging quality. A shallow dendrite angle preserving most of the dendritic trees requires to either patch cell bodies well below the slice surface or to measure dendritic calcium dynamics relatively deep into the slice.

     Note: The cutting angle can be optimized using animals that express a fluorescent marker in the neuron type of interest. This can be achieved using viral constructs, or in the case of mice, using a mouse line with a genetically encoded fluorescent marker (24).
   - (c) Slicing procedure

     For proper mounting of the brain onto the platform of the vibratome, it is important to use the right amount of glue. Apply a thin and uniform layer of glue in the center of the platform (the platform has to be completely dry). The brain has to be well glued so that it does not move during the slicing procedure. Too much glue, however, may cover some of the tissue facing the blade, thus causing tissue damage. Several settings of the vibratome are crucial to obtain good slice quality. These include the angle, speed, and lateral amplitude of the blade. As a rule of thumb, use an angle of 14–17°, a slow forward speed and relatively large vibration amplitude. The specific values need to be optimized under visual control and depend on the brain region and the age of the animals. During the slicing, the blade should not push the tissue but rather cut through it in a smooth manner. Ice-cold cutting solution helps to make the brain firmer providing better cutting conditions. For older animals, in addition to the *dura mater* also the *pia mater* may need to be removed before slicing. This has to be done with fine forceps by peeling it off in a very careful manner.

2. Choice of fluorescent dye
To visualize thin dendritic structures during Ca^{2+}-imaging experiments, we recommend combining a Ca^{2+} indicator with a highly fluorescent Ca^{2+}-insensitive dye whose emission spectra can be clearly separated from each other. For our recordings, we use a combination of the red Ca^{2+}-insensitive dye Alexa 594 (20–50 μM) and a green Ca^{2+}-indicator, either OGB-1 (200 μM) or Fluo-5F (200 μM). Today, a large number of different synthetic Ca^{2+}-indicators, such as Oregon Green BAPTA (OGB), Calcium Green, Fluo, Rhod and Fura are available, thereby covering a wide range of Ca^{2+}-binding affinities (for more Ca^{2+}-indicators please see Table 1). Usually these indicators combine a fluorophore that is directly coupled to a Ca^{2+} buffer (e.g., BAPTA). Ca^{2+}-binding to this buffer directly changes the fluorophore properties resulting in an increasing or a decreasing fluorescence signal, which indirectly reports an alternating Ca^{2+} concentration. The binding properties of the dye are usually given as *Kd* which directly indicates its binding affinity to Ca^{2+}. While low, affinity dyes (high *Kd*) are mostly used to monitor large changes in Ca^{2+} concentration, high affinity dyes with a low *Kd* can be used to measure small changes in the Ca^{2+} concentration on a fast time scale (5). More recently, promising genetic Ca^{2+} indicators have been successfully demonstrated in vitro and in vivo (21, 23, 25).

In order to choose the appropriate Ca^{2+} indicator, one should consider the experimental conditions and aims. Usually, experimenters are interested in maximizing the SNR during their recordings, but dynamical ranges of Ca^{2+} indicators differ depending on the underlying change in the Ca^{2+} concentration. Since indicators with a high Ca^{2+} affinity usually saturate at smaller Ca^{2+} concentrations, large concentration changes can be better obtained with low affinity dyes. For example, changes in the somatic Ca^{2+} concentration induced by single APs or short AP trains are expected to largely increase compared to Ca^{2+} changes induced by subthreshold events. Table 1 provides an overview of standard Ca^{2+} indicators and their dissociation

Table 1
Properties of fluorescent Ca^{2+} indicators

Fluorescent Ca^{2+} indicator	*Kd* nM (or mM where indicated)	ABS	EM	Distributor	AM dye available
Oregon green 488 BAPTA-1	170	494	523	IN	✓
Oregon green 488 BAPTA-2	580	494	523	IN	✓
Oregon green 488 BAPTA-6F	3 μM	494	523	IN	✓
Orange green 488 BAPTA-5N	20 μM	494	521	IN	✓
Calcium green-1	190	506	533	IN	✓
Calcium green C_{18}	280	509	530	IN	–
Calcium green-5N™	14 μM	506	532	IN	✓
Calcium green-2™	550	503	536	IN	✓
Calcium orange™	185	549	576	IN	✓
Calcium orange 5N™	20 μM	549	582	IN	
Calcium crimson™	185	589	615	IN	✓
Fluo-3	390	506	526	IN	✓
Fluo-3FF	41 μM	526	526	IN	✓
Fluo-4	345	494	516	IN	✓
Fluo-4FF	9.7 μM	494	516	IN	✓
Fluo-5F	2.3 μM	494	518	IN	✓
Fluo-5N	90 μM	493	518	IN	✓
Rhod-2	570	552	571	IN	✓
Rhod-FF	19 μM	552	577	IN	✓
Rhod-5N	320 μM	551	577	IN	✓
X-rhod-1	700	580	602	IN	✓
X-rhod-5F	1.6 μM	580	602	IN	✓
X-rhod-FF	17 μM	578	602	IN	✓
Magnesium green™	6 μM	506	531	IN	✓

Data taken from www.invitrogen.com
K_d determined in 100 mM KCl, pH 7.2 at 22°C. *ABS* absorption; *EM* emission; *AM* acetoxymethyl ester; *IN* Invitrogen

constants (Kd_s). For the best linear proportionality between the fluorescence intensity and the Ca^{2+} concentration, we recommend to choose an indicator for which $[Ca^{2+}] >> Kd_,$ as defined by:

$$\frac{F - F_{min}}{F_{max} - F_{min}} = \frac{[Ca^{2+}]}{K_d}$$

where Fm_{in} and Fm_{ax} indicate the minimum or maximum possible fluorescence, respectively. To allow a fast and accurate measurement of the fluorescence signals, standard microscope detectors use one or more photomultiplier tubes (PMTs). Combining multiple PMTs with dichroic beam splitters and filters permits simultaneous detection of different fluorescence indicators at different wavelengths. For in vitro slice preparations we recommend the use of two PMTs for each channel; two to collect the epi-fluorescence and two to detect the transmitted fluorescence (see also Fig. 1).

3. Loading of the dye

   Fluorescent dyes are loaded through the recording pipette. A low access resistance (<20 MΩ) is critical for a fast and efficient dye loading. To achieve low access resistance, use pipettes with low open tip resistance (4–6 MΩ). However, be aware that if the pipette resistance is too low, it can be difficult to maintain long-lasting recordings and washout of cytoplasmic elements may impair the calcium measurements. Calcium measurements should start 20–40 min after establishing the whole-cell configuration to ensure homogeneous dye concentration throughout the dendritic arbors. As a rough guide, one should wait >15 min for very proximal dendrites, >30 min for more distal ones, and >45 min for calcium measurements in the apical dendritic tuft. Other methods can be used for dye loading such as bolus injection of membrane permeant dyes (AM-dyes).

4. No response/small response

   If there are no responses following stimulation when you start to image, first try to increase the stimulation (number of AP in the train/frequency) and/or change slightly imaging location. Try also slightly increasing the laser power. If you still do not see any change in $[Ca^{2+}$, even when imaging close to the soma, check your Ca^{2+}-sensitive dye efficiency. Compare the fluorescence of the recording pipette at different Ca^{2+} concentrations (i.e., add 2 mM Ca^{2+}, measure fluorescence before and after). If there is still no difference in the fluorescence intensity, use a new batch of dye.

   Small fluorescence responses in response to a strong stimulation can be due to a washout of cytoplasmic components due to a too low access resistance resulting in a degradation of the Ca^{2+} channel-mediated responses. It can also be caused by photodamage leading to an increase in baseline fluorescence. TPLSM permits good tissue penetration, thus allowing one to image deeper in the slice, as well as a good SNR and reduced photobleaching in the region of interest. However, photodamage and photobleaching can occur in the focal volume (26) and should be avoided by minimizing the laser power to the lowest setting that still allows a good measurement of the fluorescence change.

Trace amounts of heavy metals in the experimental solutions may severely affect fluorescence properties of the dyes, resulting in an inaccurate response to stimulation. Always use highly purified water for preparing the solutions to avoid this problem (27). In addition, because voltage-gated ion channels are highly sensitive to temperature (16), a wrong setting of the bath temperature could also affect the amplitude of the signal. Be careful to always set the temperature between 32 and 36°C.

Acknowledgments

The authors wish to thank Drs. B. Kampa and M. Ginger for valuable comments on the manuscript.

References

1. Stuart G, Sakmann B (1994) Active propagation of somatic action potentials into neocortical pyramidal cell dendrites. Nature 367(6458):69–72
2. Larkum M, Zhu J (2002) Signaling of layer 1 and whisker-evoked Ca^{2+} and Na^{+} action potentials in distal and terminal dendrites of rat neocortical pyramidal neurons in vitro and in vivo. J Neurosci 22(16):6991–7005
3. Frick A, Magee J, Koester H, Migliore M, Johnston D (2003) Normalization of Ca^{2+} signals by small oblique dendrites of CA1 pyramidal neurons. J Neurosci 23(8):3243–3250
4. Waters J, Larkum M, Sakmann B, Helmchen F (2003) Supralinear Ca^{2+} influx into dendritic tufts of layer 2/3 neocortical pyramidal neurons in vitro and in vivo. J Neurosci 23(24):8558–8567
5. Grewe B, Bonnan A, Frick A (2010) Back-propagation of physiological action potential output in dendrites of slender-tufted L5A pyramidal neurons. Front Cell Neurosci 4:13
6. Waters J, Schaefer A, Sakmann B (2005) Backpropagating action potentials in neurones: measurement, mechanisms and potential functions. Prog Biophys Mol Biol 87(1):145–170
7. Frick A, Johnston D (2005) Plasticity of dendritic excitability. J Neurobiol 64(1):100–115
8. Sjöström P, Rancz E, Roth A, Häusser M (2008) Dendritic excitability and synaptic plasticity. Physiol Rev 88(2):769–840
9. Sabatini B, Maravall M, Svoboda K (2001) Ca(2+) signaling in dendritic spines. Curr Opin Neurobiol 11(3):349–356
10. Konur S, Ghosh A (2005) Calcium signaling and the control of dendritic development. Neuron 46(3):401–405
11. Lohmann C, Wong R (2005) Regulation of dendritic growth and plasticity by local and global calcium dynamics. Cell Calcium 37(5):403–409
12. Grewe B, Helmchen F (2009) Optical probing of neuronal ensemble activity. Curr Opin Neurobiol 19(5):520–529
13. Mainen Z, Maletic-Savatic M, Shi S, Hayashi Y, Malinow R, Svoboda K (1999) Two-photon imaging in living brain slices. Methods 18(2):231–239

14. Majewska A, Yiu G, Yuste R (2000) A custom-made two-photon microscope and deconvolution system. Pflugers Arch 441(2–3):398–408
15. Koester HJ, Baur D, Uhl R, Hell SW (1999) Ca^{2+} fluorescence imaging with pico- and femtosecond two-photon excitation: signal and photodamage. Biophys J 77(4):2226–2236
16. Tsubokawa H, Ross W (1997) Muscarinic modulation of spike backpropagation in the apical dendrites of hippocampal CA1 pyramidal neurons. J Neurosci 17(15):5782–5791
17. Stuart GJ, Dodt HU, Sakmann B (1993) Patch-clamp recordings from the soma and dendrites of neurons in brain slices using infrared video microscopy. Pflugers Arch 423(5–6):511–518
18. Grewe B, Langer D, Kasper H, Kampa B, Helmchen F (2010) High-speed in vivo calcium imaging reveals neuronal network activity with near-millisecond precision. Nat Methods 7(5):399–405
19. de Kock C, Bruno R, Spors H, Sakmann B (2007) Layer- and cell-type-specific suprathreshold stimulus representation in rat primary somatosensory cortex. J Physiol 581:139–154
20. Helmchen F, Svoboda K, Denk W, Tank D (1999) In vivo dendritic calcium dynamics in deep-layer cortical pyramidal neurons. Nat Neurosci 2(11):989–996
21. Tian L, Hires S, Mao T, Huber D, Chiappe M, Chalasani S et al (2009) Imaging neural activity in worms, flies and mice with improved GCaMP calcium indicators. Nat Methods 6(12): 875–881
22. Murayama M, Larkum M (2009) Enhanced dendritic activity in awake rats. Proc Natl Acad Sci USA 106(48):20482–20486
23. Lütcke H, Murayama M, Hahn T, Margolis D, Astori S, Zum Alten Borgloh S et al (2010) Optical recording of neuronal activity with a genetically-encoded calcium indicator in anesthetized and freely moving mice. Front Neural Circuits 4:9
24. Gong SC, Doughty M, Harbaugh CR, Cummins A, Hatten ME, Heintz N, Gerfen CR (2007) Targeting Cre recombinase to specific neuron populations with bacterial artificial chromosome constructs. J Neurosci 27(37):9817–9823
25. Wallace D, Meyer zum Alten Borgloh S, Astori S, Yang Y, Bausen M, Kügler S et al (2008) Single-spike detection in vitro and in vivo with a genetic Ca^{2+} sensor. Nat Methods 5(9): 797–804
26. Svoboda K, Yasuda R (2006) Principles of two-photon excitation microscopy and its applications to neuroscience. Neuron 50(6): 823–839
27. Grynkiewicz G, Poenie M, Tsien R (1985) A new generation of Ca^{2+} indicators with greatly improved fluorescence properties. J Biol Chem 260(6):3440–3450

Chapter 9

Juxtacellular Labeling in Combination with Other Histological Techniques to Determine Phenotype of Physiologically Identified Neurons

Ruth L. Stornetta

Abstract

This chapter summarizes the extracellular recording and juxtacellular labeling method and its application to the characterization of cardiorespiratory brainstem neurons, although this technique could be applied to any neuron. The combination of this method with immunohistochemical and in situ hybridization techniques is emphasized in detail.

Key words: Juxtacellular, Biotinamide, Phenotype, In situ hybridization, Immunohistochemistry

1. Introduction

1.1. History of Juxtacellular Labeling

The ability to label a single physiologically characterized neuron recorded from an intact live animal has many obvious applications for a wide variety of neuroscience studies. The juxtacellular approach has some advantages over the intracellular approach as the intracellular technique requires fine-tipped electrodes with high impedance that are difficult to introduce through many millimeters of brain tissue in the intact whole animal. The juxtacellular method is much more productive and many neurons can be recorded and labeled from a single animal over the course of several hours.

This technique was pioneered by Pinault (1) and described in detail by this author in 1996. Several updates on the technique are available in combination with other histological methods including immunohistochemistry and in situ hybridization (2) as well as anterograde/retrograde tracing and electron microscopy (3). The juxtacellular labeling technique has been used by many researchers

Emilio Badoer (ed.), *Visualization Techniques: From Immunohistochemistry to Magnetic Resonance Imaging*, Neuromethods, vol. 70, DOI 10.1007/978-1-61779-897-9_9, © Springer Science+Business Media, LLC 2012

for determining phenotype and structure of physiologically identified neurons from cortex (4) to spinal cord (5). The technique may also be applied in other systems where cells may be recorded with extracellular pipettes such as in vitro brain slices (6) and the perfused working heart brain preparation (7). In this chapter, I will illustrate how to record a cell using the juxtacellular technique and to process brain tissue containing biotinamide juxtacellularly labeled neurons in combination with multiple histochemical labeling techniques to further refine the phenotype of a physiologically identified cell in vivo.

1.2. Fundamentals of Juxtacellular Labeling

Although the exact mechanism of juxtacellular labeling is not known, it is most likely a form of electroporation. Electroporation relies on a relatively large (5–10 V) electrical current disturbing the lipid bilayer and opening pores that allow entry of extracellular material (8). Classically, electroporation has been used to transfer DNA or RNA into cells using negative current (8) and more recently has been used for other substances including siRNA (9) and calcium indicator dyes (10, 11). However, in the case of juxtacellular labeling, much lower current pulses are used, in the neighborhood of 1.5–9 nA (equivalent of ~100 mV).

The other factor essential to juxtacellular labeling that is quite different from traditional electroporation is that the subject cell's electrical activity must be entrained to the current pulses from the extracellular recording pipette. Without this entrainment, juxtacellular labeling is not observed (1). Because a cell's activity is most easily manipulated by depolarizing (i.e., positively charged) currents, the substances with the greatest likelihood of success are positively charged small molecules such as biotinamide that can be expelled from the pipette iontophoretically using the same current that is capable of activating the cells.

1.3. Summary of Protocol Steps

The labeling is accomplished by the juxtacellular recording and entraining a cell with positive pulses of current through a glass pipette containing a solution of biotinamide. The deeply anesthetized animal is then perfused and the brain is postfixed and sectioned. This method can also be applied to brain slices, in which case, the slices are merely placed into 4% paraformaldehyde for up to 4 days. The sections are processed for the specific histological method to reveal the labeled cell as well as other histological markers. The cell is imaged and the location of the cell is plotted.

2. Materials

All materials stored at room temperature (RT) unless otherwise noted.

2.1. Juxtacellular Labeling

1. K-series borosilicate glass with interior fine filament (WPI #K200F-4)
2. Biotinamide (*N*-(2-aminoethyl)biotinamide, hydrobromide (biotin ethylenediamine)) (Molecular Probes/Invitrogen, catalog #A-1593) stored at 4°C
3. MicroFil syringe needle (WPI #MF34G-5)
4. Fiberglass strands
5. 1 cc Luer lock syringe with 0.45-μm syringe filter
6. Intracellular amplifier with bridge mode capability (Axoclamp, Molecular Devices, Inc., Sunnyvale, CA)
7. Stimulator to step activate the amplifier (e.g., Grass 588, Grass Instruments, Quincy, MA)

2.2. Brain Perfusion

1. Peristaltic pump
2. Paraformaldehyde prills (EM Sciences)

2.3. Brain Sectioning

1. Vibrating microtome (Leica).
2. 24-Well tissue culture dish with lid.
3. Cryoprotectant solution (50 mM phosphate buffer (autoclaved), 30% ethylene glycol, 20% glycerol) stored at 4°C.

2.4. Histological Processing

2.4.1. For Solely Revealing Biotinamide Labeled Neuron for Optimal 3D Reconstruction

1. Mesh well dishes for rinsing tissues (Nason Fabrications, Fort Bragg, CA)
2. 24-Well tissue culture dish with lid
3. Tris-buffered saline (TBS, 100 mM Tris–Cl pH 7.4 in 154 mM NaCl in water)
4. 100 mM sodium phosphate buffer, pH 7.4
5. Triton-X-100
6. Vectastain Elite ABC kit from Vector Labs stored at 4°C
7. Tyramide amplification kit from NEN stored at 4°C
8. Diaminobenzidine (DAB) stored at −20°C
9. 1% Nickel ammonium sulfate (Ni) w/v in H_2O (Ni stock solution stored at RT)

2.4.2. For Combination with Immunohistochemistry

1. 24-Well tissue culture dish with lid
2. Mesh well dishes for rinsing tissues (Nason Fabrications, Fort Bragg, CA) Tris-buffered saline (TBS, 100 mM Tris–Cl pH 7.4 in 154 mM NaCl in water)
3. 100 mM sodium phosphate buffer, pH 7.4
4. Normal horse serum (undiluted serum stored at −20°C, diluted serum stored at 4°C)
5. Triton-X-100

6. Primary antibodies (stored as recommended by manufacturer)
7. Secondary antibodies (stored as recommended by manufacturer)
8. Avidin tagged with appropriate marker to reveal biotinamide labeled cell (stored as recommended by manufacturer)

2.4.3. For Combination with In Situ Hybridization

1. All materials listed under Sect. 2.4.2
2. Sheep anti-digoxigenin antibody, Fab fragments, alkaline phosphatase tagged (Roche# 11093274910)
3. Water baths at 37 and 55°C
4. EDTA
5. Sodium pyrophosphate (NaPPi)
6. Dextran sulfate
7. Yeast total RNA
8. Yeast transfer RNA (yeast tRNA)
9. Denhardt's bovine serum albumin (BSA)
10. Deionized (d.i.) formamide
11. Poly A+
12. 4 Ribonucleotide triphosphates (rNTPs)
13. Dithiothreitol (DTT)
14. Herring (or salmon) sperm (presheared and stored at 4°C)
15. Sodium citrate
16. Sodium thiosulfate (NaTS)
17. Ribonuclease (RNAse) A (pancreatic)
18. Digoxigenin-labeled riboprobe (made according to e.g., Promega Riboprobe Systems kit using) stored at –80°C

3. Methods

3.1. Juxtacellular Labeling

Record neurons with an extracellular pipette prepared from K-series borosilicate glass with interior fine filament pulled with a standard electrode puller to produce the taper ranging in length from 1 to 1.5 cm. The length of the taper is determined by the depth of brain tissue to be penetrated with longer tips necessary for deeper penetrations. Break the pipette tip to ~1 μm under a microscope or such that the resistance in vitro is 18–25 MΩ after filling with biotinamide. Fill the pulled pipette from the top using a MicroFil syringe needle with a 0.45-mm syringe filter and 1-cc syringe. The filling solution is 2% biotinamide in 0.5 M sodium acetate (see Note 1). After brain penetration, acceptable resistances will range

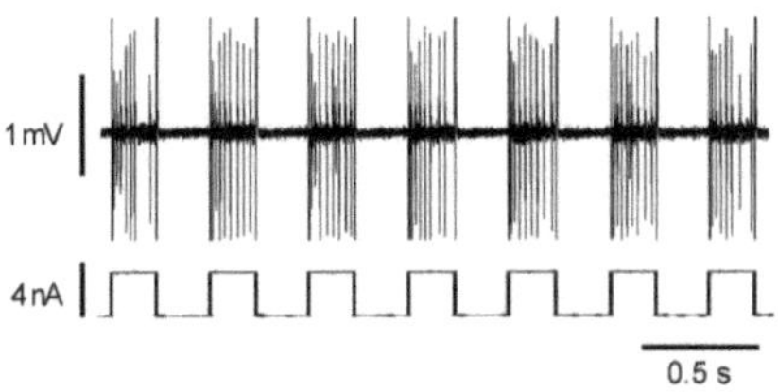

Fig. 1. Entrainment of a single neuron by extracellular application of 250-ms pulses of positive current (4 nA).

from 25 to 40 MΩ. Recordings are made with an intracellular amplifier in bridge mode so that current can be injected through the electrode while monitoring action potentials. The signals are subsequently filtered (100 Hz–3 kHz) (12). When the recording pipette is close to a cell with well discriminated action potentials, pulse anodal current (200 ms at 2.5 Hz) using a gradually increasing amount of current (0.5–9 nA) until the current injection produces bursts of action potentials from the recorded cell (see Fig. 1). At this point the current should be adjusted to the smallest possible to continue to entrain the cell's activity as this will minimize any possible damage to the cell. If another cell's activity is detected while you are attempting to entrain your neuron of interest, stop and try another cell. If two cells are close together and both cells' action potentials are noticeable or if the cells are connected by gap junctions, it is possible that two cells will be labeled. This is not usually a problem in the brainstem reticular formation where we have the most experience with this method. The cell should be entrained for as long as possible. While there is no upper limit to the duration, 2–3 min is usually adequate to fill soma and some dendrites. We have recovered neurons that have been entrained for as little as 10 s. Certainly the longer a cell is entrained, the better the filling of dendrites and initial axonal processes. After the entrainment, if possible, continue recording the cell to show recovery of normal cell activity.

3.2. Tissue Fixation

If recording from a live animal or a perfused, reduced preparation (e.g., heart-brain preparation), after successful recording session, perfuse transcardially, first with heparinized saline (~100 mL for adult rat) followed by freshly prepared 4% paraformaldehyde in 100 mM sodium phosphate buffer, pH 7.4 (~400–500 mL for adult rat). If recording from a brain slice, immerse slice in the same fixative for 24–48 h (see Note 2).

3.3. Brain Sectioning

Prepare the brain for sectioning by first removing the dura and pia if possible. Although the brain could be cryoprotected and sectioned frozen, the histological preservation is better with brains cut at RT on a vibrating microtome. Mark the side of the brain opposite to where the labeled cell is located (see Note 3). Fill a

24-well tissue culture dish with 500 μL to 1 mL of cryoprotectant per well. Section the area of the brain where the cell was labeled putting individual sections in sequential wells of a 24-well tissue culture dish. Section thickness may be 30–50 μm. Thicker sections may be possible as we have recovered labeled neurons in 300 μm brain slices from neonates. Sections may be stored at −20°C for up to several years (see Note 4).

3.4. Histological Processing of Sections for Immuno-histochemistry

All steps are performed on rotating shaker (at RT unless noted).

Transfer sections from cryoprotectant to 24-well mesh bottom dish and rinse 3×5 min in TBS (see Notes 5 and 6).

3.4.1. Revealing Biotinamide Labeled Neuron Only for Optimal 3D Reconstruction

1. Transfer sections and rinse as above.
2. Incubate 30 min in 1% H_2O_2 made in TBS (Meanwhile, prepare A/B solution from Vectastain Elite kit by adding 4 drops A+4 drops B in 20 mL TBS with 0.1% Triton, set aside at RT).
3. Rinse 3×5 min with TBS.
4. Incubate in A/B solution for 3 h. (Prepare a second A/B solution during this time).
5. Rinse 3×5 min with TBS.
6. Fill clean 24-well tissue culture dish with biotinylated tyramide (bt) solution (300–500 μL/well). Each milliliter of bt solution requires 5 μL of bt and 2 μL 5% H_2O_2.
7. Transfer sections into bt solution and incubate for 8–10 min.
8. Transfer sections back to mesh bottom dish and rinse 3×5 min in TBS.
9. Incubate 1 h in second A/B solution.
10. Rinse 3×5 min with TBS.
11. Fill clean 24-well tissue culture dish with Ni-DAB solution (4% w/v DAB with 200 μL Ni stock solution/1 mL DAB solution in TBS). Add 2 μL 5% H_2O_2 per mL of DAB solution immediately prior to use (see Note 7).
12. Add sections and react for 5–10 min under visual inspection. See Notes 8 and 9.
13. Transfer sections back to mesh bottom dish and rinse 3×5 min in TBS.
14. Transfer sections to PO_4 buffer, sort into serial order and mount onto gelatin-coated slides.
15. Dehydrate through a series of graded alcohols and xylene.
16. Coverslip with DPX mountant.

3.4.2. Processing Sections for Fluorescent Immunohistochemical Label(s) in Addition to Biotinamide Labeled Neuron

1. Transfer sections into mesh bottom dish and rinse as above (Sect. 3.4).
2. Incubate in blocking solution (10% NHS, 0.1% Triton-X-100 in TBS) for 30–60 min.
3. Add primary antibody at desired concentration made in blocking solution to new 24-well tissue culture dish (300–500 μL per well). Transfer tissue sections to primary antibody and incubate overnight (16–18 h) at 4°C.
4. Transfer sections from 24-well tissue culture dish to 24-well mesh bottom dish and rinse 3 × 5 min in TBS.
5. Add secondary antibody and avidin conjugate (for detecting biotinamide labeled neuron) at desired concentration made in 1% NHS in TBS to new 24-well tissue culture dish (300–500 μL per well).
6. Transfer sections into secondary antibodies and avidin conjugate and incubate 45 min. See Note 10.
7. Transfer sections into mesh bottom dish and rinse 3 × 5 min in TBS.
8. Transfer sections to 100 mM sodium phosphate buffer, sort into serial order and mount onto gelatin-coated slides.
9. Dehydrate through a series of graded alcohols and xylene.
10. Coverslip with DPX mountant.

3.4.3. Processing Sections for In Situ Hybridization for Specific mRNA in Addition to Biotinamide Labeled Neuron

1. Rinse sections briefly in sterile phosphate buffered saline (PBS) by transferring sections to sterile petri dish filled with PBS and then immediately transferring sections to 24-well sterile tissue culture dish filled with hybridization solution (300 μL/well). See Note 11 for hybridization recipe.
2. Incubate for 30–60 min shaking at RT, then 1 h at 37°C.
3. Add riboprobe directly to wells at ~1 μL/well. Incubate shaking at RT for 15 min.
4. Put dish with sections in hybridization solution at 55°C overnight. See Note 12.
5. Transfer sections to mesh bottom dish and rinse through the following solutions:
6. 2 × 20 min with 4× SSC/10 mM NaTS in 37°C bath.
7. 1 × 30 min with RNAse A (20 μL of RNAse A stock/5 mL RNAse buffer) in 37°C bath. See Note 13 for RNAse stock and RNAse buffer recipe.
8. 1 × 20 min with RNAse buffer in 37°C bath.
9. 1 × 20 min with 2× SSC/10 mM NaTS in 37°C bath.
10. 1 × 20 min with 0.5× SSC in 37°C bath.
11. 1 × 30–60 min with 0.1× SSC in 55°C bath. See Note 14 for SSC recipe.

12. Rinse 3 × 5 min in TBS.
13. Incubate in blocking solution (10% NHS, 0.1% Triton-X-100 in TBS) for 30–60 min.
14. Add sheep anti-digoxigenin alkaline phosphatase-tagged antibody at 1:1,000 made in blocking solution to new 24-well tissue culture dish (300–500 μL per well). Transfer tissue sections to primary antibody and incubate overnight (16–18 h) at 4°C. See Note 15.
15. Transfer sections from 24-well tissue culture dish to 24-well mesh bottom dish and rinse 3 × 5 min in TBS.
16. Rinse 1 × 10 min in NMT (0.1 M NaCl/50 mM $MgCl_2$ in 0.1 M Tris, pH 9.5).
17. Prepare alkaline phosphatase substrate colorization reaction on ice and filter with 0.45-μm syringe filter before use (Nitro Blue Tetrazolium 4.5 μL/mL NMT and BICP 3.5 μL/mL NMT). Add 300–500 μL/well to sterile 24-well tissue culture dish. Transfer sections into this solution and incubate for 1–4 h. See Note 16.
18. Transfer sections to 10 mM Tris pH 8.5/1 mM EDTA (TE 8.5) in mesh bottom dish. Rinse 3 × 10 min in TE 8.5.
19. Rinse 3 × 5 min in TBS.
20. Add avidin conjugate (for detecting biotinamide labeled neuron) at desired concentration made in 1% NHS in TBS to new 24-well tissue culture dish (300–500 μL per well).
21. Transfer sections into avidin conjugate and incubate 45 min.
22. Transfer sections into mesh bottom dish and rinse 3 × 5 min in TBS.
23. Transfer sections to 100 mM sodium phosphate buffer, sort into serial order and mount onto gelatin-coated slides.
24. Cover with aqueous mounting media, e.g., VectaShield. Seal edges with nail polish. See Note 17.

3.5. Application of the Technique

3.5.1. Juxtacellularly Labeled Neurons Optimally Labeled for Reconstruction

Reconstruction of a CO_2 sensitive neuron in the retrotrapezoid nucleus (RTN).

Neurons were recorded in anesthetized rats in the area of the RTN. The neurons that responded to increasing end tidal CO_2 with vigorous increases in activity were targeted. After juxtacellular labeling of one neuron, tissues were processed according to the protocol in Sect. 3.4.1. Figure 2a–c illustrates one such recovered neuron and one dendritic process traced over two adjacent sections and reconstructed. By using this technique, we were able to demonstrate that RTN neurons recorded in vivo and in vitro (slices of neonates age P6–P8) had both similar physiological responses to CO_2 or pH and had similar structures (Fig. 2d–f) (6).

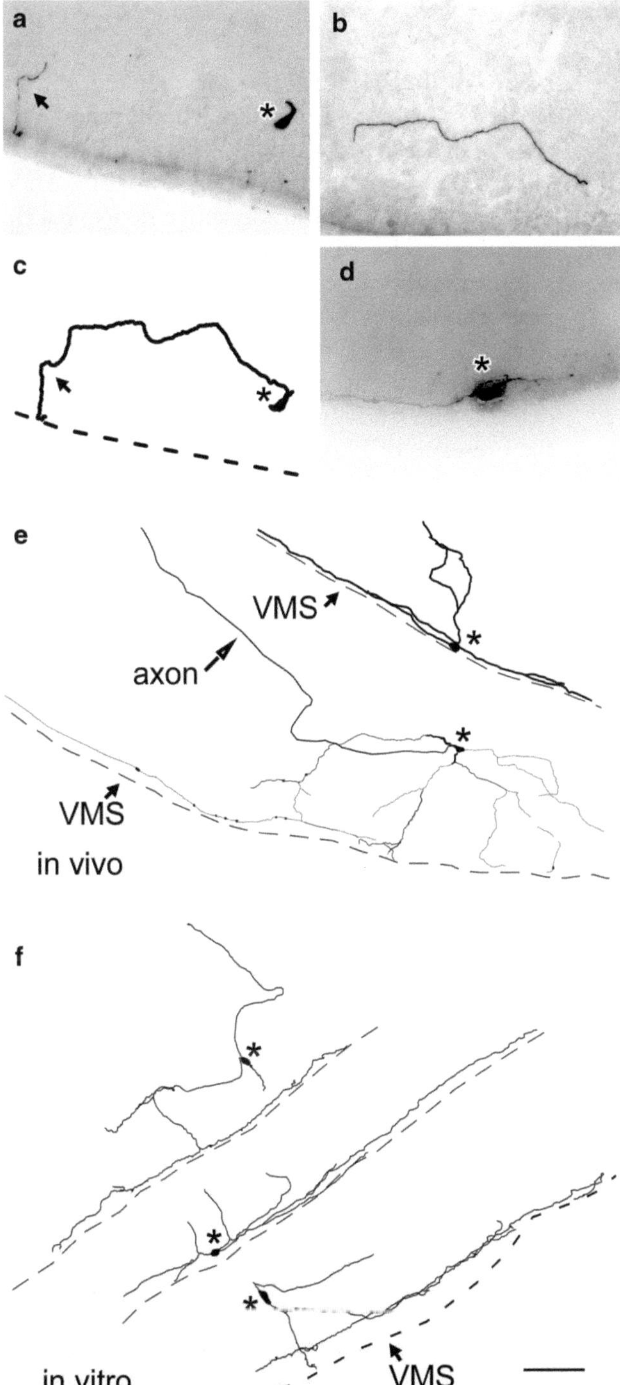

Fig. 2. Examples of juxtacellular labeling of neurons recorded in vivo and in vitro visualized for three dimensional reconstruction. (**a**, **b**) Neuron sensitive to CO_2 recorded in vivo in anesthetized rat with the juxtacellular technique and visualized with the Nickel-DAB reaction detailed in protocol (Sect. 3.4.1). The cell body in (**a**) is marked with an *asterisk*, an *arrow* points to a piece of dendrite. (**b**) Dendrite from neuron recorded in (**a**). (**c**) Drawing reconstructed from (**a**) and (**b**). The *dashed line* represents the ventral surface of the brain. (**d**) CO_2 sensitive neuron recorded and juxtacellularly labeled from transverse brain slice in neonate rat visualized by the Nickel-DAB reaction. Cell body marked with *asterisk*. (**e**) Examples of reconstruction drawings of CO_2 sensitive neurons recorded in vivo in anesthetized rats. (**f**) Examples of reconstruction drawings of CO_2 sensitive neurons recorded in vivo in transverse slices of neonate rats. For both (**e**) and (**f**) cell bodies are marked with *asterisk*. *VMS* ventral medullary surface. Scale bar is 100 μm.

3.5.2. Identifying the Phenotype of Physiologically Identified Neurons by Combining Immunohistochemistry with Juxtacellular Labeling

CO_2 Sensitive Neurons Immunoreactive for Phox2b

Neurons in the RTN in anesthetized rats were juxtacellularly labeled in order to determine whether the CO_2-senstive RTN neurons first described in 2004 (6) and illustrated in Fig. 2 also contained the transcription factor Phox2b. An example of a CO_2 sensitive Phox2b immunoreactive neuron of the RTN is shown in Fig. 3a–c. The presence of Phox2b in these RTN neurons is significant because the human disease, Congenital Central Hypoventilation Syndrome is associated with a mutation in the *PHOX2B* gene. When the same mutation is introduced in mice, the RTN is absent and the mice die at birth with major respiratory defects. The study by Stornetta et al. (13) provides more details on the presence of Phox2b in the RTN.

CO_2 Insensitive Neurons Immunoreactive for Tryptophan Hydroxylase

Serotonergic neurons in the parapyramidal area were recorded from anesthetized rats and tested for CO_2 sensitivity. All the neurons tested and juxtacellularly labeled were immunoreactive for tryptophan hydroxylase, a marker of serotonergic cells, but did not respond to CO_2 in an anesthetized rat (6). An example of one labeled neuron is illustrated in Fig. 3e, f.

pH Sensitive eGFP Immunoreactive Neurons In Vitro

eGFP-positive neurons were recorded in a 300 μm transverse slice from P6 to P8 neonate eGFP-Phox2b transgenic mouse. All eGFP-positive neurons recorded in the area of the RTN in this transgenic mouse were found to be pH sensitive. An example of one neuron recorded with juxtacellular labeling is illustrated in Fig. 3g, h. These transgenic mice are described in detail in the study by Lazarenko et al. (14).

3.5.3. Identifying the Phenotype of Physiologically Identified Neurons by Combining In Situ Hybridization with Juxtacellular Labeling

Glycinergic Neurons in Bötzinger Complex

Expiratory augmenting neurons were recorded and juxtacellularly labeled in the Bötzinger complex in anesthetized rats. The procedure for in situ hybridization in combination with detection of the labeled neurons was performed. An example of a neuron whose firing rate increased during the expiratory phase and then switched its firing pattern to preinspiratory/postinspiratory during hypercapnic hypoxia is shown in Fig. 4a from the study by Fortuna et al. (15). This neuron was positive for glycine transporter 2 mRNA revealed by in situ hybridization (Fig. 4b).

CO_2 Sensitive Neurons in RTN Contain a Marker for the Peptide Galanin

We tested whether the galanin peptide is expressed by the RTN neurons by recording and juxtacellular labeling of CO_2 sensitive neurons in the RTN of anesthetized rats. An example of such a neuron is shown in Fig. 4c. This tissue was reacted for in situ hybridization using a riboprobe for preprogalanin and the labeled neuron contained the mRNA that encodes the galanin peptide (Fig. 4d) (for more details see (16)).

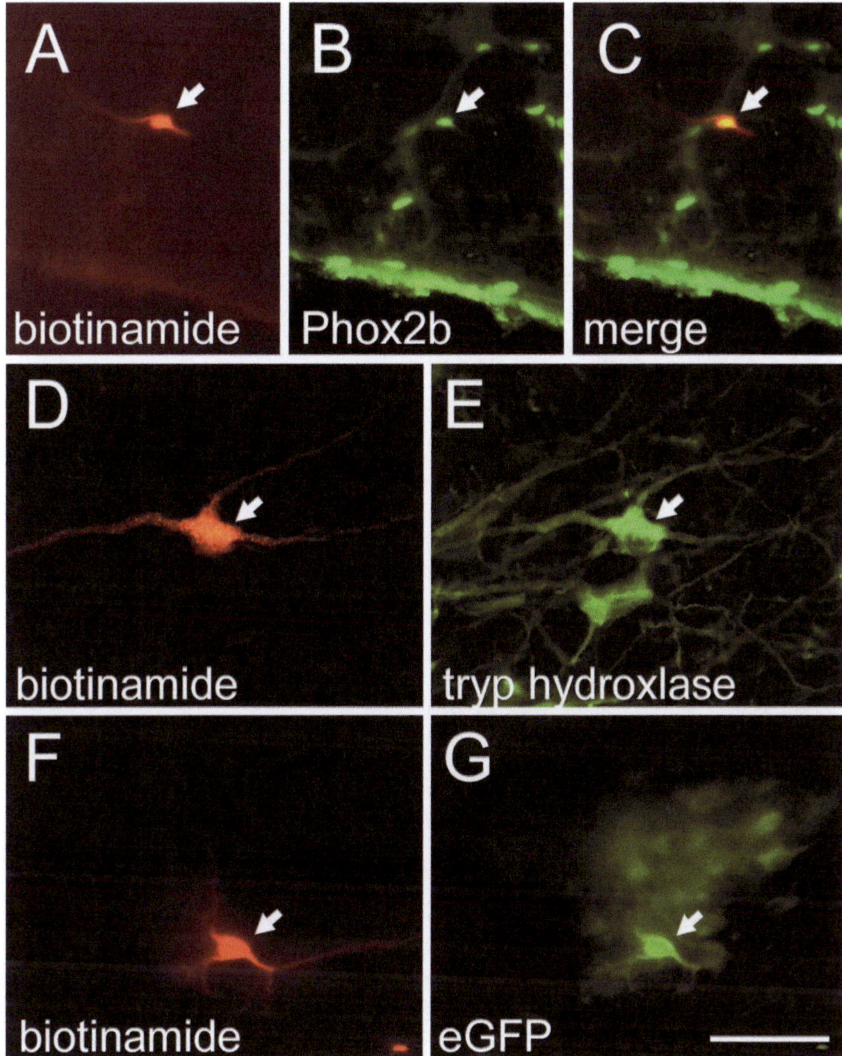

Fig. 3. Examples of juxtacellular labeling of neurons recorded in vivo and in vitro also labeled for immunoreactivity to illustrate various phenotypes. (**a–c**) Phox2b immunoreactive CO_2 sensitive neuron in RTN with (**a**) biotinamide revealed by avidin tagged with Cy3, (**b**) same neuron showing Phox2b immunoreactivity revealed with a rabbit Phox2b antibody (provided by J.-F. Brunet) and further reacted with anti-rabbit IgG tagged with Alexa 488. (**c**) Merge of (**a**) and (**b**). *Arrow* points to labeled neuron. (**d, e**) CO_2 insensitive serotonergic neuron in ventrolateral medulla. (**d**) CO_2 insensitive neuron recorded and juxtacellularly labeled in anesthetized rat. Biotinamide is revealed by avidin tagged with Cy3. (**e**) Same area of tissue showing tryptophan hydroxylase immunoreactivity revealed with mouse tryptophan hydroxylase antibody (Chemicon/Millipore) further reacted with anti-mouse IgG tagged with Alexa 488. *Arrow* points to labeled neuron. (**f, g**). CO_2 sensitive eGFP-positive neuron in vitro. (**f**) CO_2 sensitive neuron recorded in vitro in transverse slice of Phox2b-eGFP transgenic neonate mouse. Biotinamide revealed by avidin tagged with Cy3. (**g**) Same field as (**f**) with tissue reacted with chicken antibody against eGFP (AVES) and further reacted with anti-chicken IgY tagged with Cy2. *Arrow* points to labeled neuron. Scale bar in (**a–c**) = 50 μm, in (**e–h**) = 25 μm.

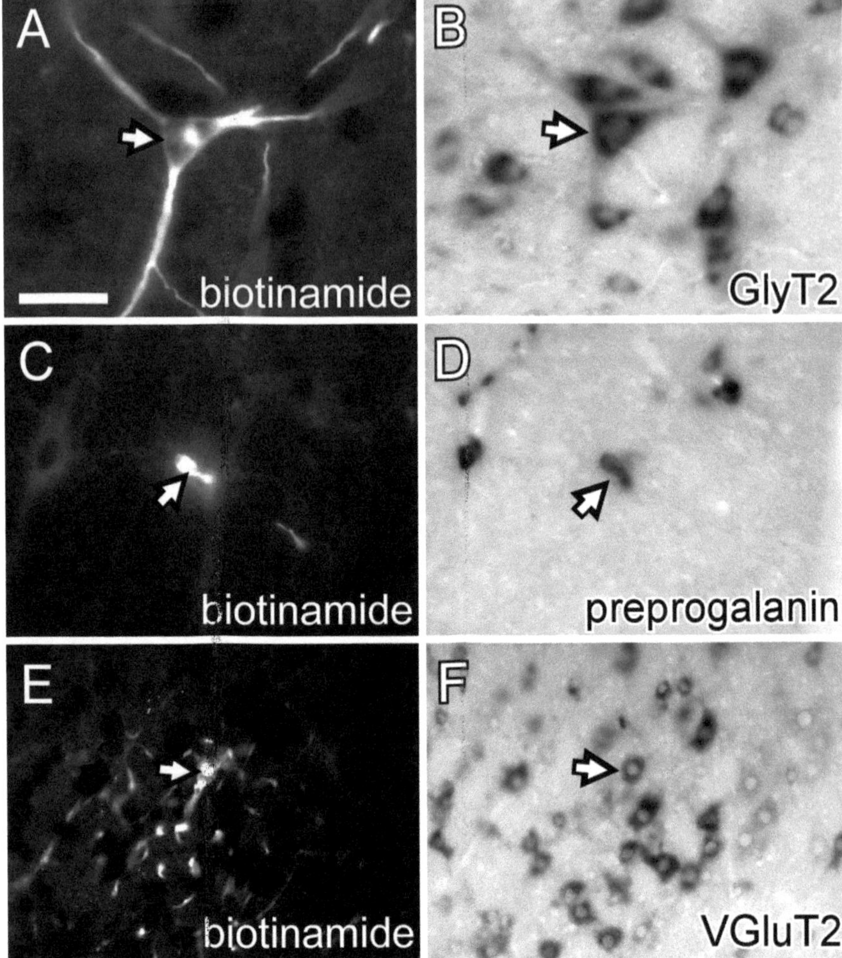

Fig. 4. Examples of juxtacellular labeling of neurons recorded in vivo and in vitro also reacted for in situ hybridization to illustrate various phenotypes. (**a**, **b**) Glycinergic neuron in Bötzinger complex. (**a**) Neuron recorded in anesthetized rat in the Bötzinger complex displayed expiratory augmenting activity (cell activity increased during expiration) whose activity became preinspiratory/postinspiratory during hypercapnic hypoxia. Biotinamide revealed by avidin tagged with Cy3. (**b**) *Arrow* points to same neuron as in (**a**) reacted to reveal glycine transporter-2 (GlyT2) mRNA. (**c**, **d**) Galanin-containing neuron in RTN. (**c**) CO_2 sensitive neuron recorded in anesthetized rat in the RTN. Biotinamide revealed by avidin tagged with Cy3. (**d**) Same neuron as C reacted to reveal preprogalanin mRNA. (**e**, **f**) Glutamatergic inspiratory bursting neuron in caudal ventral medulla in vitro. (**e**) Neuron with inspiratory bursting activity recorded in the in situ arterially perfused preparation in the caudal ventral respiratory area. Biotinamide revealed by avidin tagged with Cy3. Note the blood vessel labeling (artifact described in Sect. 4). (**f**) Same neuron as (**e**) reacted to reveal vesicular glutamate transporter 2 (VGluT2) mRNA. Scale bar in (**a**) = 50 μm for (**a–d**) and 100 μm for (**e**, **f**).

Bursting Neurons in the Area of the Pre-Bötzinger Complex Are Glutamatergic

Bursting neurons with inspiratory discharges were recorded and juxtacellularly labeled in the pre-Bötzinger complex and caudal ventral respiratory group in the in situ arterially perfused preparations of neonate and juvenile rats. An example of such a neuron is shown in Fig. 4e. This neuron expressed the mRNA for the vesicular glutamate transporter 2 (VGluT2, a marker for glutamatergic neurons) shown by in situ hybridization in Fig. 4f.

4. Notes

One of the most common artifacts seen in juxtacellular labeling is that of the micropipette track used for the recording (see Fig. 5a). Occasionally this will interfere with the ability to see the labeled neuron (Fig. 5b). The best practice is to make as few penetrations as possible in order to avoid this problem. Another artifact seen consistently in the in situ arterially perfused preparation (working heart-brain preparation pioneered by Paton (17)) is the labeling of blood vessels close to the recording pipette (see Figs. 4e and 5c). This seems unavoidable with this preparation. We assume that biotinamide gains access to the capillaries because of the edematous nature of this preparation.

4.1. Juxtacellular Labeling

Note 1: It is important to dislodge any air bubbles in the pipette by twirling a thin strand of fiberglass into the pipette (care is needed not to touch the fiberglass- if you insert with your bare fingers, the oils from your fingers can cause problems).

4.2. Tissue Fixation

Note 2: It seems to make a difference that the fixative is freshly prepared (within 24–48 h). If electron microscopic (EM) analysis is desired, the brain may be fixed with acrolein and/or glutaraldehyde fixatives and processed as per usual EM practices.

4.3. Brain Sectioning

Note 3: Marking the side of the brain helps to sort and mount the sections in the correct serial order. Use a luer stub adapter between 18 and 21 gauge to punch a hole in an inconspicuous place on the opposite side of the brain from the labeled cell.

Note 4: It is important to keep the sections in serial order to facilitate location of the labeled neuron and to do three-dimensional reconstructions if desired. By taking sequential sets of sections into the same wells, the order of the sections can be easily reconstructed. Keep sections from areas more caudal and rostral to the labeled area in separate trays.

4.4. Histological Processing

Note 5: For most tissue transfers, use glass rods made from glass Pasteur pipettes that have had the tips melted and curled using a Bunsen burner flame. These glass rods produce minimal tissue damage, do not transfer large amounts of solution and are easy to keep clean.

Note 6: Care must be taken to transfer all sections from each well every time. The labeled cell is usually only seen on one section!

4.4.1. Revealing Biotinamide Labeled Neuron

Note 7: Do not use the mesh bottom dish for DAB reactions as the DAB will adhere to and damage the plastic mesh over time.

Note 8: This timing is determined by when the reaction is judged to be optimal based on signal to noise. It may be necessary

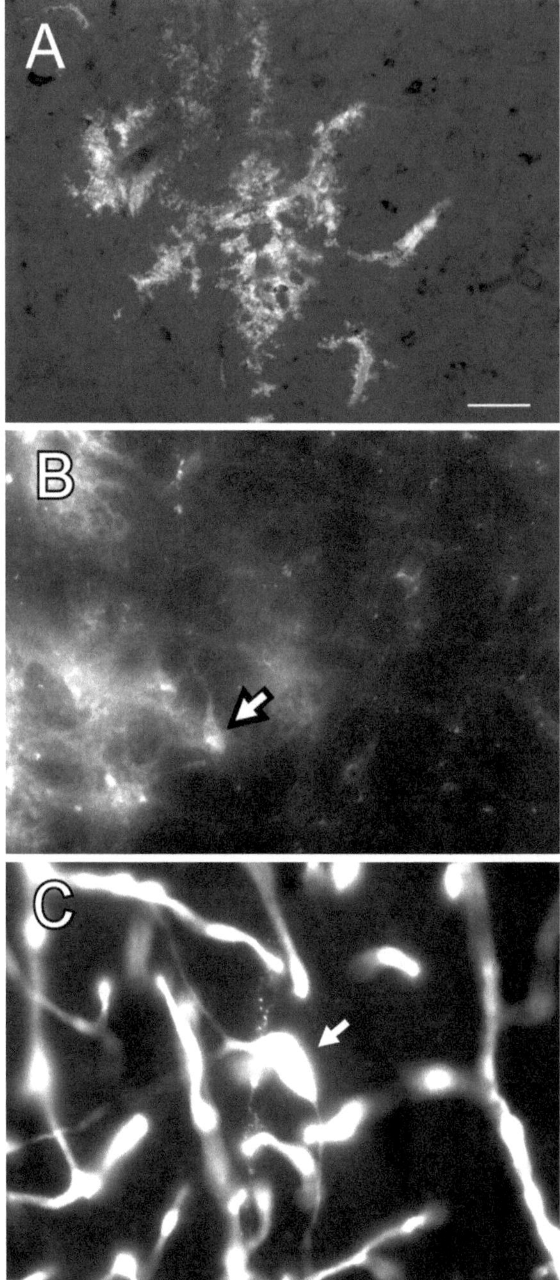

Fig. 5. Common artifacts associated with juxtacellular labeling. (**a**) Pipette track artifact. (**b**) Juxtacellularly labeled cell (*arrow*) next to pipette track artifact. (**c**) Blood vessel artifact in working heart brain (a.k.a. in situ arterially perfused) preparation. *Arrow* points to labeled cell amid the artifact. Scale bar in (**a**) = 100 μm for (**a**), 50 μm for (**b**) and 25 μm for (**c**).

to observe the tissue under a dissecting microscope. Use a light box to illuminate the tissue under a dissecting scope to prevent the direct illumination of the material by bright light. The tissue should NOT be inspected at this point under a regular microscope as the concentrated light source will cause the DAB to overreact.

Note 9: It may be necessary to process the tissue in "batches" to prevent some of the tissue from overreacting.

4.4.2. Processing Sections for Fluorescent Immunohistochemical Label(s)

Note 10: For best results, all secondary antibodies should be raised in the same species, e.g., donkey. Good fluorescent combinations are Alexa 488 with Cy3 and/or Cy5.

4.4.3. Processing Sections for In Situ Hybridization

Note 11: Recipe for 20 mL hybridization solution (may be aliquoted and frozen at –80°C until use) is shown in the Table 1.

Note 12: Covering the dish with provided lid is sufficient to keep wells from drying out if dish is set in water bath.

Table 1
Recipe for 20 mL hybridization solution

Ingredient	Final concentration	Stock	Amount (mL)
NaCl	0.60	5.0 M	2.400
Tris–Cl (7.5)	0.10	1.0 M	2.000
EDTA	0.01	0.5 M	0.400
NaPPi	0.05%	5.00%	0.200
Yeast total RNA	0.50	10.00 mg/mL	1.000
Yeast tRNA	0.05	10.00 mg/mL	0.100
Denhardt's BSA	1.00	50.00	0.400
d.i. formamide	50.00%	100.0%	10.00
Poly A+	0.05	5.00 mg/mL	0.200
4 rNTP's	0.00001	0.010 M	0.020
DTT	0.010	1.00 M	0.200
Sterile H_2O			2.080
Dextran sulfate (add this last, put in 37°C bath to aide in dissolving)	5.00%	100.00%	1.000 g
Sheared herring sperm boiled for 10 min and quenched on ice immediately prior to adding	50 μL/mL	10 mg/mL	1.000
Total (mL)			20.000

Note 13: RNAse stock is 5 mg/mL RNAse A in Tris-EDTA pH 7.6. Aliquots stored at –20°C. RNAse buffer is 500 mM NaCl, 10 mM Tris pH 8.0, 5 mM EDTA pH 8.0.

Note 14: Recipe for stock 20× SSC: Dissolve 175.3 g NaCl and 88.2 g sodium citrate in 800 mL H_2O. Adjust pH to 7.0 with concentrated HCl. Q.S. to 1 L with H_2O.

Note 15: For anti-digoxigenin antibody, centrifuge antibody before using and take only from the top of the tube.

Note 16: This timing is determined when the reaction is judged to be optimal based on signal to noise. It may be necessary to observe the tissue under a dissecting microscope. Use a light box to illuminate the tissue under a dissecting scope to prevent the direct illumination of the material by bright light. The tissue should NOT be inspected at this point under a regular microscope as the concentrated light source will cause the substrate/enzyme to overreact. This process is light sensitive, keep sections covered with a box to minimize exposure to light when not directly observing the sections for reaction product.

Note 17: Do not dehydrate the tissue sections as the alkaline phosphatase reaction product is soluble in alcohols and xylene.

Acknowledgments

This work was supported by National Institutes of Health Grants HL74011 and HL28785.

References

1. Pinault D (1996) A novel single-cell staining procedure performed in vivo under electrophysiological control: morpho-functional features of juxtacellularly labeled thalamic cells and other central neurons with biocytin or Neurobiotin. J Neurosci Methods 65: 113–136
2. Guyenet PG, Stornetta RL, Weston MC, McQuiston T, Simmons JR (2004) Detection of amino acid and peptide transmitters in physiologically identified brainstem cardiorespiratory neurons. Auton Neurosci 114:1–10
3. Duque A, Tepper JM, Detari L, Ascoli GA, Zaborszky L (2007) Morphological characterization of electrophysiologically and immunohistochemically identified basal forebrain cholinergic and neuropeptide Y-containing neurons. Brain Struct Funct 212:55–73
4. Manns ID, Alonso A, Jones BE (2000) Discharge profiles of juxtacellularly labeled and immunohistochemically identified GABAergic basal forebrain neurons recorded in association with the electroencephalogram in anesthetized rats. J Neurosci 20:9252–9263
5. Tang X, Neckel ND, Schramm LP (2003) Locations and morphologies of sympathetically correlated neurons in the T10 spinal segment of the rat. Brain Res 976:185–193
6. Mulkey DK, Stornetta RL, Weston MC, Simmons JR, Parker A, Bayliss DA, Guyenet PG (2004) Respiratory control by ventral surface chemoreceptor neurons in rats. Nat Neurosci 7:1360–1369
7. St-John WM, Stornetta RL, Guyenet PG, Paton JF (2009) Location and properties of respiratory neurones with putative intrinsic bursting properties in the rat in situ. J Physiol 587:3175–3188
8. Neumann E, Kakorin S, Toensing K (1999) Fundamentals of electroporative delivery of drugs and genes. Bioelectrochem Bioenerg 48:3–16

9. Boudes M, Pieraut S, Valmier J, Carroll P, Scamps F (2008) Single-cell electroporation of adult sensory neurons for gene screening with RNA interference mechanism. J Neurosci Methods 170:204–211
10. Nevian T, Helmchen F (2007) Calcium indicator loading of neurons using single-cell electroporation. Pflugers Arch 454:675–688
11. Nagayama S, Zeng S, Xiong W, Fletcher ML, Masurkar AV, Davis DJ, Pieribone VA, Chen WR (2007) In vivo simultaneous tracing and Ca(2+) imaging of local neuronal circuits. Neuron 53:789–803
12. Brown DL, Guyenet PG (1985) Electrophysiological study of cardiovascular neurons in the rostral ventrolateral medulla in rats. Circ Res 56:359–369
13. Stornetta RL, Moreira TS, Takakura AC, Kang BJ, Chang DA, West GH, Brunet JF, Mulkey DK, Bayliss DA, Guyenet PG (2006) Expression of Phox2b by brainstem neurons involved in chemosensory integration in the adult rat. J Neurosci 26:10305–10314
14. Lazarenko RM, Milner TA, Depuy SD, Stornetta RL, West GH, Kievits JA, Bayliss DA, Guyenet PG (2009) Acid sensitivity and ultrastructure of the retrotrapezoid nucleus in Phox2b-EGFP transgenic mice. J Comp Neurol 517:69–86
15. Fortuna MG, West GH, Stornetta RL, Guyenet PG (2008) Botzinger expiratory-augmenting neurons and the parafacial respiratory group. J Neurosci 28:2506–2515
16. Stornetta RL, Spirovski D, Moreira TS, Takakura AC, West GH, Gwilt JM, Pilowsky PM, Guyenet PG (2009) Galanin is a selective marker of the retrotrapezoid nucleus in rats. J Comp Neurol 512:373–383
17. Dutschmann M, Wilson RJA, Paton JFR (2000) Respiratory activity in neonatal rats. Auton Neurosci 84:19–29

Chapter 10

Visualization of Activated Neurons Involved in Endocrine and Dietary Pathways Using GFP-Expressing Mice

Rim Hassouna, Odile Viltart, Lucille Tallot, Karine Bouyer, Catherine Videau, Jacques Epelbaum, Virginie Tolle*, and Emilio Badoer*

Abstract

Neuropeptide Y (NPY) and growth hormone releasing hormone (GHRH) neurons in the hypothalamic arcuate nucleus play an important role in the neuroendocrine control of energy homeostasis and growth hormone secretion. These neuropeptides are synthesized in such small quantities that the common practice was to use colchicine, a neuronal transport inhibitor, to enhance the levels of neuropeptides in the cell body in order to visualize them by immunochemistry. However, colchicine also induces a marked increase in c-Fos levels, suggesting that it affects other mechanisms than solely axonal transport. The development of transgenic mice expressing fluorescent proteins under the control of specific neuropeptide promoters allows direct visualization of specific cell population within the central nervous system. The use of these genetic tools alleviates the need for colchicine pretreatment and can thus be combined with techniques aimed at detecting activated neurons, using the protein c-Fos, as a marker.

In the present article we discuss different variants of green fluorescent protein (GFP), which vary in brightness, cellular localization and can be differentially expressed through the use of different promoters, to visualize GFP-positive neurons. We also address methods to ensure that the GFP fluorescent signal is optimized during additional immunohistochemical procedures. We compared different GFP constructs in which the fluorescence exhibits variable levels of intensity or cellular localization. Recent developments with new GFP variants expressed under the control of strong promoters may provide enhanced and increased stability of fluorescence. Two examples are mice expressing an enhanced GFP protein in GHRH neurons and the NPY-Renilla-GFP mouse, in which a long bacterial artificial chromosome sequence as a promoter and a free cytoplasmic GFP renders the signal more stable and brighter than other variants of GFPs. In these examples GFP expression is localized mostly in cell bodies, and these two models are advantageous for studies investigating the phenotypical identity of neurons activated by physiological stimuli, like ghrelin. Differential subcellular expression of GFP is highlighted by the NPY-Tau-Sapphire-GFP mice in which the blue-shifted GFP variant (Sapphire-GFP) is fused to the tau sequence, allowing its cellular targeting to axon terminals thereby making this model suitable for studies investigating neuronal terminals.

Key words: c-Fos, Green fluorescent protein, Antibody, Fluorescence, Immunohistochemistry, NPY-Tau-Sapphire-GFP, NPY-Renilla GFP

*The authors Virginie Tolle and Emilio Badoer contributed equally for this chapter.

Emilio Badoer (ed.), *Visualization Techniques: From Immunohistochemistry to Magnetic Resonance Imaging*, Neuromethods, vol. 70, DOI 10.1007/978-1-61779-897-9_10, © Springer Science+Business Media, LLC 2012

1. Introduction

Due to its localization and the proximity of the fenestrated capillaries of the median eminence, the arcuate nucleus of the hypothalamus is a key target for peripheral signals that provide the central nervous system with information concerning the body's nutritional needs (1). The vast majority of these signals, originating either from adipocytes such as leptin or from endocrine cells along the gastrointestinal tract, such as PYY or CCK, are anorexigenic (2). In contrast, ghrelin is the only orexigenic hormone synthesized in the gastrointestinal tract and is a long-term regulator of energy homeostasis (3). Originally ghrelin was isolated from the stomach as the endogenous ligand for the growth hormone secretagogue receptor GHS-R1a (4, 5) and its ability to stimulate Growth hormone (GH) secretion. Several populations within the arcuate nucleus, including Growth Hormone Releasing Hormone (GHRH) and Neuropeptide Y (NPY) neurons, express GHS-R1a receptors (6, 7) and are activated by ghrelin (8–10). NPY neurons in the hypothalamic arcuate nucleus play an important role in the neuroendocrine control of energy homeostasis. NPY neurons are also present within the dorsal vagal complex, more specifically in the nucleus tractus solitarius (NTS), a region that is also considered as a satiety center and receives information from peripheral signals, partly mediated by afferent nerves traveling through the vagal nerve bundle.

Usually, neuropeptides are synthesized as prepropeptides in small quantities and immediately targeted to the terminals and thus cannot be visualized in the cell bodies using conventional immunohistochemistry. Therefore, it was common practice to use colchicine, an inhibitor of microtubule polymerization to block neuropeptide transport and artificially enhance its expression in the soma. This requirement has been overcome by the development of transgenic mice expressing fluorescent proteins under the control of specific neuropeptide promoters that allow direct visualization of specific cell populations within the central nervous system. Several transgenic mouse models allow the identification of GHRH and NPY neuronal populations through the expression of green fluorescent protein (GFP) reporting tool. GFP-expressing mice alleviate the need for colchicine treatment and can be combined with conventional immunohistochemical techniques to identify activated neurons by detecting the protein c-Fos, a marker of increased neuronal activity (11).

In the present article the visualization of GFP-positive neurons is described. As not all GFP-expressing neurons may express sufficient GFP to be easily detected above background fluorescence, novel developments with GFP has allowed enhanced signal intensity and increased fluorescence stability. One recent example is the development of the strong Renilla-GFP-expressing mouse under the control of the NPY promoter (12).

2. Materials

2.1. Drug Administration to GFP-Expressing Mice

2.1.1. Animals

1. GHRH-eGFP transgenic mice (13) expressing the jellyfish aequorea victoria enhanced green fluorescent protein (eGFP) under the transcriptional control of a rat GHRH hypothalamic promoter.
2. NPY-Tau-Sapphire-GFP mice expressing the blue-shifted GFP variant fused to the protein Tau under the transcriptional control of the NPY genomic sequence (14).
3. NPY-Renilla-GFP transgenic mice expressing humanized *Renilla reniformis* green fluorescent protein (hrGFP) under control of the mouse NPY promoter fused to a long bacterial artificial chromosome (BAC) sequence driving expression of a bright and stable fluorescent signal that is more resistant to fading compared to other GFP's (12).

2.1.2. Treatments

1. Rat ghrelin was obtained in powder form from *NeoMPS Peptide* (Strasbourg, France) and stored dessicated at 4°C. When ready to use, it is recommended not to weigh out specific amounts but to reconstitute the peptide directly in the original bottle. If the peptide is reconstituted directly in saline solution, it is recommended to use the solution as soon as possible and avoid freezing/thawing cycles. If smaller quantities are to be used, the powder can be resuspended in sterile water, aliquoted according to the amount needed and can be relyophilized and stored as before.
2. Prepare a working solution of 30 nmol of ghrelin in 0.2 mL of sterile saline solution 0.9% (i.e., 99 μg in 0.2 mL).

2.1.3. Colchicine Administration

1. Anesthetic: isoflurane (*Abbot Laboratory, Centravet, Plancoet, France*)
2. Pump to deliver isoflurane (*Univentor 400 anaesthesia unit, Phymep, Paris, France*)
3. Colchicine: prepare a working solution of 50 mg in 1 mL of sterile saline solution 0.9% (*Sigma, Saint-Quentin Fallavier, France*)
4. Diazepam (*Valium, Roche Laboratories, Meylan, France*)
5. Stereotaxic frame

2.2. Perfusion

1. Phosphate buffer (PB) (stock solution) 1 M (NaH_2PO_4 1 M, K_2HPO_4 1 M) further diluted to 0.2 M or 0.1 M (working solutions). This buffer is stable for several months if stored at 4°C.
2. Paraformaldehyde (PFA) 4% is prepared in PB 0.1 M (working solution) : dissolve 40 g of PFA in 500 mL of heated water at 55°C under a fumed hood and neutralize with about 1 mL of NaOH 4% and stir until solution is clear. Add 500 mL of PB 0.2 M to get the PFA in a final solution of PB 0.1 M. Filter, cool the solution, and adjust the final pH to 7.4. The PFA should be freshly prepared (the same day or the day before the experiment) and stored at 4°C until use.
3. Sodium pentobarbital (*Ceva santé animal, Libourne, France*).
4. Sucrose 30% in PB 0.1 M (working solution) should be prepared freshly and stored at 4°C.
5. 2-Methyl-butane (*Sigma, Saint-Quentin Fallavier, France*) is flammable and should be kept at 4°C.
6. Saline solution.

2.3. Preparation of the Brain Slices

1. Freezing microtome (*Frigomobile, Leica, Wetzlar, Germany*).
2. Tris-buffered saline (TBS) 1 M (stock solution) : Tris base 1 M, NaCl 1.5 M, and HCl until pH is at 7.4, further diluted to 0.1 M (working solution). This buffer is stable for several months if stored at 4°C.
3. Cryoprotectant : TBS 0.1 M (working solution), glycerol 0.3 M, ethylene glycol 0.3 M. Can be stored at room temperature for several months (see Note 1).

2.4. Immuno-histochemistry

1. TBS 0.1 M (working solution).
2. Triton 0.3% in TBS 0.1 M (working solution). This solution can be stored for several months at 4°C.
3. NDS = Normal donkey serum (*BioWest, Miami, FL, USA*).
4. NGS = Normal goat serum (*BioWest, Miami, FL, USA*).
5. Rabbit anti-c-Fos antibody (Ab-5, *Jackson Laboratories, West Grove, PA, USA*). The stock solution can be diluted 1:2 in a final concentration of 50% glycerol and stored at −20°C.
6. Chicken anti-GFP antibody (*Invitrogen, Molecular Probes, Carlsbad, CA, USA*). The stock solution is in water and is stored directly at 4°C.
7. Rabbit anti-GHRH antiserum (L0851) raised against the 25-amino-acid C-terminal part of the mouse GHRH sequence (15).
8. Donkey anti-rabbit antibody coupled with CY3 (*Jackson Laboratories, West Grove, PA, USA*). The powder stock solution can be resuspended in a final concentration of glycerol

50% to obtain a 1:2 dilution and stored at −20°C, protected from light (see Note 2).

9. Goat anti-chicken coupled with Alexa 488 (*Invitrogen, Molecular Probes, Carlsbad, CA, USA*). The stock solution is in water and can be diluted in a final concentration of glycerol 50% to obtain a 1:2 dilution and stored at −20°C, protected from light (see Note 2).
10. Biotinylated Rabbit Anti-c-Fos (*Jackson Laboratories, West Grove, PA, USA*). The powder stock solution can be resuspended in a final concentration of glycerol 50% to obtain a 1:2 dilution and stored at −20°C.
11. Avidin biotin complex (ABC) (*Vectastain Elite Kit, Vector Laboratories, Burlingame, CA, USA*).
12. H_2O_2 30% (Stock solution) (*Sigma, Saint-Quentin Fallavier, France*).
13. Diaminobenzidine (DAB) 10 mg pellets stored at −20°C (*Sigma, Saint-Quentin Fallavier, France*).
14. Gelatin coated or superfrost slides.
15. Coverslip.
16. Staining sets (6×4 wells) containing a 1/8″ leakproof laminated bottom, 1/8″ lid and a 5/8″ insert with 6×4 array 500 µm mesh (*Nason Machine, Fort Bragg, CA, USA*).
17. Small paintbrush (N°4).
18. Fluoromount (*SouthernBiotech, Birmingham, AL, USA*) (See Note 3).
19. Permount (*Sigma, Saint-Quentin Fallavier, France*).
20. Vectashield (*Vector Laboratories, Burlingame, CA*) stored in the dark and at 4°C (See Note 3).
21. Zeiss Axioplan microscope (*Carl Zeiss, Le Pecq, France*) and Leica SP2 or SP5 confocal microscope (*Leica, Wetzlar, Germany*).

3. Methods

3.1. Drug Administration to GFP-Expressing Mice

3.1.1. Animals

1. Animals are housed on a 12/12 h light/dark cycle at 22°C with *ad libitum* access to regular chow and water.
2. Intraperitoneal (ip) injections were performed between 8.00 and 11.00 am (see Note 4). Lights-on at 7.00 am.

3.1.2. Treatments

1. Each animal receives intraperitoneal injections of either:
   - Ghrelin (0.2 mL of working solution)/ 30 g BW
   - Saline (0.2 mL)

2. After injection, return the animals into their home cage.
3. Ninety minutes after treatment, perfuse the animal.

3.1.3. Colchicine Administration

1. Pretreat mice with diazepam (100 μg/30 g BW, ip) to avoid convulsion induced by colchicine.
2. Deeply anesthetize the mouse with isoflurane (200 mL/min, 2% isoflurane).
3. Place the mouse on the stereotaxic frame.
4. Insert a micropipette filled with the colchicine solution into the lateral ventricle (See Note 5).
5. Slowly inject 1 μL of the colchicine solution delivered with a microinjecting pump (0.007 μL/min).
6. Leave the micropipette in place for an additional 10-min period in order to reduce backflow.
7. Forty-eight hours after recovery, perfuse the animal.

3.2. Perfusion

1. Anesthetize the animals by intraperitoneal injection of sodium pentobarbital (~1.25 mg per 25 g of body weight) (see Note 6).
2. Check by tail pinching that the mouse is fully asleep and perfuse the animals through the aorta with saline solution (1 min) followed by 4% PFA in PB 0.1 M kept in ice, for 9 min at 20 mL/min under a fumed hood.
3. Dissect the brains and postfix in 4% PFA in PB 0.1 M for 2 h at room temperature.
4. Leave the brains for a minimum of 2 days in a 30% sucrose solution, at 4°C.
5. Freeze the brains by immersing them in 2-methyl-butane at −30 to −35°C (see Note 7) and store them at −80°C until further usage.

3.3. Preparation of the Brain Slices

1. Cut 25 μm coronal brain sections using a freezing microtome (see Note 8).
2. Slices can be kept in the cryoprotectant solution at −20°C or collected in TBS and immediately processed for immunochemistry (see Note 1).

3.4. Immuno-histochemistry

3.4.1. Immuno-histochemistry with Fluorescence

1. Immunohistochemistry is performed on free-floating sections (see Note 8). Sections should be protected from light since they contain GFP.
2. Select the sections to be stained and transfer them into the staining sets with a small paintbrush (see Note 9).
3. Wash the sections overnight in TBS 0.1 M under agitation, at room temperature (see Note 1).
4. The following day, rinse the slices four times, 10 min each time, in TBS 0.1 M under agitation, at room temperature.

5. Incubate for 1 h in the blocking solution (TBS 0.1 M, Triton 0.3%, NDS 5% and/or NGS 5%) under agitation, at room temperature (see Note 10).
6. Incubate overnight with the primary antibody solution (TBS 0.1 M, Triton 0.3%, NDS 0.5% and/or NGS 0.5%, anti-c-Fos 1/20,000 and/or anti-GFP 1/2,000 or anti-GHRH 1/3,000) under agitation, at room temperature (see Note 10).
7. The following day, rinse the slices four times, 10 min each time, in a solution of 0.1 M TBS, under agitation, at room temperature.
8. Incubate for 1 h with the secondary antibody solution (TBS 0.1 M, NDS 0.5%, NGS 0.5%, donkey antirabbit Cy3 at 1/800, and goat antichicken Alexa 488 at 1/400) under agitation, at room temperature (see Notes 10–12).
9. Then, rinse four times for 10 min under agitation, at room temperature.
10. Counterstain the sections using DAPI or SYTOX nuclear stainings in order to visualize the anatomical landmarks of the brain (see Note 13).
11. Mount the sections on superfrost or gelatin-coated slides. Air-dry and coverslip sections using Fluoromount or Vectashield. Store at 4°C in the dark (see Note 2).

3.4.2. Immunohistochemistry with Diaminobenzidine

1. Immunohistochemistry is performed on free-floating sections as described above (see Note 8).
2. Select the sections and transfer them into the staining sets with a small paintbrush (see Note 9).
3. Wash the sections overnight in TBS 0.1 M under constant agitation, at room temperature (see Note 1).
4. The following day, rinse the sections four times, 10 min each time, in TBS 0.1 M under agitation, at room temperature.
5. Incubate the sections for 30 min in H_2O_2 0.3% in TBS 0.1 M under agitation, at room temperature to inactivate endogenous peroxidase (should be prepared freshly).
6. Wash the sections four times, 10 min each time, in a solution of 0.1 M TBS, under agitation, at room temperature.
7. Incubate for 1 h in the blocking solution (TBS 0.1 M, Triton 0.3%, NDS 5%) under agitation, at room temperature (see Note 10).
8. Incubate overnight with the primary antibody solution (TBS 0.1 M, Triton 0.3%, NDS 0.5% and/or NGS 0.5%, anti c-Fos 1/20,000) under agitation, at room temperature (see Note 10).
9. The following day, rinse the sections four times, 10 min each time, in a solution of 0.1 M TBS, under agitation, at room temperature.

10. Incubate for 2 h with the secondary antibody solution (TBS 0.1 M, NDS 0.5%, biotinylated donkey anti-rabbit at 1/500) under agitation, at room temperature (see Notes 10 and 11).
11. Then, rinse four times for 10 min under agitation, at room temperature.
12. Incubate the slices with the ABC solution under agitation, at room temperature: mix two drops of A and two drops of B in 5 mL of TBS 0.1 M (should be prepared 30 min ahead).
13. Wash the slices four times, 10 min each time, in a solution of 0.1 M TBS, under agitation, at room temperature.
14. During the washing step, prepare DAB 0.05% by dissolving a 10 mg pellet in 20 mL of TBS 0.1 M under a fumed hood and with gloves, filter this solution and add 33 μL of H_2O_2 30%. Discard all plastics in special trashes.
15. Incubate the sections with the DAB solution until a brown precipitate appears (about 2 min but may be variable) and wash with TBS 0.1 M to stop the reaction.
16. Mount the slices on superfrost or gelatin coated slides and air-dry the sections.
17. Dehydrate in successive alcohol baths (10 min in 70% ethanol, 10 min in 95% ethanol, 10 min in 100% ethanol) and 10 min in Xylene before mounting in Permount medium with coverslip.

3.5. Analysis/Interpretation of Data

3.5.1. Visualization and Quantification of Cells/Nuclei by Epifluorescence Using Optical Microscopy

1. The anti-GFP and the endogenous Renilla-GFP signals can be observed with a normal GFP filter set (excitation: 475/40 nm and emission: 530/50 nm) and the CY3 signal can be observed with a CY3 filter set (excitation: 545/25 nm and emission: 605/70 nm). The endogenous Sapphire-GFP can be observed with a specific sapphire filter set (excitation: 395/40 nm and emission: 510/40 nm).
2. The section is initially observed at low magnification (×5 or ×10) in order to observe the whole structure and determine in which nuclei the staining is localized. To count numbers of cells or nuclei stained, choose the ×40 magnification (see Note 14).
3. The Metaview software (or similar) is used to take pictures on the epifluorescent Axioplan microscope.
4. Mercator software (or similar) is used to outline the region of interest and count the nuclei using a pointing tool.

3.5.2. Colocalization Detection Using Confocal Microscopy

1. Three different lasers are used: an Argon/Krypton 488 nm ion laser in order to observe the green fluorescence, a He/Ne 543 nm laser in order to observe the red fluorescence, and a 405 nm diode to observe the sapphire fluorescence. The range of emission used to detect the Alexa488/endogenous GFP signal, the CY3 signal, and the sapphire signal are respectively 498–522, 568–674, and 490–530 nm.

2. Immunostained sections through the arcuate nucleus of the hypothalamus and the dorsal vagal complex of the brainstem are analyzed using a ×40 objective (numerical aperture 1.25). Adjacent image planes through the *z*-axis are collected over a 10 μm thickness (where staining is the strongest) using a 1 μm *z*-step and scanned three times at a frequency of 200 Hz. A Pinhole size of 1 Airy unit is used for each channel. In case of multiple labelings, acquisitions are performed using a sequential set-up to maximize spectral separation.
3. All images are saved and stored for further analysis.

3.5.3. Analysis of Confocal Data

1. For each staining (Alexa 488/GFP and CY3), the ten images collected from each section using the confocal microscope are stacked in a single file and the two stacks are merged.
2. The Image J. software (or similar) is then used to count the colocalized cells using a pointing tool in each section of the *Z*-stack to ensure that the same cell is not counted twice in adjacent *Z* planes.

3.6. Application of the Technique/Interpretation of Data

3.6.1. Visualization of Neuropeptides Using GFP-Expressing Mice Alleviates the Need for Colchicine

Many neuropeptides are synthesized in such small quantities that they cannot be visualized in the cell body using conventional immunohistochemistry. As alluded in the introduction, before the development of transgenic mice expressing fluorescent proteins under the control of specific neuropeptide promoters, colchicine, a blocker of axonal transport, was commonly used to enhance the levels of neuropeptides in the cell body. For example, GHRH neurons in the arcuate nucleus of the hypothalamus could not be visualized in cell bodies without colchicine pretreatment (data not shown). Initially shown to block axonal transport of noradrenaline (16), colchicine also induces a marked increase in c-Fos levels, suggesting that it affects other mechanisms than solely axonal transport (17). Most importantly, colchicine is toxic to the animal, which rules out its use in studies aimed at measuring neuronal activity using early gene markers. In mice expressing the eGFP under the control of the GHRH promoter, GHRH neurons can easily be visualized without any pretreatment. The localization of the endogenous eGFP signal in neurons coincides with that of endogenous GHRH visualized using an anti-GHRH antibody in animals that were treated with colchicine (Fig. 1).

3.6.2. Visualization of Activated Neurons: Comparison of Chromogenic Versus Fluorescence-Based Immunodetection

c-Fos protein can be detected using immunofluorescence or using DAB as a chromogen. Immunofluorescence is good for confocal use and unequivocal colocalization studies but the signal can bleach with time, whereas DAB is good for permanent staining. As shown in Fig. 2, the detection of DAB may be slightly more sensitive than fluorescence as the number of c-Fos positive nuclei in serial sections of the same animal is slightly higher with DAB than with fluorescence. However, using DAB as a chromogen is more time-consuming and

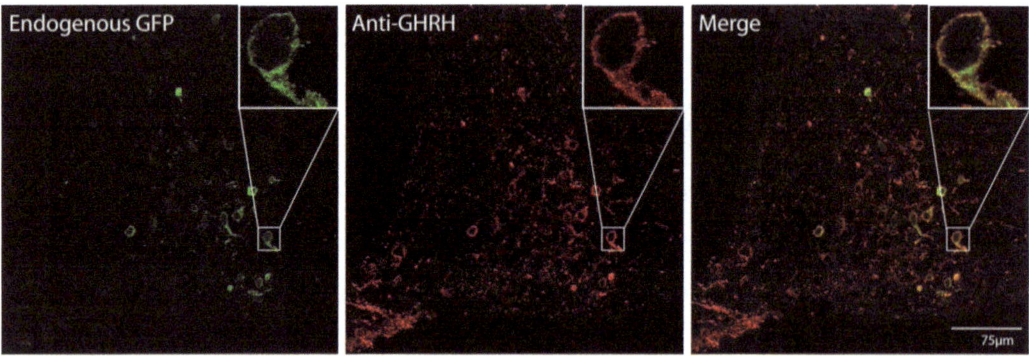

Fig. 1. Visualization of endogenous GFP fluorescence (*left panel*), growth hormone releasing hormone (GHRH) positive neurons detected using an anti-GHRH antibody (*middle panel*) and following the merger of both signals (*right panel*) in 30 μm-thick coronal sections of arcuate nucleus in a GHRH-eGFP mouse injected with colchicine.

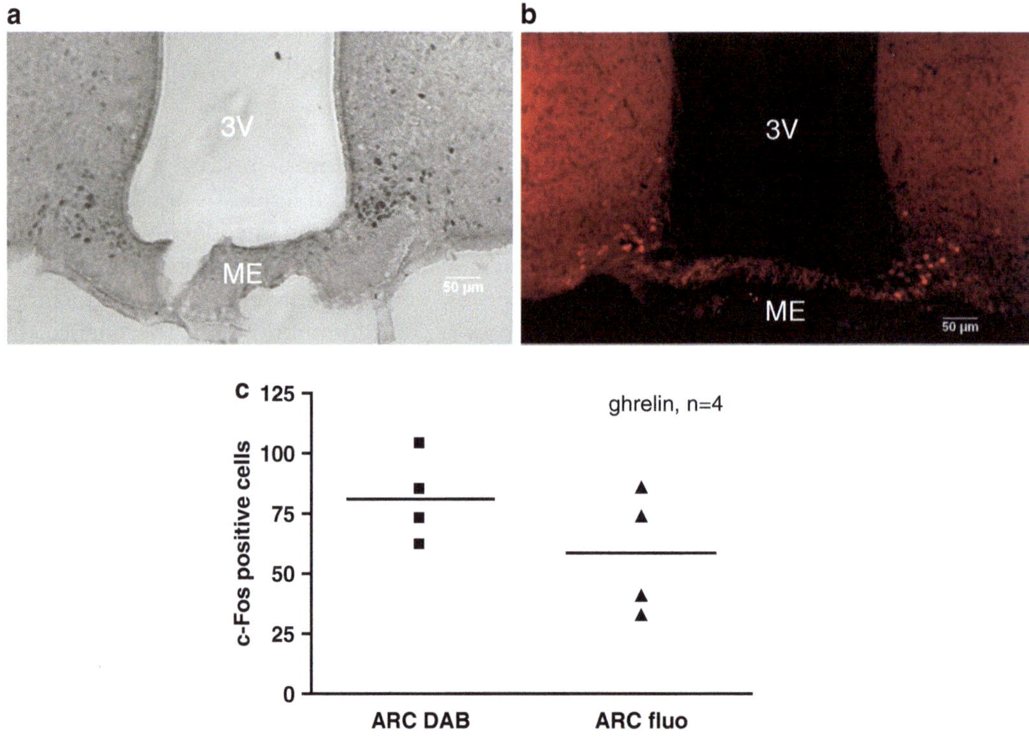

Fig. 2. Visualization of c-Fos positive cells using a chromogenic visualization method (**a**) or fluorescence immunohistochemistry (**b**) and quantification of c-Fos positive nuclei in serial sections of the arcuate nucleus (ARC) (**c**) from mice injected with 30 nmol ghrelin intraperitoneally. *ME* median eminence; *3V* third ventricle.

colocalization studies using DAB may provide more equivocal evidence of colocalization than immunofluorescence.

3.6.3. Weak GFP Signal Can Be Improved by the Utilization of an Anti-GFP Antibody

Under perfused conditions, not all GFP-expressing neurons may express sufficient GFP to be easily detected above background fluorescence. In such instances, it is possible to "enhance" the detection of GFP by using an antibody against GFP and to visualize

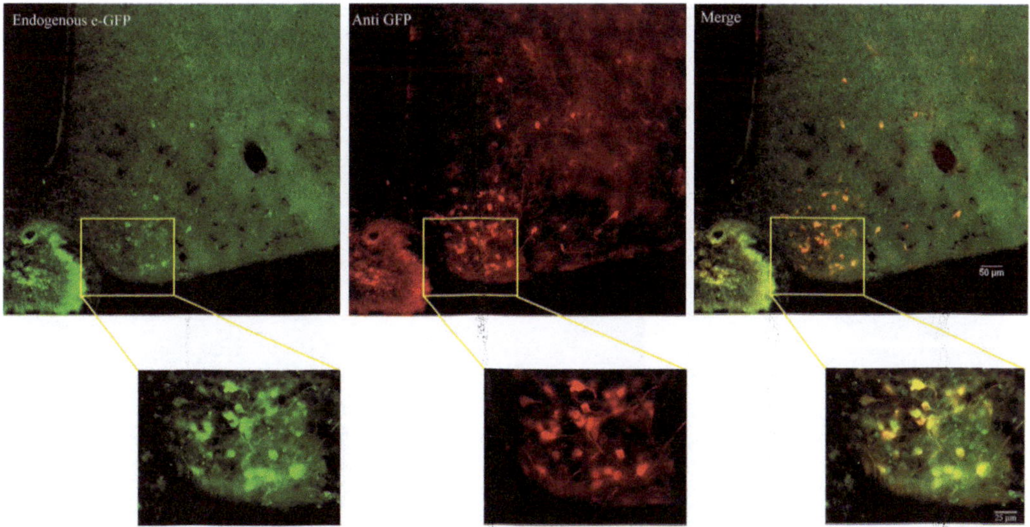

Fig. 3. Visualization of endogenous GFP (*left panel*), anti-GFP signal using an anti-GFP antibody (*middle panel*) and the merger of both signals (*right panel*) in 25 μm-thick coronal sections of the ARC in a GHRH-eGFP mouse. Insets in the *top panels* are shown at higher magnification in the *lower panels*.

it using standard immunofluorescence histochemistry (Figs. 3 and 4). Antibodies can be raised against a common epitope found in different GFP isoforms and, therefore, will recognize eGFP or Sapphire-GFP. As illustrated in Fig. 3, eGFP in GHRH-eGFP mice has a strong signal but the signal to noise ratio is improved when using an anti-GFP antibody (Fig. 3).

A further example is the NPY-Tau-Sapphire-GFP mouse in which the GFP is blue-shifted and can be visualized with a confocal microscope using a 405 diode. As the signal from the Sapphire-GFP was weak and bleached quickly in the arcuate nucleus of the hypothalamus, an anti-GFP antibody was used to improve the signal (Fig. 4). Using the anti-GFP antibody, more GFP-positive cells were visible than with the endogenous GFP fluorescence. However, in the brainstem, the endogenous Sapphire-GFP signal was much stronger and could be visualized under an epifluorescence microscope equipped with a classical GFP filter set without the need for signal amplification.

3.6.4. Visualization of Activated Neurons (c-Fos) in NPY-Tau-Sapphire-GFP or NPY-Renilla-GFP Mice

As mentioned previously, colchicine is toxic and this prevents studies aimed at detecting the neuropeptide content in cell bodies in combination with functional or physiological experimentation. GFP-expressing mice alleviate the need for colchicine treatment and can be combined with techniques aimed at detecting activated neurons, using the protein, c-Fos, as a marker.

As not all GFP-expressing neurons may express sufficient GFP to be easily detected above background fluorescence levels, new developments with modified GFP may provide enhanced and increased stability of fluorescence. One example is the development of the Renilla-GFP-expressing mouse under the control of the NPY promoter. In this mouse, the use of a long BAC sequence

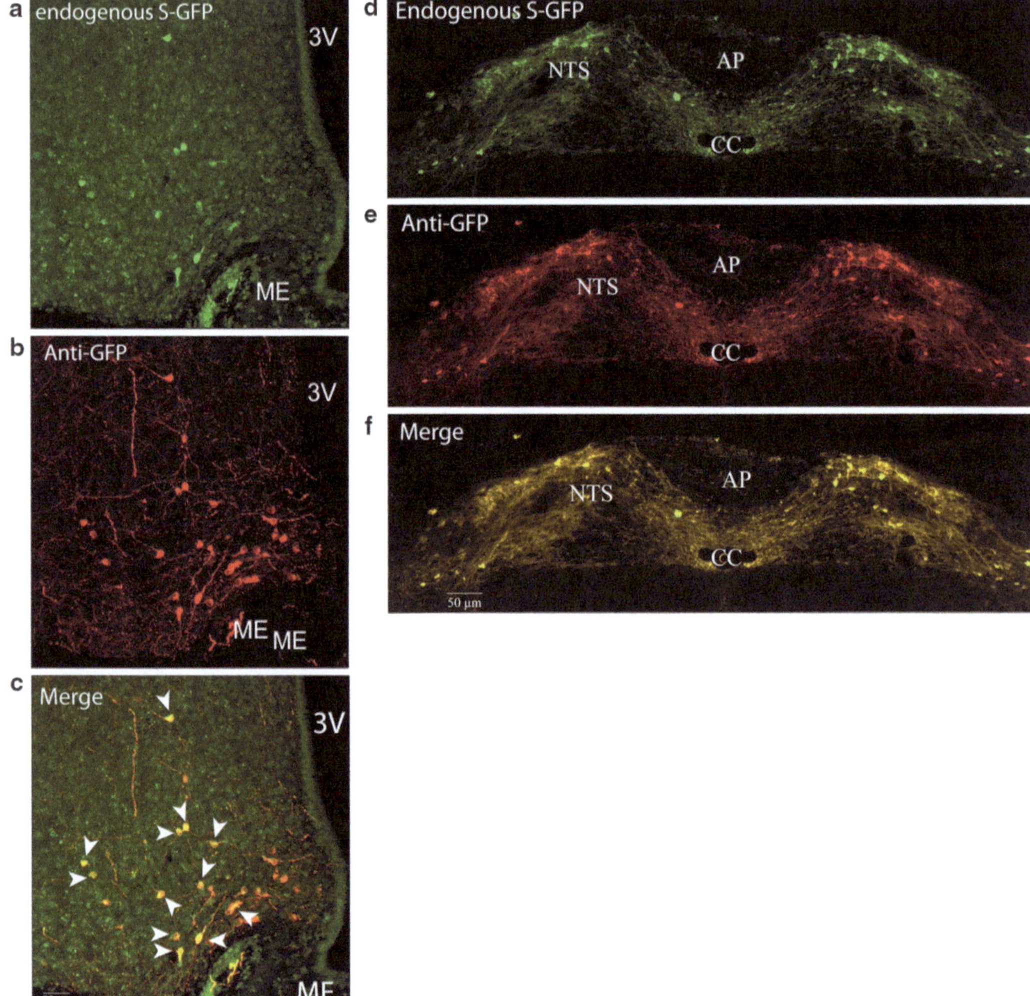

Fig. 4. Visualization of endogenous GFP and anti-GFP signal using an anti-GFP antibody in 25 μm-thick coronal sections of the ARC (*left panel*, confocal microscope) (**a**-**c**) and dorsal-vagal complex (*right panel*, epifluorescence microscope) (**d**-**f**) in a NPY-Tau-Sapphire-GFP mouse. (**a,d**) Endogenous Sapphire-GFP, (**b,e**) anti-GFP, (**c,f**) Merge of both signals. *ME* median eminence; *3V* third ventricle; *AP* area postrema; *NTS* nucleus tractus solitarius; *cc* central canal.

(clone name = RP24-177I12) (12) fused to the mouse NPY promoter sequence and combined to a free cytoplasmic GFP renders the signal more stable and brighter than other variants of GFPs and localized mostly in cell bodies. In contrast to the Sapphire-GFP signal, which is addressed mostly in terminals in non colchicine-treated animals (12, 14) and, thus, is suitable for studies investigating axon terminals (18), the Renilla-GFP-expressing mouse is an appropriate model for studies investigating neuropeptide content of neurons activated by physiological stimuli. For instance, after intraperitoneal administration of ghrelin, we observed c-Fos expression in only 8% of GFP-positive neurons using NPY-Tau-Sapphire-GFP mice, whereas c-Fos was activated in 27% of arcuate GFP-positive neurons using Renilla-GFP mice (Figs. 5 and 6).

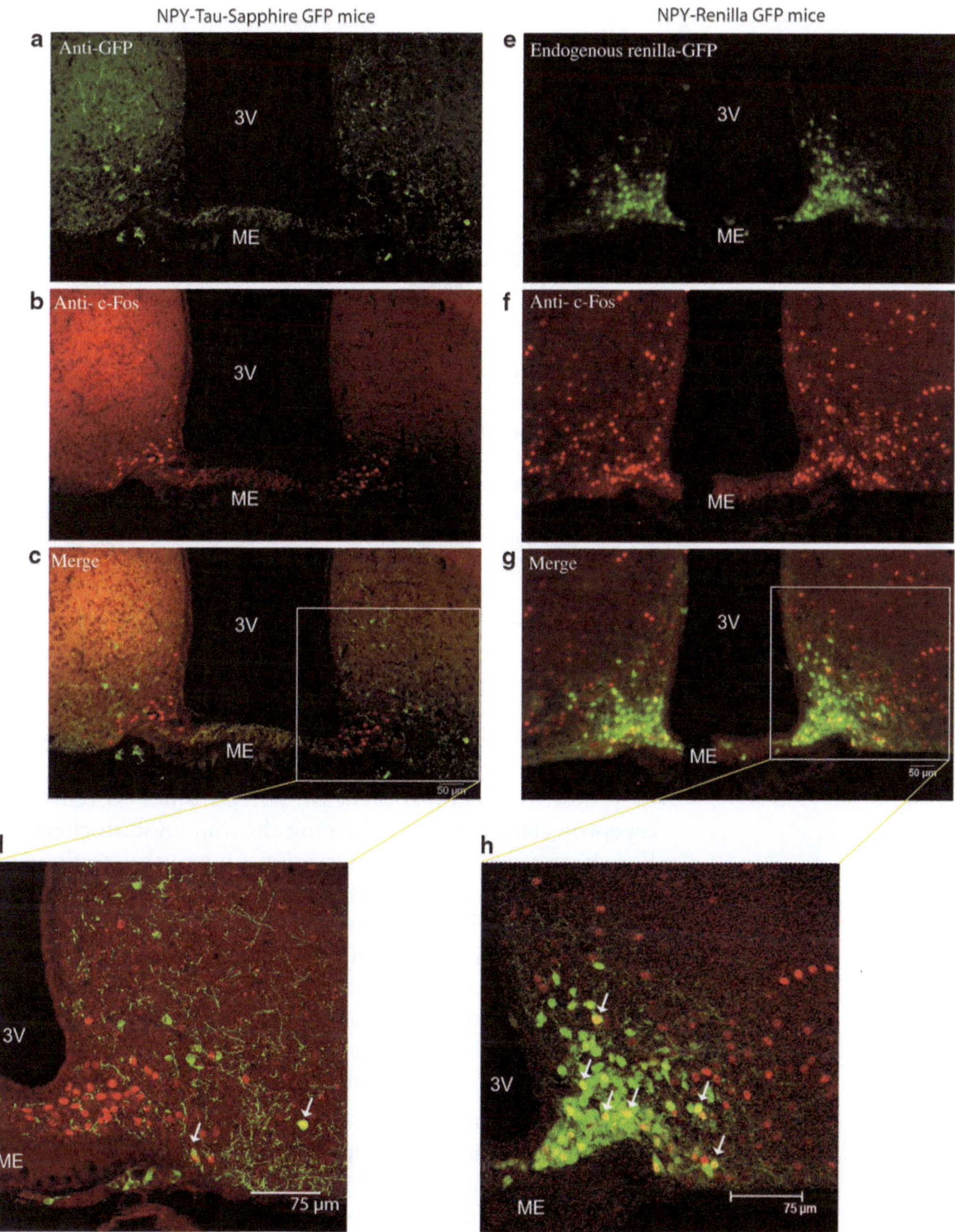

Fig. 5. Visualization of GFP and c-Fos in 25 μm-thick coronal sections of the ARC in a NPY-Tau-Sapphire-GFP mouse and in a NPY-Renilla-GFP mouse following intraperitoneal injections of ghrelin. *Left panel* shows a NPY-Tau-Sapphire-GFP mouse: (**a**) GFP signal using an anti-GFP antibody, (**b**) c-Fos staining using an anti-c-Fos antibody, (**c**) merge of both signals, (**d**) higher confocal magnification of (**c**). *Right panel* shows a NPY-Renilla-GFP mouse: (**e**) Endogenous GFP signal, (**f**) c-Fos staining using an anti-c-Fos antibody, (**g**) merge of both signals, (**h**) higher confocal magnification of (**g**). *White arrows* indicate cells expressing both GFP and c-Fos.

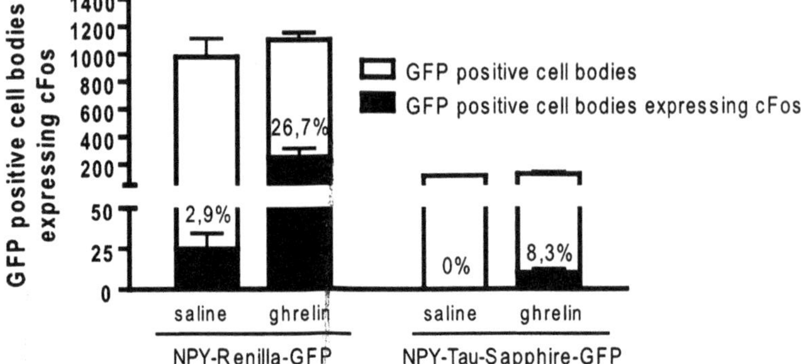

Fig. 6. Total number of GFP-positive cell bodies expressing c-Fos in the ARC (2.2–1.6 mm from interaural line) after intraperitoneal injections of saline vs. ghrelin in NPY-Tau-Sapphire-GFP mice (saline: $n=5$, ghrelin: $n=5$) and in NPY-Renilla-GFP mice (saline: $n=3$, ghrelin: $n=2$). Percentages represent the proportion of GFP-positive cells expressing c-Fos. GFP-positive cell bodies expressing c-Fos: $P<0.05$ ghrelin vs. saline (in both transgenic mice) and $P<0.05$ NPY-Renilla-GFP mice treated with ghrelin vs. NPY-Tau-Sapphire-GFP mice treated with ghrelin. GFP-positive cell bodies: $P<0.05$ NPY-Renilla-GFP vs. NPY-Tau-Sapphire-GFP mice.

4. Notes

1. Cryoprotectant is used to store slices at −20°C to preserve their antigenicity for several months. It is important to carefully wash the sections (at least 1 h) in order to remove all cryoprotectant before performing the immunohistochemistry. For shorter periods of conservation (1 month), sections can also be stored at 4°C in TBS 0.1 M containing sodium azide (50 mg/100 mL). Sodium azide, used to prevent bacterial contamination, is toxic and should be prepared with gloves. Slices should also be washed before immunohistochemistry.

   Although tissues are stored at 4°C, antigenicity may still deteriorate over time and it is better to perform immunohistochemistry as soon as possible.
2. The secondary antibodies coupled to CY3 or Alexa 488 as well as the slices incubated with these antibodies should be protected from light to avoid bleaching. Once the sections are mounted on slides, they should also be stored in the dark and preferably at 4°C to preserve fluorescence.
3. Fluoromount and Vectashield are mounting mediums containing anti-bleaching agents that help maintain the fluorescence. Fluoromount is a semi-hardest mounting medium that polymerizes and cannot easily be removed, whereas coverslips mounted with Vectashield can be removed.
4. As c-Fos protein is a transcription factor that can be activated by a variety of stimuli, including stress, it is recommended to

manipulate and habituate the animals to intraperitoneal injections several days before the experiments (19, 20).

5. Stereotaxic coordinates for the lateral ventricle were chosen according to the atlas of Franklin and Paxinos (21): 0.07 mm posterior, 0.10 mm lateral, and 0.21 mm deep from bregma.
6. c-Fos is an immediate early gene that is induced very rapidly but transiently (19, 20). The protein c-Fos reaches maximum levels approximately 60–120 min after stimulation. Ghrelin has been shown to increase c-Fos production in several brain nuclei, including the arcuate nucleus, after intraperitoneal injection (8, 9, 22).
7. For freezing a mouse brain in 2-methyl-butane, temperature should not go below –35°C because the brain tissue may be damaged.
8. Twenty-five micrometer-thick brain sections should be handled with care because of their fragility. This thickness is more adequate for colocalization studies when using epifluorescent microscope. However, if using a confocal microscope, thicker sections (over 30 μm) are also adequate. Using free-floating sections allows the antibody to penetrate into the tissue on both sides. On thicker sections, the antibody will not stain the whole tissue and it should be taken into consideration when analyzing the staining.
9. Staining sets consist of an array of 6×4 wells with a 500 μm mesh at the bottom. Sections can be placed in these wells and the array can easily be transferred into a leakproof laminated container. This system is very convenient to wash all the sections without too much handling of the brain slices. However, the container's volume in which the meshed wells fit is big and consumes a lot of reagents. In order to save antibodies, the tissue sections can be transferred into individual 48 well-plates for the incubation steps.
10. The serum for the blocking solution will be chosen depending on the nature of the secondary antibody (e.g., NDS with a secondary antibody produced in donkey or NGS with a secondary antibody produced in goat). If two primary antibodies are used in the same incubation (e.g., Rabbit Anti-c-Fos and Chicken Anti-GFP), two secondary antibodies directed against the species of the first antibodies should be used (e.g., Donkey Anti-Rabbit and Goat Anti-Chicken, respectively). However, it is recommended to use secondary anti-bodies raised in the same specie to minimize background (e.g., Goat anti-rabbit and Goat anti-chicken). In this case, a single blocking solution (e.g., NGS) should be used.
11. Dilutions of primary and secondary antibodies should be optimized. The best signal to noise ratio in our conditions was

obtained by testing serial dilutions of the rabbit anti-c-Fos (1:5,000–1:20,000) and chicken anti-GFP (1:1,000–1:20,000). Dilutions should be done extemporaneously.

12. Fluorochrome that emits light into the green (e.g., Alexa 488) and orange/red (e.g., CY3) are classical tools used for colocalization studies, as their emission spectra are easily distinguishable. For more details on fluorochromes, visit the Abcam website: http://www.abcam.com/fluorochromes.
13. DAPI (4′,6′-diamidino-2-phenylindole) and SYTOX are fluorescent nuclear stainings that are used to visualize brain structures and define anatomical landmarks. DAPI can penetrate both alive and dead cells. DAPI is excited by UV light and emits at 449 nm. SYTOX is a high-affinity nucleic acid stain that easily penetrates dead cells. Different SYTOX dyes exist (e.g., SYTOX green that emits at 523 nm) (Invitrogen). As these markers incorporate into the DNA, they should be manipulated with caution (gloves). Concentrations of these dyes should be tested empirically.
14. When observing a section at a high magnification (×40), the signal can fade. Consequently, this magnification should be used for short period of time and/or only when performing the quantification or acquiring images.

Acknowledgments

We are grateful to Dr ICAF Robinson (National Institute for Medical Research, Mill Hill, London, England) for providing the GHRH-GFP mice, to Dr Anne Lorsignol (UMR 5241 CNRS-UPS, Toulouse, France) for the NPY-Sapphire-GFP mice and to Dr Odile Viltart (INSERM U837, Lille, France) for the NPY-Renilla-GFP mice.

References

1. Morton GJ, Cummings DE, Baskin DG, Barsh GS, Schwartz MW (2006) Central nervous system control of food intake and body weight. Nature 443:289–295
2. Goldstone AP (2006) The hypothalamus, hormones, and hunger: alterations in human obesity and illness. Prog Brain Res 153:57–73
3. Tschöp M, Smiley DL, Heiman ML (2000) Ghrelin induces adiposity in rodents. Nature 407:908–913
4. Kojima M, Hosoda H, Date Y, Nakazato M, Matsuo H, Kangawa K (1999) Ghrelin is a growth-hormone-releasing acylated peptide from stomach. Nature 402:656–660
5. Tomasetto C, Karam SM, Ribieras S, Masson R, Lefèbvre O, Staub A, Alexander G, Chenard MP, Rio MC (2000) Identification and characterization of a novel gastric peptide hormone: the motilin-related peptide. Gastroenterology 119:395–405
6. Tannenbaum GS, Lapointe M, Beaudet A, Howard AD (1998) Expression of growth hormone secretagogue-receptors by growth hormone-releasing hormone neurons in the mediobasal hypothalamus. Endocrinology 139:4420–4423
7. Willesen MG, Kristensen P, Rømer J (1999) Co-localization of growth hormone

secretagogue receptor and NPY mRNA in the arcuate nucleus of the rat. Neuroendocrinology 70:306–316

8. Kobelt P, Wisser A, Stengel A, Goebel M, Inhoff T, Noetzel S, Veh R, Bannert N, Vandervoort I, Wiedenmann B (2008) Peripheral injection of ghrelin induces Fos expression in the dorsomedial hypothalamic nucleus in rats. Brain Res 1204:77–86
9. Wang L, Saint-Pierre DH, Taché Y (2002) Peripheral ghrelin selectively increases Fos expression in neuropeptide Y—synthesizing neurons in mouse hypothalamic arcuate nucleus. Neurosci Lett 325:47–51
10. Lawrence CB, Snape AC, Baudoin FM-H, Luckman SM (2002) Acute central ghrelin and GH secretagogues induce feeding and activate brain appetite centers. Endocrinology 143:155–162
11. Cham JL, Badoer E (2007) Exposure to a hot environment can activate spinally projecting and nitrergic neurones in the lower brainstem in the rat. Exp Physiol 92:529–540
12. van den Pol AN, Yao Y, Fu L-Y, Foo K, Huang H, Coppari R, Lowell BB, Broberger C (2009) Neuromedin B and gastrin-releasing peptide excite arcuate nucleus neuropeptide Y neurons in a novel transgenic mouse expressing strong Renilla green fluorescent protein in NPY neurons. J Neurosci 29:4622–4639
13. Balthasar N, Mery P-F, Magoulas CB, Mathers KE, Martin A, Mollard P, Robinson ICAF (2003) Growth hormone-releasing hormone (GHRH) neurons in GHRH-enhanced green fluorescent protein transgenic mice: a ventral hypothalamic network. Endocrinology 144:2728–2740
14. Pinto S, Roseberry AG, Liu H, Diano S, Shanabrough M, Cai X, Friedman JM, Horvath TL (2004) Rapid rewiring of arcuate nucleus feeding circuits by leptin. Science 304:110–115
15. Bouyer K, Loudes C, Robinson ICAF, Epelbaum J, Faivre-Bauman A (2006) Sexually dimorphic distribution of sst2A somatostatin receptors on growth hormone-releasing hormone neurons in mice. Endocrinology 147:2670–2674
16. Dahlström A (1968) Effect of colchicine on transport of amine storage granules in sympathetic nerves of rat. Eur J Pharmacol 5:111–113
17. Ceccatelli S, Villar MJ, Goldstein M, Hökfelt T (1989) Expression of c-Fos immunoreactivity in transmitter-characterized neurons after stress. Proc Natl Acad Sci USA 86:9569–9573
18. Rancillac A, Lainé J, Perrenoud Q, Geoffroy H, Ferezou I, Vitalis T, Rossier J (2010) Degenerative abnormalities in transgenic neocortical neuropeptide Y interneurons expressing tau-green fluorescent protein. J Neurosci Res 88:487–499
19. McClung CA, Ulery PG, Perrotti LI, Zachariou V, Berton O, Nestler EJ (2004) DeltaFosB: a molecular switch for long-term adaptation in the brain. Brain Res Mol Brain Res 132:146–154
20. Sagar SM, Sharp FR, Curran T (1988) Expression of c-fos protein in brain: metabolic mapping at the cellular level. Science 240:1328–1331
21. Franklin BJ, Paxinos G (1997) The mouse brain in stereotaxic coordinates. Academic, San Diego
22. Rüter J, Kobelt P, Tebbe JJ, Avsar Y, Veh R, Wang L, Klapp BF, Wiedenmann B, Taché Y, Mönnikes H (2003) Intraperitoneal injection of ghrelin induces Fos expression in the paraventricular nucleus of the hypothalamus in rats. Brain Res 991:26–33

Chapter 11

Use and Visualization of Neuroanatomical Viral Transneuronal Tracers

J. Patrick Card and Lynn W. Enquist

Abstract

Methods to define circuit organization in the brain are largely based upon the axonal transport capabilities of neurons. Numerous tracers have been developed since the 1970s that are sequestered by neurons and transported through axons in either the anterograde or retrograde directions. These tracers have been widely applied and are integral to the literature that has defined the functional organization of the nervous system. Nevertheless, with few exceptions these tracers do not cross synapses and therefore cannot provide insight into the polysynaptic organization of neurons devoted to a particular function. Viral transneuronal tracers introduced in the 1980s have been developed to fill this void. These tracers exploit the abilities of neurotropic viruses to invade neurons and generate infectious progeny that cross synapses to infect other neurons within a circuit. The method has become increasingly popular with the development of recombinant strains of virus that are reduced in virulence and express unique reporters. In this chapter we provide an overview of guidelines for the use of pseudorabies virus for transneuronal tracing. This swine alpha herpesvirus is certainly the most widely used virus for circuit analysis and it has benefited extensively from technological developments that have increased its power as a tracer.

Key words: Neurotropic, Alpha herpesvirus, Pseudorabies virus, Reporter genes

1. Introduction

1.1. General Overview

The principal means of information exchange in the nervous system is through synaptic connections between polarized neurons. The circuits formed through these contacts contribute to all aspects of nervous system function. Thus, analysis of circuit architecture is fundamental to understanding normal brain function as well as the mechanisms that contribute to loss of function in injury and disease.

Emilio Badoer (ed.), *Visualization Techniques: From Immunohistochemistry to Magnetic Resonance Imaging*, Neuromethods, vol. 70, DOI 10.1007/978-1-61779-897-9_11, © Springer Science+Business Media, LLC 2012

Anatomical methods for the characterization of synaptic relationships among neurons have evolved substantially over the past half-century. Transmission electron microscopy (TEM) provides the most definitive means of identifying synaptic contacts (1), but the technical demands of the method when applied alone limit the ability to define synaptic relations of a polysynaptic network. Since the 1970s tracing methods based upon the axonal transport capabilities of neurons have evolved as the primary means of characterizing neural circuits. Early studies combined the uptake and transport of tritiated amino acids with autoradiography to demonstrate connections between regions of the brain. The resolution of this approach at the light microscopic level is low but, when combined with TEM, the method effectively defines the synaptology of monosynaptic projections (2). The use of enzyme histochemistry to localize transported molecules such as horseradish peroxidase (HRP) dramatically improved the cytoarchitecture revealed at the light microscopic level in tracing studies (3) and served as a springboard for the subsequent identification and development of numerous other molecules for circuit analysis. It is now commonplace for investigators to use a variety of tracers that are transported through axons selectively in either the retrograde (e.g., FluoroGold; (4)) or anterograde (e.g., PHAL; (5)) directions to provide high-resolution morphological detail of cells contributing to projections between cell groups. When combined with TEM and immunocytochemistry these tracers provide a powerful means of defining synaptic connections between neurochemically defined populations of neurons (6, 7) but do not define the polysynaptic organization of neural circuits.

Viral transneuronal tracing has added another dimension to modern circuit analysis. With few exceptions (e.g., the C-fragment of tetanus toxin; (8, 9)), the aforementioned "classical" tracers do not cross synapses. In contrast, viral tracers replicate within neurons to produce infectious progeny that cross at or near synapses to infect synaptically linked neurons. With increasing time after infection, additional neurons contributing to a circuit become infected. In effect, neurotropic viruses are self-amplifying circuit tracers that have the potential to label all components of a polysynaptic circuit. Thus, transneuronal viral tracers provide investigators with a unique and powerful tool that can be effectively exploited to define neural circuits at a systems level.

In this chapter we discuss important guidelines for the use of pseudorabies virus (PRV) as a transneuronal tracer. Importantly, we also discuss the factors that should be considered in interpreting data derived from viral tracing studies. We focus upon PRV for the following reasons. First, it is among the earliest of the neuroinvasive viruses developed for transneuronal tracing and it remains the most widely applied (10, 11). As a result, there is considerable literature to draw upon in describing the strengths and limitations of the method. Second, PRV is amenable to genetic modification; consequently, a variety of modification have been made to facilitate dissection of neural circuit organization. For example, the construction of PRV recombinants that express different reporter genes enables dual infection analyses of complex circuits and the use of Cre-lox technology provides controlled gene expression with conditional replication and spread (12, 13). Finally, the guiding principles for the use of PRV are largely applicable to other viral tracers, such as herpes simplex virus (HSV) and rabies virus (14).

1.2. PRV Overview

PRV is an alpha herpesvirus that causes Aujeszky's disease, a severe neurological disorder, in a broad range of mammalian species. The natural host of PRV is the pig and it is the only animal able to survive a productive infection and carry the viral genome as a latent infection. Infected adult pigs usually show mild clinical signs of respiratory distress, weight loss, or reproductive failure. Baby pigs, in contrast, quickly succumb to infection, often with a fatal encephalitis (15). Pruritus (violent scratching and self-mutilation) is a common symptom in nonnatural hosts. These species represent classical "dead-end" hosts because they play no obvious role in transmission to other species (15–22).

PRV is a member of the alpha herpesvirus family that includes HSV and varicella-zoster virus (23, 24). As noted above, the natural host for PRV is the swine, but it has an extensive host range that includes essentially all mammals, except higher primates. Higher primates are refractory to infection by PRV, which is a benefit from the standpoint of the safety of investigators but places obvious limitations on the use of the virus for tracing in higher species. In general, infection of laboratory animals can happen only by injection, scarification or biting, and not by aerosol or inadvertent or casual contact. PRV is not shed in feces or urine. Wild type isolates of PRV infect the nervous system through *both* anterograde and retrograde transneuronal transport and often produce pronounced cytopathic changes in infected neurons. These viruses cause an infection with characteristic symptoms of violent pruritus (the "mad itch"), in which death is rapid (3–4 days) (23, 25, 26). In contrast, attenuated strains of PRV such as PRV-Bartha were developed originally as live vaccine strains (27). Animals infected with this strain exhibit markedly reduced symptoms after peripheral or central infection (reduced virulence), a property that facilitates its use in circuit tracing studies. Another property of PRV-Bartha and its recombinants is that they move through circuits only by retrograde routes (28, 29).

The structure of the PRV virion (virus particle), the organization of the viral genome, and the life cycle of the virus are illustrated in Fig. 1 and discussed in detail in a number of prior reports (reviewed in (23)). The genome of this DNA virus is sequestered within an icosahedral protein capsid, which in turn is covered by a tegument

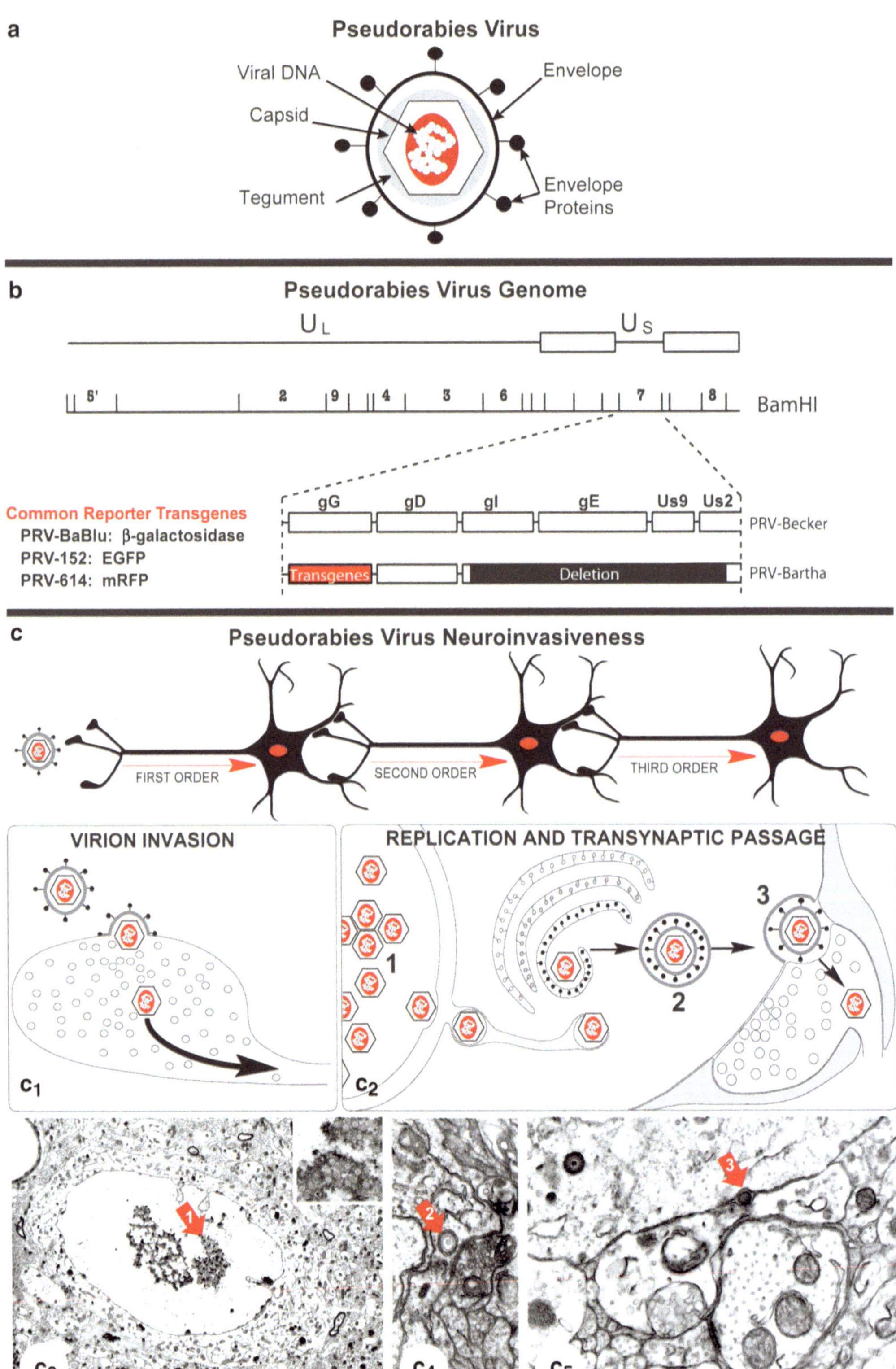

a
Pseudorabies Virus
Viral DNA
Capsid
Tegument
Envelope
Envelope Proteins
b
Pseudorabies Virus Genome
UL
US
BamHI
gG
gD
gI
gE
Us9
Us2
PRV-Becker
Transgenes
Deletion
PRV-Bartha
Common Reporter Transgenes
PRV-BaBlu: β-galactosidase
PRV-152: EGFP
PRV-614: mRFP
c
Pseudorabies Virus Neuroinvasiveness
FIRST ORDER
SECOND ORDER
THIRD ORDER
VIRION INVASION
REPLICATION AND TRANSYNAPTIC PASSAGE
c1
c2
c3
c4
c5

layer containing proteins important for initiation of viral replication. A lipid membrane bi-layer (envelope) surrounds the capsid and tegument. This envelope contains viral encoded membrane proteins essential for target cell attachment and receptor-mediated invasion. The viral genome contains about 74 open reading frames that encode the RNAs and proteins necessary for synthesis of progeny genomes and virions. The genome is arranged in two portions termed the unique long (U_L) and unique short (U_S) segments. The U_S segment is bracketed by large inverted repeat sequences (30). After considerable work by many laboratories, we now know that the reduced virulence characteristic of PRV-Bartha and the restriction of the virus to travel only retrogradely through circuits is largely the result of a small deletion in the U_S segment that eliminates genes encoding the gE, gI, and Us9 proteins (reviewed in (23, 25)). Deletion of these genes individually or in combination from a wild type PRV genome gives rise to the PRV-Bartha phenotype. Conversely, restoring these genes in the PRV-Bartha genome results in wild type virulence and spread of the virus. The PRV-Bartha genome carries other point mutations that have modest effects on virulence and replication, including those in the U_L21 gene that reduce the rate of retrograde transport of virus through neural circuits compared to wild type virus (31).

PRV-Bartha was among the earliest virus strains employed routinely for transneuronal tracing of neural circuits (32). In recent years, recombinant PRV strains have been developed with markedly improved utility as transneuronal tracers. PRV-Bartha is the parent for the majority of these recombinants. Typically, construction of the recombinants has involved insertion of transgenes into the gG locus of the viral genome (Fig. 1b), since deletion of gG

Fig. 1. The structure of a PRV virion (**a**), the organization of the PRV genome (**b**), and fundamental features of PRV life cycle (**c**) that lead to uptake, replication, and transneuronal passage are illustrated. The PRV virion consists of a central icosahedral capsid that houses the viral genome (**a**). The capsid is surrounded by a protein matrix known as the tegument and a lipid membrane (envelope) that contains virally encoded membrane proteins. The viral genome consists of unique long (UL) and short regions (US) flanked by internal and terminal repeat sequences (**b**). The BamH1 restriction map highlights a region in the unique short region of the wild type (PRV-Becker) genome that contains a deletion in the PRV-Bartha genome. This deletion eliminates genes that encode envelope proteins necessary for anterograde transport of the virus. In addition, loss of these gene products reduces virulence. PRV virions have a particularly high affinity for axon terminals (C_1), but also are capable of infecting neurons by invading the somatodendritic compartment. In either case the virion envelope fuses with the plasma membrane of the target cell, releasing the capsid and tegument into the target cell cytoplasm (C_1). Capsids are transported to the cell body where they deliver the viral genome to the cell nucleus. Replication and transcription of the viral genome produces all of the proteins necessary to assemble progeny virus. Capsids assemble in the cell nucleus (C_2 & *red block arrow* and *inset* of C_3) and bud through the nuclear envelope to gain access to the cell cytoplasm (C_2). The mature envelope of progeny virus is acquired from membrane of the trans-Golgi network or a late endosomal compartment, providing each virion with two membranes containing mature viral membrane proteins (C_2 & *red block arrow* of C_4). The outer membrane of these newly enveloped particles fuses with the plasma membrane of the cell to release the virion from the parent cell and the virion membrane proteins facilitate fusion and delivery of the capsid to synaptically connected neurons (C_2 & *red block arrow* of C_5). Adapted from (49, 88) with permission.

has little if any effect on replication and spread. These transgenes act as unique reporters that enable direct visualization of infected cells and thereby permit the analysis of complex circuits (e.g., dual infection paradigms using viruses expressing unique reporters), or restrict the replication of virus to phenotypically defined populations of neurons (e.g., conditional replication). The genomic organization of a subset of these recombinants is illustrated in Fig. 1b and a list of viruses most commonly used for circuit analysis is provided in Table 1.

Table 1
Strains of PRV available from the CNNV

Virus	Attributes
PRV-Becker	Wild type laboratory strain of PRV. Anterograde and retrograde, but highly virulent
PRV-Bartha	Attenuated vaccine strain of PRV most widely used for retrograde transneuronal tracing. "Attenuated" means that animals live several days longer after infection without significant symptoms compared to PRV-Becker. While much less virulent than PRV-Becker, PRV-Bartha does kill animals after infection
PRV-NIA3	Wild type, highly virulent PRV isolate. It is transported bidirectionally through circuits and has been used in tracing experiments in animal species (e.g., cats) that are refractory to infection with other PRV strains
PRV-BaBlu	PRV-Bartha containing a β-galactosidase reporter gene (lacZ gene) inserted into gG locus. The gG promoter drives expression of the lacZ gene and the virus is selectively transported retrogradely
PRV-91	PRV-Becker with a deletion of the gE gene. The virus is attenuated compared to PRV-Becker and is only transported retrogradely through neural circuits
PRV-98	PRV-Becker with a deletion of the gI gene. The virus is attenuated compared to PRV-Becker and is only transported retrogradely through neural circuits
PRV-151	PRV-Becker containing the CMV-EGFP reporter gene cassette inserted into the gG locus of the viral genome. The CMV promoter drives expression of the EGFP gene and virus is transported bidirectionally
PRV-152	PRV-Bartha containing the CMV-EGFP reporter gene cassette inserted into the gG locus of the viral genome. The CMV promoter drives expression of the EGFP gene and the virus is selectively transported retrogradely
PRV-154	PRV-Bartha that expresses Us9-EGFP fusion protein from the gG locus. Us9-GFP is a protein that concentrates in the trans-Golgi network in infected cells and also is incorporated into virion envelopes
PRV-180	PRV-Becker expressing mRFP-VP26 (capsid) fusion protein. The fusion protein is incorporated into capsids; strongly expressed in nuclei of infected cells; transported both in the anterograde and retrograde directions
PRV-GS443	PRV-Becker expressing EGFP-VP26 (capsid) fusion protein. The fusion protein is incorporated into capsids; strongly expressed in nuclei of infected cells; transported both in the anterograde and retrograde directions

(continued)

Table 1 (continued)

Virus	Attributes
PRV-181	PRV-Becker recombinant expressing mRFP-VP26 (capsid) and GFP-VP22 (tegument) fusion proteins. Both fusion proteins are incorporated into dual colored viral particles. Transported both in the anterograde and retrograde directions
PRV-614	PRV-Bartha containing the CMV-mRFP reporter gene cassette inserted into the gG locus of the viral genome. Isogenic with PRV-152. Often used with PRV-152 in dual infection studies
PRV-760	PRV-181 lacking the gE envelope glycoprotein gene. Less virulent than PRV 181 and only transported retrogradely
PRV-813	PRV-Becker lacking the Us3 gene. This virus is modestly less virulent than PRV Becker, but not as attenuated as PRV Bartha. Anterograde and retrograde tracer
PRV-823	PRV-Becker lacking the Us3 gene expressing mRFP-VP26 (capsid) fusion protein. This virus is less virulent than PRV-813 but not as attenuated at PRV Bartha. Anterograde and retrograde tracer
PRV-833	PRV-Becker lacking the Us3 gene and expressing mRFP-VP26 (capsid) fusion protein and eGFP-VP22 (tegument) fusion protein. This virus is modestly less virulent than PRV-823 and is an anterograde and retrograde tracer
PRV-2001	PRV-Bartha whose replication and expression of tau-EGFP is conditional upon Cre recombinase mediated recombination of the viral genome. A retrograde only tracer
PRV-263	PRV-152 carrying the Brainbow 1.0L cassette in place of the EGFP reporter. This virus is replication competent and expresses a default red dTomato fluorescent reporter unless in the present of Cre recombinase. In the presence of Cre the red reporter gene is excised, liberating the expression of cyan (mCerulean) or yellow (EYFP) reporters

2. Materials

Here we define the materials and facilities essential to viral tracing studies. Our description is tightly constrained by the unique requirements of viral tracing and focuses upon reagents that are essential either for the conduct of experiments or for detection of virus infected cells. Whenever possible we provide sources for these reagents and equipment. We do not go into detail on equipment and supplies that are commonly applied in circuit studies (e.g., immunohistochemistry and stereotaxic surgery) and refer the reader to other sources for this information (33).

2.1. Biosafely Level 2 Certified Facilities

PRV is an NIH/CDC class 2 infectious pathogen that requires special facilities for growing the virus and conducting animal experiments. Most institutions require that these facilities conform to regulations stipulated in Health and Human Services Publication

#88-8395 (*Biosafety in Microbiological and Biomedical Laboratories*). These regulations mandate that the experiments be conducted in a dedicated facility with restricted access. If the planned experiments involve growing quantities of virus, the facility should also contain a biosafety cabinet (Class II, type A/B3 biohood) and the necessary equipment (detailed in Sect. 2.3) for tissue culture. The biohoods can be self-venting, pumping HEPA-filtered air back into the room and are not vented to the outside. In general, biosafety level 2 conditions require lab coats, gloves, restricted access with appropriate signage, and other details as defined by the NIH guidelines.

The need for a biohood when producing stocks of PRV is twofold: first, to keep contaminants from getting into virus cultures, and second, to keep aerosols from large volumes of infected liquid from occurring. Biohoods have relevance to PRV safe handling only when growing the virus in quantity (e.g., a volume of more than 10 mL). Since animal infections involve very small quantities of infectious virus (usually less than 100 μL of liquid in a stock vial, with less than 5 μL of liquid being injected) the biohood is superfluous. In fact, doing injections in the confines of a biohood often creates more hazards to the animal and investigator. Thus, the most appropriate method of injecting animals is outside of a hood on a dedicated surgical table equipped with a surgical microscope in a quiet room with all of the facilities necessary to ensure a safe working environment.

The facility we use to inject animals for circuit-tracing experiments is self-contained. It comprises a surgical suite for virus injections as well as animal rooms that the animals live within throughout the experiment. Live animals are delivered to the facility, cared for in the animal housing room, and only leave the facility after they have been euthanized. Carcasses are autoclaved as medical waste and, depending upon the regulations of the institution, incinerated. All personnel involved in the studies are trained in the handling of virus and experimental animals. Only personnel certified for participation in the experiments are permitted to access the facility.

The equipment necessary for injections is detailed in Sect. 2.4. Our institutions mandate that animals living within the facility be housed in HEPA-filtered units. Animals are caged individually in plexiglass tubs with free access to food and water. Cages are placed on shelves on the front of the unit, which draws air across the shelves and through the HEPA-filter before ventilating it back into the room. Filters must be replaced every 3 months and the draw of the unit must be certified annually.

Our facility also contains a chemical fume hood that is used for transcardiac perfusion fixation of deeply anesthetized animals at the conclusion of each experiment. The hood is ventilated to the exterior and is certified for the use of paraformaldehyde. It is inspected annually to certify that there is adequate draw. The aldehyde-based perfusion fluid inactivates PRV virions quickly such that the hazard from infectious aerosols during perfusion is nonexistent when conducted in the chemical fume hood.

2.2. Virus Stocks

Table 1 provides a list of well-characterized strains of PRV that have been used extensively in transneuronal tracing studies. These strains of virus are currently available through the *Center for Neuroanatomy with Neurotropic Viruses* (CNNV). This center is funded by the National Institutes of Health through the National Center for Research Resources (Grant P40 RR018604). The mission of this center is to develop viral tracing technology and make reagents available to investigators interested in applying the method. More information regarding the reagents and services available through the center can be found at the following website: http://www.cnnv.pitt.edu. However, it is important to note that other prominent investigators have also produced well-characterized strains of PRV for transneuronal tracing (13, 20, 34).

Selecting a strain of virus for viral tracing involves several considerations. These include (a) the overall goals of the investigation, (b) experimental design, and (c) the methods used for detecting infected neurons. Each of these issues is discussed in later sections of this chapter.

2.3. Reagents, Supplies, and Equipment for Growing Virus

PRV can be grown in a variety of cell lines, but the PK15 cell line (transformed pig kidney epithelial cells) is commonly employed. The following sections detail the basic requirements for preparing purified stocks of PRV for circuit analysis. We assume a basic knowledge and expertise in tissue culture procedures. The specific protocols for growing and purifying the stocks as well as determining their concentration are provided in Sects. 3.1 and 3.2.

Equipment

1. Biosafety cabinet (Class II, type A/B3 biohood)
2. 37°C incubator
3. Cup sonicator
4. Rocking platform

Supplies

1. Sterile 6-well tissue culture plates
2. Sterile 1.7-mL microcentrifuge tubes, snap cap
3. Sterile 50-mL screw-cap plastic tubes
4. Plastic cell scraper

Reagents

1. PK15 cells grown in 100-mm dishes containing DMEM/10%FBS/penicillin–streptomycin (see recipe below). Unless stated otherwise, all tissue culture media contains penicillin and streptomycin. PK15 cells are commercially available (American Type Culture Collection) but also can be obtained from the CNNV (see Sect. 2.1).

   DMEM (Dulbecco's modified Eagles medium) is available through a variety of vendors. Store 100 and 500 mL aliquots of DMEM in sterile screw-cap bottles and add the following *just prior to use.*

   100 U/mL penicillin G

   100 μg/mL streptomycin

   Sodium bicarbonate to 7/5% (w/v) (This reagent can be added or media can be purchased that already contains it in the required concentration)

   Heat-inactivated fetal bovine serum (FBS) to 10% (Use only qualified, performance-tested, mycoplasma-, virus-, and endotoxin-tested FBS and use the same lot number as much as possible)

   The above solution can be stored at 4°C for up to 3 months.

2. Virus stock.
   Available through the CNNV (see Sect. 2.1) or researchers that routinely use PRV.

3. Trypsin–EDTA
   Purchase the stock solution containing 0.5% trypsin and 5.3 mM EDTA (e.g., Invitrogen) and store in 2 mL aliquots at –20°C. Prepare the final working concentrations for tissue culture by diluting 1:10 to 0.1% trypsin and 1.06 mM EDTA with Phosphate buffered saline (PBS).

4. Unsupplemented DMEM with 2% FBS and penicillin/streptomycin (pen/strep).

5. DMEM/10%FBS/pen-strep (see above for recipe, adjust FBS concentration).

6. PBS modified with calcium and magnesium (e.g., HyClone; SH30028.03).

7. DMEM/1% methocel/sodium bicarbonate/2%FBS/pen-strep.
   Weigh out a 2% (w/v) solution of methylcellulose (methocel) in distilled water. Mix the solution using a stir bar and agitation to disperse the methocel. Autoclave the solution (include the stir bar) for 30 min using a liquid cycle. Resuspend the

methocel by stirring overnight at 4°C. Mix equal volumes of 2× DMEM and the methocel suspension in a sterile 1-L bottle. Add 5 mL of 100× pen-strep per 500 mL (Life Technologies) and 25 mL of a 7.5% sodium bicarbonate solution per 500 mL of the methocel solution. The solution can be stored at 4°C for up to 3 months.

8. 1 M HEPES buffer, sterile. (e.g., GIBCO catalog number 15630).
9. 20% D-Sorbitol buffer for virus concentration.
   In 50 mL sterile water, add 10 g D-sorbitol, 2.5 mL of 1 M Tris–HCL (pH 7.2), and 50 μL of 1 M $MgCl_2 \cdot 6H_2O$.
10. Methelene blue for staining plaques. 0.5% in 70% methanol.

2.4. Reagents, Supplies and Equipment for Virus Injection

Viral tracing uses the same basic approaches as conventional tracing with the obvious exception being the use of infectious pathogens that require Biosafety level 2 facilities (Sect. 2.1) and procedures. The equipment necessary for a particular experiment will vary (e.g., stereotaxic frame for intracerebral injections). The following lists focus on what is necessary to support the use of viruses in tracing studies rather than the specific needs of a particular experiment. We assume a basic understanding of the procedures and equipment necessary for conducting tract-tracing experiments.

Equipment

1. Microisolator cages or hepafiltered housing units for housing experimental animals.

   As noted in Sect. 2.1, we conduct our experiments in a dedicated facility equipped to both perform the experiments and house the experimental animals. This is not available to all investigators and institutions vary in the procedures that they employ to address the BSL2 regulations mandated for viral tracing experiments. It is necessary to consult with the Division of Laboratory Animal Resources and/or the Institutional Animal Care and Use Committee (IACUC) at your institution to determine the regulations relevant to housing of experimental animals in viral tracing studies.
2. Chemical fume hood for transcardiac perfusion of infected animals with buffered aldehyde solutions.
3. Peristaltic pump for transcardiac perfusion of infected animals with buffered aldehyde solutions.
4. Autoclave. An autoclave should be available for sterilization of surgical packs and decontamination of animal cages and bedding.

5. Freezing microtome for sectioning of fixed tissue. We use a Leica microtome equipped with a freezing stage (Physitemp Instruments, Inc., Clifton, NJ).
6. Cryovials (The Lab Depot, Inc., Dawsonville, GA) for storage of tissue in cryoprotectant at –20°C.
7. Staining Nets & Dishes (Brain Research Laboratories; Newton, MA) for washing tissue during immunocytochemical processing.
8. Clinical Rotator table (Fisher Scientific, Pittsburgh, PA) for agitation of tissue during immunocytochemical processing.

Supplies

1. Anesthetic for animals during surgery. Gas anesthesia (e.g., isoflurane) requires appropriate disperser for administration and an effective scavenging system.
2. Surgical instruments. Instruments will vary depending upon the requirements of the experimental design. All experiments should use sterilized instrument packs. A bead sterilizer provides a useful means of maintaining sterile conditions for instruments within the confines of a single surgery.
3. Injection apparatus. This also will vary depending upon the requirements of the experiment. Intracerebral injection of small aliquots of virus (1 μL or less) is most effectively accomplished using a PicoPump (World Precision Instruments, Sarasota, FL). Hamilton microliter syringes provide an effective vehicle for reproducible injection of peripheral organs or tissues. The most localized injections are achieved using syringes with small gauge (26 gauge) beveled needles.
4. Disinfectant. All equipment and unused virus solutions should be disinfected at the conclusion of the experiment and disposed of according to regulations defined by your institution. A 5% Clorox solution or commercially available disinfectants (e.g., Envirocide) are effective for this purpose.
5. Biohazard disposal bags. After exposure to disinfectants, all disposables (e.g., cotton swabs, Q-tips, unused suture) should be placed in clearly labeled biohazard bags and disposed of according to institutional regulations. Our institution requires that unused tissues and animal carcasses be placed in these bags and stored frozen until disposal by incineration.

Reagents

1. Virus stock

Available through the CNNV (see Sect. 2.1) or researchers that routinely use PRV

2. 0.1 M sodium phosphate buffer

0.2 M sodium phosphate buffer (Na_2HPO_4–NaH_2PO_4)			
Sodium phosphate dibasic			
Na_2HPO_4 anhydrous	MW = 141.96	0.2 M Solution	28.39 g in 100 mL
$Na_2HPO_4 \cdot 2H_2O$	MW = 178.05	0.2 M Solution	35.61 g in 100 mL
$Na_2HPO_4 \cdot 7H_2O$	MW = 268.07	0.2 M Solution	53.64 g in 100 mL
$Na_2HPO_4 \cdot 12H_2O$	MW = 358.22	0.2 M Solution	71.64 g in 100 mL
Sodium phosphate mono			
NaH_2PO_4 anhydrous	MW = 120.00	0.2 M Solution	24.00 g in 100 mL
$NaH_2PO_4 \cdot H_2O$	MW = 138.00	0.2 M Solution	27.60 g in 100 mL
$NaH_2PO_4 \cdot 2H_2O$	MW = 156.03	0.2 M Solution	31.21 g in 100 mL

PH	**Dibasic (mL)**	**Monobasic (mL)**
5.8	8.0	92.0
6.0	12.3	87.7
6.2	18.5	81.5
6.4	26.5	73.5
6.5	31.5	68.5
6.6	37.5	62.5
6.8	49.0	51.0
7.0	61.0	39.0
7.2	72.0	28.0
7.4	81.0	19.0
7.6	87.0	13.0
Dilute to 200 mL with distilled water for 0.1 M solution		

3. Fixative.

   We use the paraformaldehye-lysine-periodate fixative developed by McLean and Nakane (35). This solution should be made the day that it is used according to the following recipe and procedure.

	100 mL	250 mL	500 mL	1,000 mL
Paraformaldehyde (g)	4.00	10.00	20.00	40.00
Sodium meta-periodate (g)	0.22	00.54	01.10	02.14
L-Lysine (g)	1.37	03.43	06.85	13.70

(a) Heat desired volume of 0.1 M sodium phosphate buffer to 80°C while stirring.

(b) Add paraformaldehyde and stir vigorously until solution clears.

(c) Cool paraformaldehyde solution in ice water.

(d) Add periodate-meta periodate and lysine and stir until it goes into solution.

(e) Filter solution, cool on ice, and store at 4°C until perfusion.

4. Immunohistochemical reagents.
The method for detecting infected neurons will vary according to the goals of the study and the strain of virus used for infection. Localization of viral antigens using polyclonal antibodies generated against acetone-inactivated wild type virus produces a robust signal in both immunoperoxidase and immunofluorescence localizations. Neurons infected with

Table 2
Antibodies for detection of infected neurons

Antibody	Species	Antigen	Source
Rb132, 133, & 134	Rabbit	Polyclonal antibody raised against acetone-inactivated wild type virus (PRV-Becker)	CNNV—see Sect. 2.2
PA1-081	Rabbit	Polyclonal antibody raised against acetone-inactivated attenuated virus (PRV-Bartha)	Thermo Scientific Pierce Antibodies
GFP	Rabbit	Polyclonal antibody raised against green fluorescent protein purified from the jellyfish *Aequorea victoria*	Invitrogen Molecular Probes
GFP	Chicken	Polyclonal antibody raised against recombinant full length GFP protein	Abcam
RFP	Rabbit	Polyclonal antibody raised against a RFP fusion protein corresponding to the full length amino acid sequence derived from the mushroom polyp coral *Discosoma*	Rockland
β-galactosidase (clone GAL-40)	Mouse	Monoclonal antibody raised against β-D-galactosidase purified from *E. coli*	Sigma

recombinant viruses that express unique reporters can also be detected using specific antisera and immunohistochemical techniques to localize reporter molecules (e.g., β-galactosidase, EGFP, mRFP) or simply on the basis of the native fluorescence produced by the protein product of the transgenes (EGFP, mRFP). Table 2 provides a list of primary antibodies effective for the localization of PRV and its recombinants along with potential sources.

5. Phosphate buffered sucrose solutions (20 and 30%) tissue cryotprotection.

3. Methods

In this section we describe basic protocols for preparing viral stocks and using them in viral tracing studies. Of course, well-characterized strains of PRV are the essential reagent for both of these procedures. Although we provide protocols for preparing and characterizing viral stocks it is worth reiterating that all major strains of PRV are currently available through the CNNV (Sect. 2.2 and Table 1).

3.1. Preparing Viral Stocks

Preparing purified viral stocks for circuit analysis is a two-step process involving the initial preparation of a crude stock from tissue culture infection that is subsequently partially purified to produce a "neurostock" for injection. The latter purification step is essential to reduce the cellular response that may be stimulated in animals by the debris present in the crude stocks.

3.1.1. Preparation of Crude Viral Stocks

1. Monolayers of PK15 cells should be prepared in 100-mm sterile tissue culture dishes containing DMEM and 10% FBS/pen-strep the day before they will be infected with virus. All tissue culture solutions should be at 37°C. Three plates should be prepared for each stock of virus.
2. Split the cells using trypsin-EDTA so that the monolayers reach 80% confluence the following day. Count the cells in a hemocytometer.

   It is important to have sufficient plates ready for infection and that the cells should not have reached confluence. Confluent cells produce lower amounts of virus particles. It is important to know the number of cells per plate so that the proper multiplicity of infection can be calculated.
3. Resuspend virus in DMEM containing 2% FBS/pen-strep so that 1.0 mL will provide a multiplicity of infection (MOI) of 0.01 plaque-forming units (pfu) per cell. Infection is accomplished by aspirating the medium from each dish of

PK15 cells, washing the cells once with PBS, and adding the virus in 1.0 mL of medium at the required MOI.

By infecting an 80% confluent culture so that one in a hundred cells are infected, we expect that at least one or two subsequent rounds of virus growth will occur on the plate before all cells are infected. As a result, we expect to achieve higher titers in each stock.

4. Adsorb virus to cells for 1 h in a humidified 37°C, 5% CO_2 incubator. Gently tilt the plate every 15 min to ensure that the medium is uniformly distributed over the surface of the monolayer. Better than 80% of the added virus particles will adsorb to cells during this period.
5. Remove unabsorbed virus by aspiration of the medium and replace it with 10 mL of fresh 37°C DMEM plus 2%FBS/pen-strep.
6. Incubate the infected monolayers for 2–3 days in a humidified 37°C, 5% CO_2 incubator. Monitor carefully every day (every 12 h on average) to identify cytopathic changes in the cells. A fast growing virus such as PRV-Becker will achieve 100% cytopathic effect (CPE) within 2 days, while a slower growing virus such as PRV-Bartha will achieve the same in 3 days.

   The length of time to incubate infected monolayers will depend upon the rate at which virus spreads among the cells. This can be determined by examining each plate with an inverted microscope for the appearance of rounding and sloughing of cells, margination of nuclear chromatin, and nuclear inclusions.
7. Harvest the cells and medium using a plastic scraper. Combine the contents of all infected plates in one sterile 50-mL, screw-cap conical plastic tube and mix the contents thoroughly with gentle stirring or agitation. The vast majority of newly produced virus particles remain cell-associated and therefore the best yields are obtained when both released virions (media) and cell-associated virus (cells) are harvested. Important: add 1 M HEPES buffer to a final concentration of 20 mM.

   The pH of infected cell media becomes acidic, which may have long-term adverse effects on viability of the stock. It is important to buffer the stock with HEPES buffer to neutral pH.
8. Transfer 1 mL of the solution from step 7 to each of ten 1.7-mL microcentrifuge tubes and proceed to step 2 of the next section (preparation of neurostocks). Another 1 mL should be stored frozen at –80°C for later use in determining virus concentration (Sect. 3.2). Transfer the remaining crude stock to a 50 mL screw-cap tube and store frozen at –80°C.

   Crude stocks can be stored frozen at –80°C for a year or more and do not lose viability if stored properly (see step 4 below).

3.1.2. Preparation of "Neurostocks"

1. If using a frozen crude stock from step 8 in the above protocol it should be thawed in a 37°C in a water bath with gentle agitation every 5 min. Virions tend to aggregate and associate with cell debris. Therefore, once the ice is entirely thawed, but the solution remains chilled (e.g., 4°C), sonicate the sealed tube using a cup sonicator (10 pulses for a total of 10 s at an amplitude of 80%) to disperse aggregated virus and separate virions from the cellular debris. Repeatedly invert the samples to thoroughly mix the contents and produce a homogeneous solution. Transfer 1 mL of the solution to each of ten 1.7-mL microcentrifuge tubes.

   Do not use a probe sonicator as it will inactivate virions and produce aerosols

   Do not be tempted to store smaller volume aliquots of virus. Small volumes are highly susceptible to damage and inactivation.

2. Centrifuge the microcentrifuge tubes from the previous step (or step 1 of the crude stock protocol) for 5 min at 2,000 × *g* (5,000 rpm using a Fotodyne 24 place rotor) at room temperature.

3. Pipette the supernatant from each tube into a single tube, being careful to exclude the cell debris in the loose pellet at the bottom of the tubes.

4. Divide the supernatant containing purified virus into 100- to 200-μL aliquots in screw-cap cryovials, snap-freeze on dry ice, and store at –80°C.

   *Do not aliquot in volumes less that 50*μL *since this may produce freezing artifact that reduces the titer of the virus.*

   Virus can be stored in aliquots at –80°C for extended periods of time without consequences as long at the temperature remains stable. Under no circumstances should these samples be thawed and refrozen, as this will reduce the titer of the virus. Remember, the inside of a –80°C freezer is not only cold, it is dry. Any open tube, or tube with a loose cap will set up conditions for "freeze-drying" of the frozen material inside the tube. PRV virions can be inactivated under these conditions. Avoid large surface exposure when storing aliquots in tubes as surfaces are the site of freeze-drying.

   We recommend storing the virus aliquots with Cryoguard M-40 thermal exposure indicators (Controlled Chemicals). These indicators are green at temperatures below –50°C but turn irreversibly red when the temperature rises above –40°C. If the indicator in a box of samples is red, the samples should be inactivated and discarded.

3.2. Determining Virus Concentration (Titer)

Viral concentration is measured in pfu determined on a standard cell line such as PK15 cells. A "plaque" is a site in a tissue monolayer resulting from a single infected cell that has replicated virus, spread focally, and produced cytopathology. Plaques are only produced by viable infectious virions. Thus, pfu is a measure of the viable infectious particles in a virus aliquot. The number of pfu required to infect 100% of the animals in a study is relatively constant for a given species, but such a number does vary with the viral genotype and must be determined empirically before extensive studies are initiated. In our hands, 100% infection of rodents (rat and mice) can be achieved by injecting about 10^5 pfu (typically 1 μL of a stock with a titer of 1×10^8 pfu/mL) (see Sect. 4, step 1 for further detail). In this section we describe the procedures for determining the titer of a virus stock.

1. Prepare monolayers of PK15 cells (see steps 1 and 2 of preparing crude virus stocks in Sect. 3.1). Split a 100-mm dish of cells that has just reached confluence and plate the cells in two 6-well culture plates. Cells should reach 80% confluence in each well of the plate before proceeding to step 2 of the procedure. Reaching 80% confluence typically takes 1 day, but this is only an estimate as time to confluence depends on the density of the cells that were first seeded on the dish.
2. Thaw and sonicate (cup sonicator) a 1-mL tube of crude stock (Sect. 3.1). The titer can also be determined from an aliquot of the neurostock that has been cleared of debris so it is not necessary to sonicate the sample.
3. Prepare a tenfold dilution series of the virus using the following procedure.
   (a) Prepare seven sterile tubes of 0.9 mL of DMEM/2%FBS/pen-strep.
   (b) Add 0.1 mL of virus to the first tube and mix the solution thoroughly. Take 0.1 mL of this solution and add it to the second tube.
   (c) Repeat this process until you have added virus to all seven tubes.
4. Remove the medium from each well of the 6-well plates and wash the cells with warm (37°C) PBS (Sect. 2.3). Add 0.2 mL of each of 10^{-5} (tube 5), 10^{-6} (tube 6), and 10^{-7} (tube 7) virus dilutions to duplicate wells of the 6-well plates. Return the plates to the incubator.
5. Allow virions to adsorb to the cell monolayers for 1 h in the 37°C, 5% CO_2 atmosphere of the incubator. Gently tilt the plates back-and-forth every 15 min during this period to distribute virions evenly across the monolayers.
6. Remove the plates from the incubator, aspirate the inoculum from each well, replace it with 3 mL of DMEM/1%methocel/sodium bicarbonate/2%FBS/pen-strep, and return the plates

to the incubator. The methocel solution is a thick slurry that physically impedes virion diffusion and aids in localizing virions to the area of the plaque. However, methocel is a liquid, and not a gel, so be very careful to leave the plates undisturbed for 2–3 days.

It is very important not to disturb the plates. Doing so spreads newly produced virions over the surface and thereby produces inaccurate titers.

The time for plaque formation usually takes 2 days. Some virus mutants produce very small plaques and these can sometimes be visualized if one incubates for an additional day.

7. Remove the plates from the incubator, aspirate the inoculum from each well, and then add 1 mL of 0.5% methylene blue in 70% methanol. The plates should then be placed on a rocking platform for 10 min at room temperature so that the methylene blue solution is evenly distributed over the monolayers. Remove excess stain by gently washing with tap water to visualize plaques.

   Plaques are produced by virus-induced cytopathology that can be visualized by eye in the stained cell monolayer. Viral replication in focal regions of the monolayer also produces lysis and sloughing of infected cells. Absence of cells at these sites produces unstained areas in the monolayer, each of which is a plaque.

8. The titer of the virus is calculated as plaque-forming units per mL (pfu/mL). It is determined by counting the total number of plaques on all countable wells, dividing by the total volume plated on these plates based on the lowest dilution giving countable numbers of plaques, and multiplying by the reciprocal of the lowest dilution that gave countable wells.

   For example, if there are 12 plaques in the well containing 0.2 mL of the 10^{-7} dilution and 175 plaques on the well containing 0.2 mL of the 10^{-6} dilution, the total plaques counted would be 187. The volume plated would be 0.22 mL of the 10^{-6} dilution. Thus, $(187/0.22)(10^{-6}) = 8.5\times 10^{8}$ pfu/mL.

3.3. Concentrating Viral Stocks

For direct injections of the CNS, a standard stock of PRV may be of insufficient titer to achieve reproducible infection. In these cases, it is necessary to partially purify and concentrate a standard stock to increase the pfu/mL. PRV virions can be sensitive to harsh treatments such as high-speed centrifugation. This protocol, which pellets virions through a sorbitol layer can concentrate virus stocks about 50–100-fold so that titers of about 1×10^{10} pfu/mL can be achieved.

1. Sterilize Beckman polymer tubes (size 25×89 mm) for SW28 rotor (serial # 11020). Place the tubes in a hood with the UV

light on and let them stand for 30 min to reduce surface contamination.

2. Thaw 30 mL of virus stock, sonicate using a cup sonicator to break up aggregates and then centrifuge for 20 min at 2,000 rpm to remove large aggregates and debris.
3. In the hood, add 13–14 mL of virus stock to a sterilized tube.
4. Add 7 mL of sorbitol buffer (Sect. 2.3) beneath the virus by inserting a loaded serological pipette at the bottom of the tube and slowing pipetting the sorbitol buffer so that it forms a layer below the virus solution. Make sure the layers do not mix.
5. Spin the tube in a clinical centrifuge for 1 h at room temperature at 23.5k rpm.
6. There should be a large pellet at the bottom of the tube that contains concentrated virus. Remove the supernatant above the pellet by aspiration.
7. Resuspend the pellet in 1 mL DMEM and titer.

3.4. Viral Tracing Experiments: Injection of Virus

PRV infection has been used to define neural circuits in a variety of systems and circumstances (11–13, 36–38). These include retrograde transneuronal tracing in normal adult and developing nervous system, after injection of peripheral neural targets (e.g., muscle or autonomic targets) or direct injection into the nervous system, and to define reorganization of circuitry in animal models of disease or following injury. Each system and application presents unique challenges and requires careful planning to ensure accurate definition of the circuit under study. Particularly important is recognition that rules governing the use of a live pathogen for circuit analysis differ in many respects from those that guide the use of "classical" tracers. In this section we identify variables that influence viral uptake and provide guidelines for using viruses in tracing studies (Fig. 2).

Transneuronal tracing from peripheral organs and tissues has certainly been the most widely applied application of viral tracing technology. Retrograde transport of PRV-Bartha and its recombinants from peripheral autonomic targets and muscle has provided considerable insight into the identity and functional organization of CNS circuits innervating these areas (e.g., Figs. 3, 4, and 6). Additionally, PRV has become increasingly popular for definition of circuit organization after intracerebral injection (e.g., Fig. 5). Here we identify important variables that should be considered when designing the injection strategy for any viral tracing study. Particularly important in this regard are (a) the concentration of the injected virus, (b) the distribution and density of axons innervating the injected tissue, and (c) the affinities of PRV for extracellular matrix molecules and membranes of cells within the target tissue (Fig. 2).

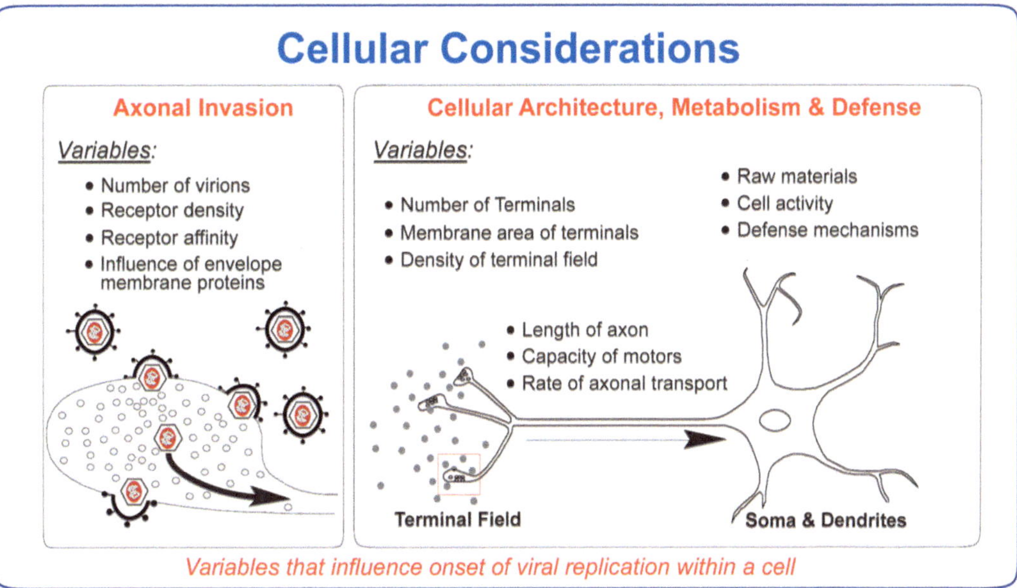

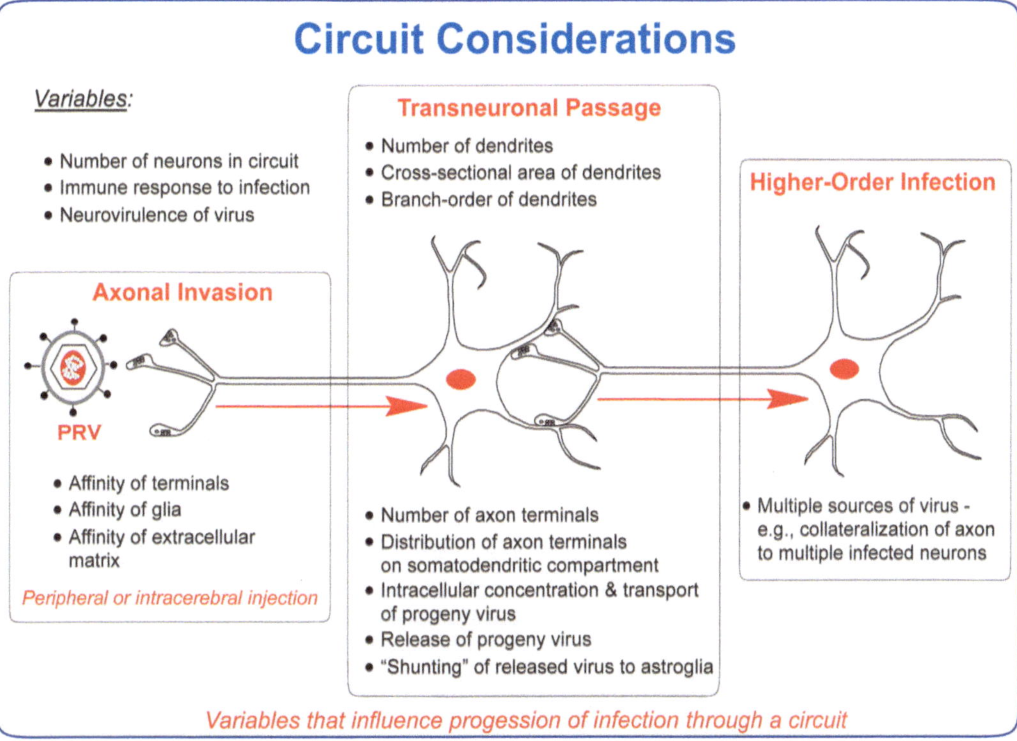

Fig. 2. The factors that influence replication and transport of virus through a polysynaptic circuit are illustrated. The *top panel* illustrates the features of individual neurons that determine the invasion and replication of PRV. The *lower panel* illustrates the features of a circuit that will determine the rate at which PRV through the component neurons of the circuit.

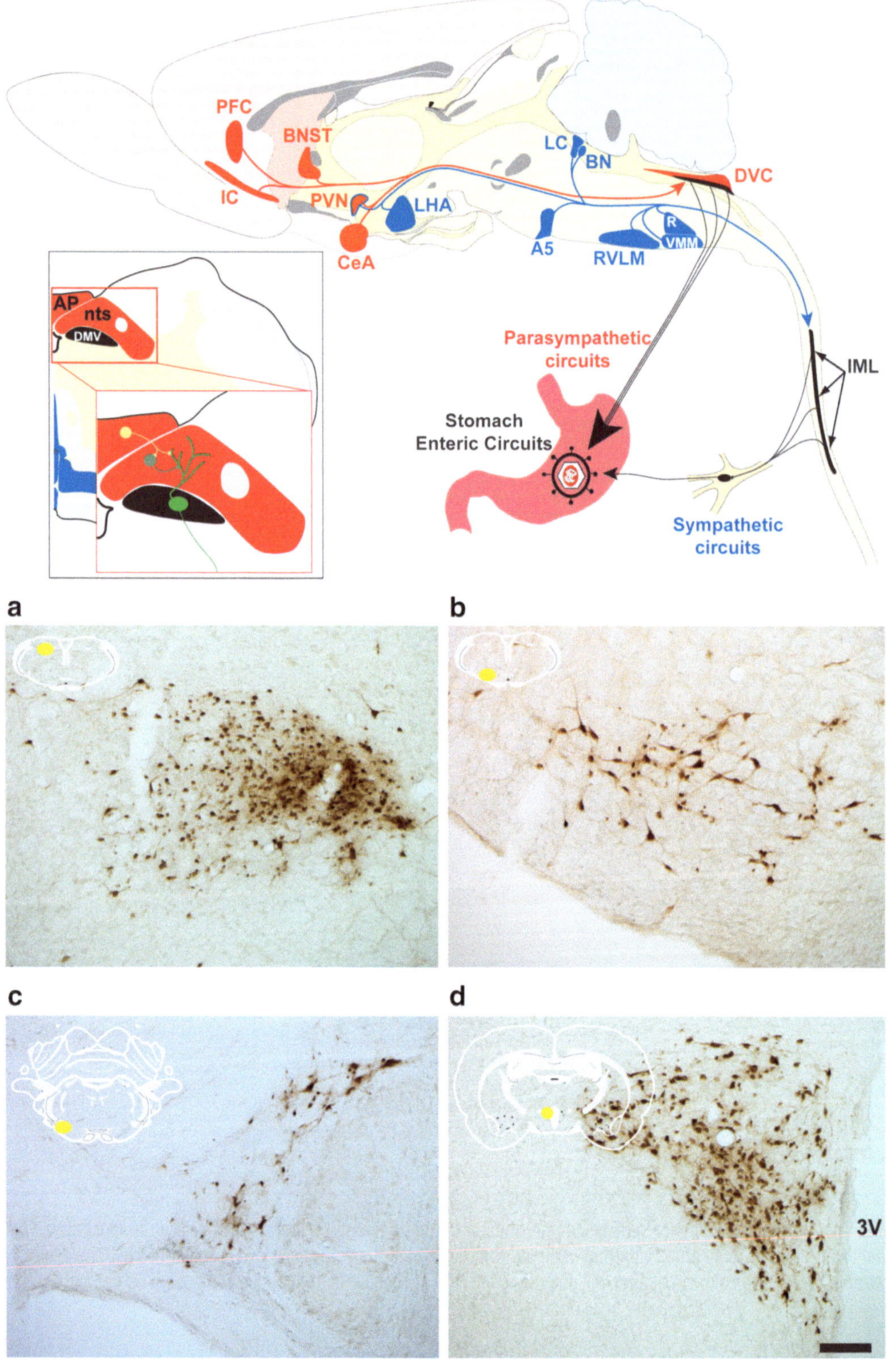

PFC
BNST
IC
PVN
CeA
LHA
LC
BN
DVC
A5
RVLM
R
VMM
AP
nts
DMV
Parasympathetic circuits
Stomach Enteric Circuits
IML
Sympathetic circuits
a
b
c
d
3V

As discussed in Sect. 4, step 4, the concentration of the injected virus has a very important influence upon the onset of viral replication and transneuronal passage through a circuit. The rate-limiting variable in this regard appears to be the number of virions that gain access to permissive cells and are able to establish a productive infection. Virions have a high affinity for charged surfaces such as the heparin and chondroitin proteoglycans of the extracellular matrix (Sect. 4, step 2). These affinities may prevent some virions from entering cells and initiating an infection, and will reduce the diffusion of virus from the site of injection. Collectively, the diffusion gradient and the density of axons within the injected tissue will influence the ability to infect the polysynaptic circuit innervating the injected tissue. Effective and informed application of the viral tracing method most often derives from parametric studies that first define the most effective measures (e.g., volume and concentration of virus, number of injection sites) that reproducibly infect circuitry linked to particular tissue or area of the brain. These control studies are particularly important in avoiding false negatives when studying the reorganization of circuitry in animal models of injury and disease.

Specific examples of the impact of tissue architecture upon the design of injection strategies are provided in the Sect. 4 (steps 5–7). Following are general guidelines for injection applicable to most systems and applications.

- The concentration of injected virus should be a minimum of 10^5 pfu (e.g., 1 μL of a 10^8 pfu/mL stock) irrespective of the injection target as lower concentrations result in variable rates of infection (Sect. 4, step 4). If smaller volumes are required, then the stock must be concentrated.
- Strategies for injection of peripheral tissues should consider the density of neural innervation and tissue architecture (Sect. 4, step 5). For example, injection of highly vascularized organs

Fig. 3. The basic organization of preautonomic circuits infected by injection of an attenuated strain of PRV into the wall of the stomach is illustrated in the schematic diagrams. Virus injected into the stomach wall infects resident enteric neurons and also invades axons from the parasympathetic and sympathetic divisions of the spinal cord. Infection of the brainstem and forebrain through parasympathetic pathways precedes infection through sympathetic pathways because of the first-order infection of neurons in the dorsal motor nucleus (DMV). The circuitry in the dorsal motor vagal complex (DVC) infected by transneuronal passage of virus from DMV neurons is shown in the *inset*. The photomicrographs below the circuit diagrams illustrate infected neurons in the nucleus of the solitary tract (nts) (**a**), rostroventrolateral medulla (RVLM) (**b**), the A5 catecholamine cell group (**c**), and the paraventricular nucleus of hypothalamus (PVN) (**d**) 72 h after injection of virus into the ventral wall of the stomach. The location of each of the illustrated regions is highlighted in *yellow* in the coronal schematics in the *upper left* of each photomicrograph. The infected neurons were identified using a rabbit polyclonal antiserum raised against acetone-inactivated virus and the immunoperoxidase procedure discussed in Sect. 3.6. Note the specific labeling of neurons within the confines of each of these regions and the absence of cytopathology in the infected neurons. *AP* area postrema; *BN* Barrington's nucleus; *BNST* bed nucleus of stria terminalis; *CeA* central nucleus of amygdala; *IC* insular cortex; *IML* intermediolateral cell column; *LC* locus coeruleus; *LHA* lateral hypothalamic area; *PFC* prefrontal cortex; *R* raphe nuclei; *VMM* ventromedial medulla. The figure is adapted from Curanovic and collaborators (31) with permission.

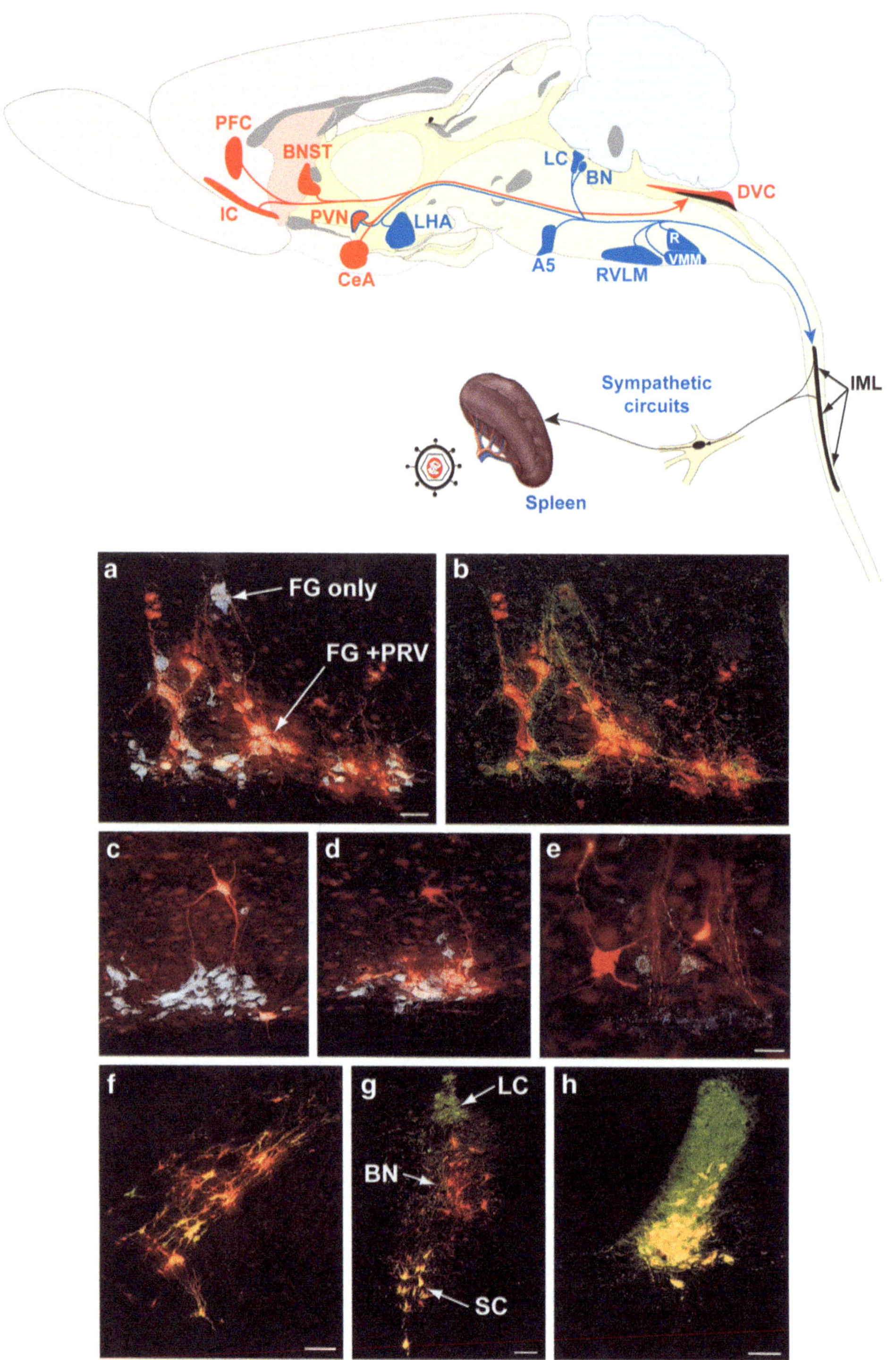
PFC
BNST
IC
PVN
LHA
CeA
LC
BN
DVC
A5
RVLM
R
VMM
IML
Sympathetic circuits
Spleen
a
FG only
FG +PRV
b
c
d
e
f
g
LC
BN
SC
h

(e.g., spleen; (39, 40)) or tissues receiving a sparse innervation (e.g., brown fat; (41)) require multiple injections and higher concentrations of virus than encapsulated parenchymal organs with a denser innervation (e.g., kidney; (42)).

- Intracerebral injections may be affected by the differential affinities of PRV for neuronal and glial membrane as well as the cytoarchitecture of the injected area since these factors exert important effects upon the diffusion gradient of virus from the injection site (Sect. 4, step 6). In this regard, injection of areas with high terminal field density (e.g., striatum; (43)) will result in very localized injections due to the high affinities of PRV for the membrane of axon terminals.
- Dual injection paradigms with viruses expressing different reporter proteins designed to explore issues of axonal collateralization should use isogenic strains with well-defined invasive profiles (Sect. 4, step 7). Failure to ensure that viruses invade circuits with similar temporal kinetics can lead to false negatives in circumstances where early replication of one virus precludes infection of neurons with a second strain (42, 44).

3.5. Viral Tracing Experiments: Temporal Considerations

Determining the rank order of neuronal infection within a circuit is a complex process that cannot be determined solely on the basis of the time after infection. As noted above, and in Sect. 4, steps 2 and 4, the progression of infection through a circuit depends upon the concentration of injected virus and the number of terminals available to serve as conduits for virion invasion. In addition, the length of axons through which capsids must be transported to gain access to neuronal cell bodies will produce additional variability in the onset of replication among neurons within a circuit. Consequently, productive infection of a long axon neuron with dense synaptic input to an infected cell will likely become infected

Fig. 4. PRV can be used in combination with other neuronal tracers and the phenotype of infected neurons can be defined with dual labeling immunofluorescence. These approaches are illustrated in this figure, which illustrates PRV infected neurons synaptically linked to the spleen. The sympathetic circuits infected by injection of virus into the spleen are illustrated in *blue* in the schematic diagram at the *top* of the figure. The photomicrographs below the schematic illustrate PRV labeling (*red*) in combination with FluoroGold (FG) labeling (*white*) (**a–e**) and dual labeling immunofluorescence localization of PRV (*red*) and dopamine-β-hydroxylase (*green*) (**f–h**). Combined use of FG and PRV is discussed in greater detail in Sect. 3.5. FG has been demonstrated to interfere with HSV replication when injected in a cocktail. Thus, in the illustrated experiment FG was injected a week prior to PRV injection and, in those circumstances, shown not to interfere with PRV replication and transport. (**a**) Through (**e**) illustrate FG labeling (*white*) of sympathetic preganglionic neurons in the intercalated nucleus and intermediolateral cell column of thoracic spinal cord. A subset of these neurons are replicating PRV (*red+ white*) and the virus has also passed transneuronally to infect local circuit interneurons (*red* only). Illumination of the area shown in (**a**) with different filters demonstrates PRV infected neurons (*red*) in relation to catecholamine containing fibers (*green*) localized with an antibody against DbH. Combined localization of PRV and DbH is also shown in (**f–h**) of the A5 catecholamine cell group (**f**), the locus coeruleus (**g, h**) and the nucleus subcoeruleus (SC) (**g**). Dual labeled neurons are *yellow*. The photomicrographs are adapted from Cano et al. (40) with permission.

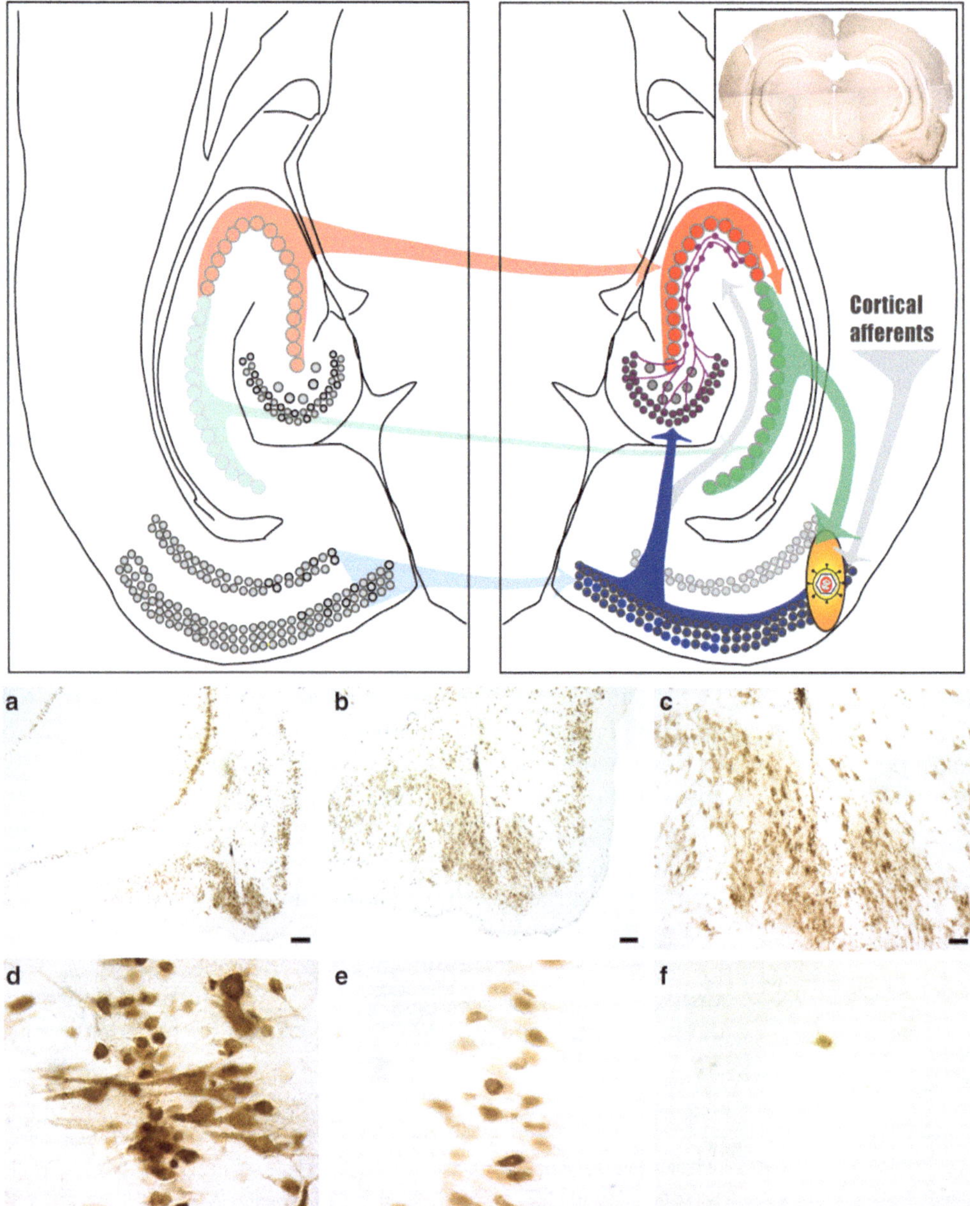

Fig. 5. Retrograde transneuronal infection of hippocampal circuitry following injection of PRV-BaBlu into the entorhinal cortex (ERC). More detailed discussion of the use of intracerebral injection of virus for circuit analysis can be found in Sect. 4, step 6. The schematic diagrams at the *top* of the figure illustrate ERC and hippocampal circuitry ipsilateral to the site of viral injection (defined by the *yellow-orange oval*). Injection sites from two animals are shown in (**a**, **b**). (**c**) Illustrates the site shown in (**b**) at higher magnification. Injection of PRV into the ERC leads to the predicted pattern of retrograde transport through these well-characterized circuits (see *inset* of circuit schematic). Differential distribution of viral antigens reflective of differing degrees of infection and the order of transneuronal transport are illustrated in (**d–f**). First-order infection of CA1 pyramidal neurons resulting from retrograde transport of PRV from ERC is illustrated in (**d**). Viral antigen is densely concentrated throughout the cell nuclei and somatodendritic compartments of infected neurons. Second-order infection of neurons in CA3 pyramidal neurons is illustrated in (**e**). Viral immunoreactivity is dense in the cell nuclei with relatively light staining in the cytoplasm of infected cell bodies; there is essentially no viral antigen apparent in dendrites. Third order infection in the dentate gyrus (**f**) is characterized by viral immunoreactivity confined almost exclusively to the cell nucleus. The figure is adapted from Card et al. (72) with permission.

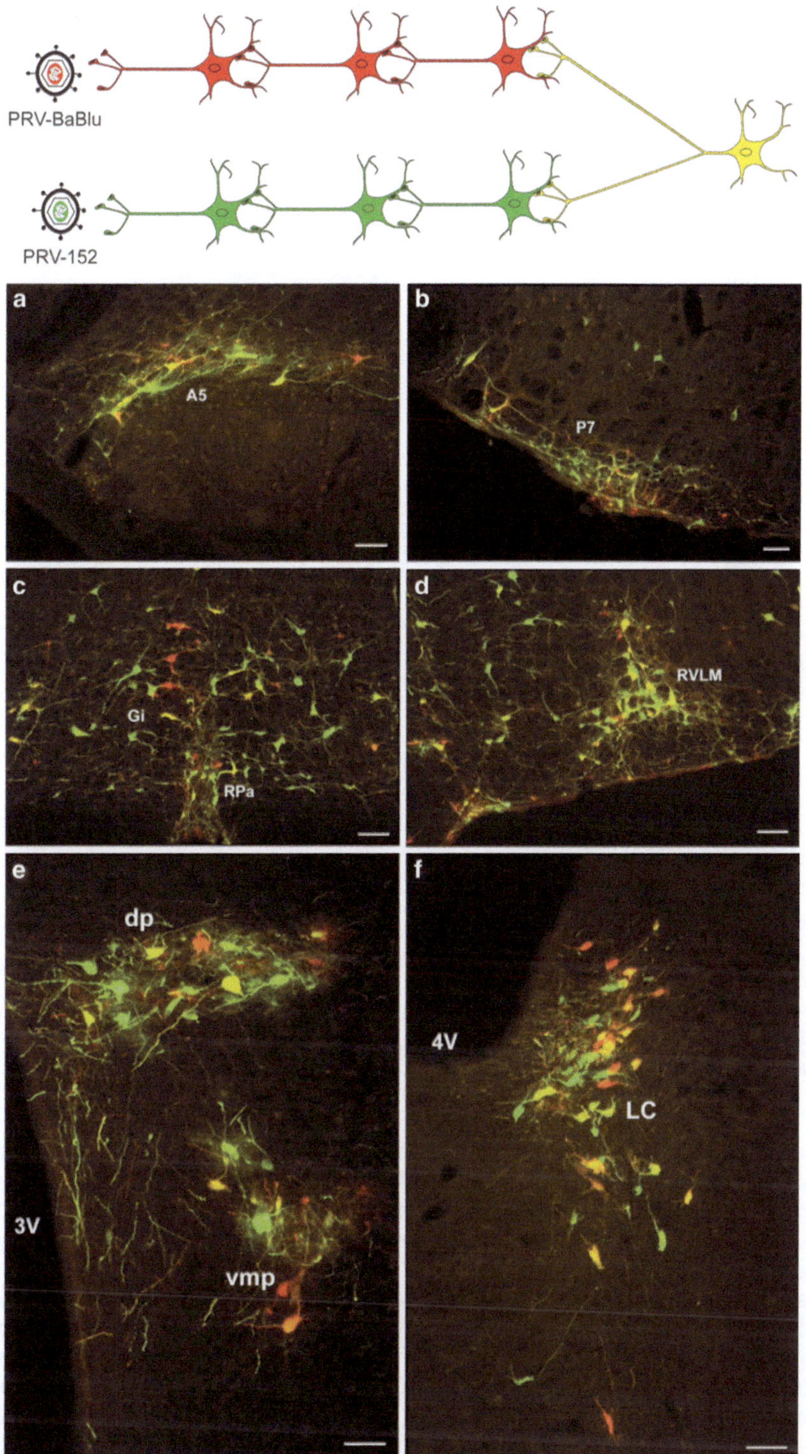

Fig. 6. The use of isogenic strains of PRV expressing unique reporters in dual infection paradigms is illustrated. Discussion of this experimental approach can be found in Sect. 4, step 7. The basic approach is diagramed schematically at the *top* of the figure. The photomicrographs illustrated in (**a–f**) illustrate data derived from applying this approach to injection one recombinant into the left kidney and the other into the right kidney. Both reporter proteins (EGFP and β-galactosidase) were localized with specific antisera and immunofluorescence. Neurons infected only with PRV-152 are *green*, those infected only with PRV-BaBlu are *red*, and those replicating both viruses are *yellow*. Neurons in each category are evident within the A5 catecholamine cell group (**a**), the P7 region of brainstem (**b**), the caudal ventromedial brainstem (**c**), the RVLM (**d**), the paraventricular nucleus of hypothalamus (**e**) and the locus coeruleus (**f**). Adapted with permission from Cano et al. (42).

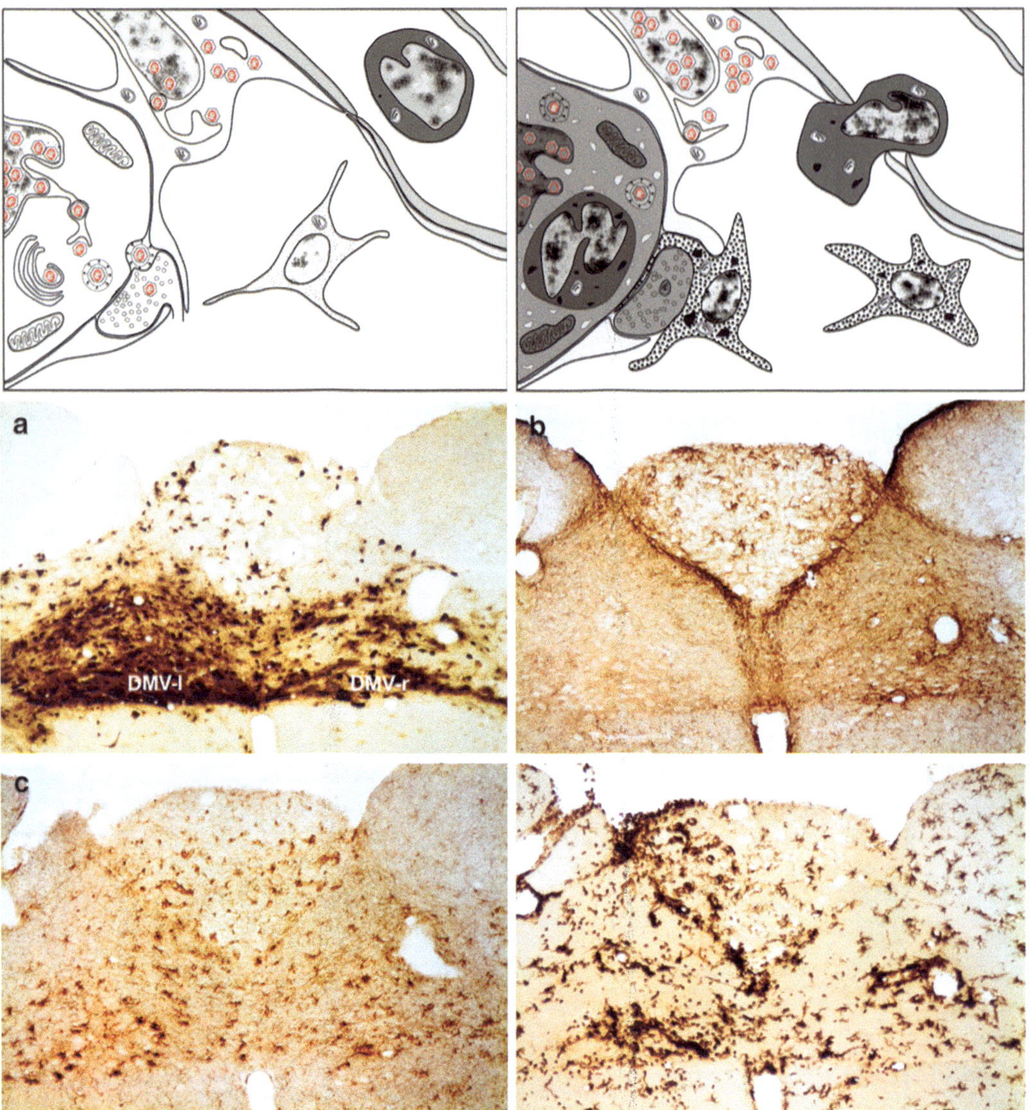

Fig. 7. A model of the nonneuronal response to viral infection that biases transport of PRV through synaptic contacts is illustrated. Data supporting this model is discussed in Sect. 4, step 3. The schematics illustrate the nonneuronal response at early (*upper left*) and late (*upper right*) postinoculation intervals. At early stages of viral replication there is no apparent pathology of infected neurons and progeny virus passes transneuronally to infect synaptically linked neurons. Reactive astrogliosis occurs at this stage, generating processes that isolate infected neurons. Virions leave infected neurons differentially at sites of synaptic contact. Those that escape transneuronal passage invade adjacent reactive astroglia, but these glial cells cannot produce infectious virions. At late stages of replication infected neurons exhibit cytopathic changes and the nonneuronal response is amplified. Hypertrophy and hyperplasia of resident astroglia and microglia isolate infected neurons and immune cells are recruited from the vasculature. The invading immune cells exhibit a directed migration to infected neurons where they appear to contribute to a killing response. The images in the *lower portion* of the figure illustrate immunocytochemical localization of PRV (**a**), GFAP (astrocytic marker) (**b**), OX-42 (microglial marker) (**c**), and ED1 (marker of cells of monocytic lineage) (**d**) in the DVC 68 h after injection of the virulent PRV-Becker strain into the ventral wall of the stomach. The ventral wall of the stomach is principally innervated by the left DMV of the vagus (DMV-l). Virus also spreads through local enteric circuitry in the stomach to the dorsal wall and gives rise to a temporally delayed infection of the right DVM (DMV-r). Thus, the left DMV is advanced in infection and the right DMV is early in infection. Note that infection of neurons is consistent with known connections of the DVC (see Fig. 3) and that the magnitude of reactive gliosis correlates with the stage of infection. It is important to note that GFAP antigenicity decreases at advanced stages of replication but that TEM analysis has demonstrated that reactive astroglia densely surround infected neurons. Adapted with permission from Card et al. (49) and Rinaman et al. (58).

before a neuron in closer proximity that has sparse synaptic input into the same cell.

Informed estimates on the temporal sequence of infection of neurons within a circuit can be obtained based upon predictions incorporating knowledge of the existing literature and basic parameters that determine viral replication and transneuronal passage. In cultured neurons, PRV replication and packaging reaches a maximum by about 10 h and virions move in the retrograde direction at approximately 1 μm/s, or 3,600 μm/h (45). On the basis of these in vitro kinetics, Granstedt et al. indicate that the time for passage of virus through a polysynaptic circuit can be estimated using the following formula:

$$T = (10)(\#S) + (\mathrm{Tr}) / 3{,}600$$

where T = transport time in hours, $\#S$ = estimated number of synapses to be crossed, and Tr = transport distance over dendrites and axons in microns (46). Obviously, precise information about the general organization and morphology of neurons that contribute to the circuit under study will not always be available. Nevertheless, this formula provides a good starting point for estimating the timing of viral neuroinvasiveness in experimental design.

Combining viral tracing with classical tracers provides a very useful means of clarifying the rank order of infection (Fig. 7). In an early study, Li et al. demonstrated the utility of this approach by combining anterograde tracing with *Phaseolus vulgaris leucoagglutinin* (PHAL) and retrograde transport of HSV to define synaptic relations of caudal brainstem neurons involved in the neural regulation of blood pressure (47). Injection of cocktails of PRV and retrograde tracers has also been used to separate first order projections from neurons in the same area with those labeled by transneuronal passage for virus. For example, coinjection of the β fragment of cholera toxin (βCT) or a CT-HRP conjugate with PRV-Bartha have been successfully employed to distinguish first-order neurons from those infected by replication and transneuronal passage of virus (48, 49). Cano et al. demonstrated that neurons previously labeled with the retrograde tracer FluoroGold are permissive to infection by PRV and can be used to distinguish sympathetic preganglionic neurons (SPNs) from interneurons infected by transneuronal passage of PRV (39). However, successful combined use of FluoroGold and PRV requires temporal separation in the injection of the tracers since LaVail et al. demonstrated that FluoroGold interferes with HSV replication when the two tracers are injected as a cocktail (50).

Lesions also provide a useful means of determining the rank order of viral infection and the routes that virus travel to infect components of a circuit. O'Donnell et al. and Aston-Jones et al. used electrolytic lesions to confirm that cell groups served as relays for transneuronal passage of virus (48, 51). Similarly, Jasmin et al.

demonstrated the utility of using ibotenic acid lesions to confirm the route of viral transport following intracerebral injection of virus into the central nucleus of amygdala (52). This approach is particularly valuable when the patterns of viral infection present alternative pathways for infection of a particular node within a circuit. However, when incorporating a lesion strategy in a viral tracing study it is important to recognize that the immune response generated by the lesion can compromise replication of virus. Thus, it is important to separate the generation of the lesion and injection of virus. Separating these aspects of the experiment by approximately a week is effective in addressing this important issue.

A good rule of thumb in designing tracing studies is to first document the invasive profile of virus in the circuit by examining short, intermediate, and long postinoculation intervals. The precise timing for each of these survival intervals will vary according to the circuit under study. General guidelines for studies involving injection of peripheral targets (e.g., autonomic tissues and skeletal muscle) are 3, 4, and 5 days. In contrast, the literature suggests that postinoculation intervals of 1, 2, and 3 days effectively define circuits labeled by intracerebral injection. Once the general pattern of infection is established one can gain further insight into the rank order of infection through analysis of more frequent survival intervals and combined injection of PRV and classical tracers.

It is useful to note that the immunocytochemical localization of viral antigens within infected neurons also provides an accurate index of the temporal course of viral infection. Viral antigen first becomes apparent within the nuclei of infected neurons as capsids assemble around replicated viral DNA. As infection advances the concentration of antigens increases within the perikarya of infected neurons and progressively invades the dendritic tree. Ultimately the entire somatodendritic compartment becomes densely labeled as progeny virus and virally encoded proteins traffic throughout the full extent of the somatodendritic compartment. Thus, a careful analysis of the intracellular distribution of viral immunoreactivity provides an informed perspective on the temporal course of infection within and between neurons.

3.6. Viral Tracing Experiments: Detection Methods & Tissue Preparation

Immunocytochemical detection of viral antigens or reporter molecules is the most widely applied means of detecting infected neurons. Well-characterized polyclonal antisera generated against acetone-inactivated PRV generate robust signals in immunoperoxidase and immunofluorescence localizations (Table 2). Currently, only one rabbit polyclonal antibody generated against acetone-inactivated PRV is commercially available (PA1-081; Thermo Scientific; Table 2). We conducted direct comparisons of the signal produced by this antiserum and demonstrated that it produces labeling equivalent to that produced by the rabbit polyclonals

(Rb132, 133, 134) produced in our laboratory. Polyclonal antisera generated in other species are not widely available.

The construction of well-characterized PRV recombinants that express unique reporters has provided another powerful means of localizing infected neurons and also expanded the way in which the method can be applied (e.g., dual infection models for analysis of collateralization). PRV-BaBlu (β-galactosidase reporter), PRV-152 (EGFP reporter), and PRV-614 (mRFP reporter) are among the most widely applied recombinants used for this purpose (Table 1). Numerous antibodies are available for effective localization of β-galactosidase in PRV-BaBlu infected circuits. We have had excellent success using the mouse monoclonal commercially available through Sigma Chemical Company. PRV-152 (EGFP) and PRV-164 (mRF) can be detected by endogenous fluorescence of the fluorophor. Nevertheless, we have had very good success localizing GFP and RFP in both immunoperoxidase and immunofluorescence applications with specific antisera. Use of these antisera for immunofluorescence enhances and stabilizes the fluorescence signal over that produced by endogenous fluorescence of the reporter (e.g., Fig. 6).

The labor-intensive nature of viral tracing experiments along with the relative lack of commercially available reagents makes it advantageous to obtain as much information as possible from each experiment. Consequently, we process the entire brain using the following protocol and store the tissue in cryoprotectant (53) until immunocytochemical processing. Direct comparisons of immunoreactivity in tissue from the same brain processed with our rabbit polyclonals (Rb132, Rb133, Rb134) immediately after sectioning and after 10 years of storage in cryoprotectant demonstrated no apparent reduction in immunoperoxidase signal. The following protocol is used in our laboratory to fix, process, and store section in cryoprotectant. A more detailed presentation of this procedure and our immunocytochemical processing procedures is available elsewhere (33).

1. Perfuse animal by transcardiac infusion of the paraformaldehyde-lysine-periodate (PLP) fixative develop by McLean and Nakane (35).
2. Postfix tissue in PLP at 4°C for 24–72 h.
3. Cryoprotect tissue by sequential immersion in 20 and 30% phosphate buffered sucrose solutions at 4°C (24 h in each sucrose concentration).
4. Section tissue in desired plane using a freezing microtome. We cut tissue sections sequentially at 35 μm/section into 6 wells of sodium phosphate buffer. Tissue not immediately processed for localization of virus is transferred to cryotubes (1 mL capacity) containing cryoprotectant and stored at –20°C.

Immunoperoxidase localization of viral antigens provides the most effective means of thoroughly characterizing the pattern of viral transport. We typically process one well of tissue for immunoperoxidase of viral antigens using one of our rabbit polyclonal antisera. With our approach, one well of tissue contains sections at a frequency of 210 μm through the brain, which provides an effective sampling of all regions of the neuraxis. A detailed protocol for immunoperoxidase localization of viral antigens has been published (33). However, it is useful to highlight specific issues as they relate to the processing of tissue for viral localization.

1. Wash cryoprotectant from sections using repeated changes of sodium phosphate buffer. It is essential to completely wash the viscous cryoprotectant from the tissue sections in order to achieve reliable immunocytochemical localizations with low background. We have found that this is easily achieved using the compartmentalized staining nets and dishes available from Brain Research Laboratories (Sect. 2.4). We use these staining nets for washing tissue after all antibody incubations and for the diaminobenzidine reaction.
2. Incubate tissue in primary antibody diluted with PBS, normal serum, and 0.3% Triton X-100. All of the polyclonal antibodies generated against PRV that are listed in Table 2 can be reliably used at a dilution of 1:20,000 when incubated for 24–48 h at 4°C.
3. We use species appropriate affinity purified antibodies for the secondary antibody incubations. When used in combination with our rabbit polyclonals we use donkey antirabbit IgG antibodies that are biotinylated. The biotinylation of the secondary antibodies is necessary for use of the Vectastain *Elite* Kit (Vector Laboratories).

Following characterization of the pattern of infected neurons derived from the immunoperoxidase localization described above we further characterize the labeled circuit using dual labeling immunofluorescence to define the phenotype of infected neurons. Immunofluorescence is also an essential component of dual infection studies employing recombinant strains of PRV that express unique reporters (Sect. 4, step 6). This procedure as it is applied in our laboratory has also been published (33).

The recent generation of recombinant strains of PRV that express fluorescent reporters also creates the need for preparing tissue for reporter localization based simply upon the fluorescence emitted by the reporter proteins. The signal produced by these reporters is typically less intense than that produced by immunofluorescence and, in some cases (e.g., the cyan reporter of the mCerulean gene), quenches under prolonged excitation. The following approach works well in our hands to reliably document the fluorophor profiles of neurons expressing fluorescing reporters.

1. As is typical of our approach, we section the brain of perfused animals into six bins of tissue at a frequency of 210 μm (35 μm/section). All of the tissue is immediately transferred to cryoprotectant and stored at −20°C to preserve antigenicity.
2. We first determine the location of infected neurons by processing a bin of tissue for immunoperoxidase localization of viral antigens using rabbit anti-PRV polyclonal antiserum. This focuses the fluorescent analysis upon specific areas and reduces the need to scan sections for fluorescence signal.
3. On the day that we plan on analyzing the tissue we rinse one bin of tissue with agitation in multiple changes of sodium phosphate buffer to remove cryoprotectant. Sections are then mounted on gelatin-coated slides and air-dried.
4. Mounted sections are coverslipped using Vectashield Hard Set Mounting Medium for Fluorescence (Vector Laboratories, Inc.; Burlingame, CA) and stored in the dark.
5. We conduct the initial analysis using an epifluorescence microscope (Olympus BX51) equipped with filters specific for the reporter proteins expressed by the PRV recombinants. The initial exposure settings are adjusted to those previously shown to be optimal for each fluorescence reporter. When cells expressing fluorescent reporters are encountered the exposure of each wavelength of light is adjusted for optimal saturation and the field is photographed.

4. Notes

1. *Storage of viral stocks.* Storage of virus aliquots is an extremely important consideration for successful viral tracing studies. The lipid envelope is easily damaged, but virions are quite stable if certain variables are controlled. Exposure to detergents, bleach, and alcohols in any concentration and ice formation should be strictly avoided. Multiple freeze-thaw cycles dramatically reduce the concentration of a virus aliquot due to ice formation. For this reason, avoid storage at −20°C (most −20°C freezers have automatic defrost cycles that are lethal to frozen virions). The −80°C or liquid nitrogen freezers will maintain virus infectivity for years, but certain precautions should be used to maintain virus titer under these conditions. Avoid storing microliter aliquots of stocks in large tubes. Use tightly sealed cryovials for long-term storage. Avoid storing virus stocks at the front of an upright freezer or at the top of a

chest freezer as these areas warm quickly when the doors are open while retrieving reagents from a freezer. The smallest volumes we aliquot for –80°C storage are 100 μL quantities in 1.5-mL tubes. Ideally, 500 μL quantities in 1.5-mL tubes should be used. At the outset of an experiment we remove a tube from the freezer and place it in an ice bucket. We thaw the tube in a warm water bath (37°C; rapid thawing avoids ice formation) until little ice remains and then return it to the ice bucket for storage during the course of the experiment. On ice, virus stocks are stable for at least 12 h. At the conclusion of the experiment we inactivate and dispose of the unused virus from the aliquot. Under no circumstances should these small aliquots of unused virus be refrozen since this action has a high probability to reduce the infectious virus concentration.

2. *Virion invasiveness.* Appropriate design of a viral tracing study is dependent upon a clear understanding of the factors that influence virus uptake, replication, and transneuronal passage (Fig. 3). First and foremost are the mechanisms that lead to virion invasion of permissive neurons. Neuroinvasive viruses have evolved to exploit the intimate synaptic relations of neurons to pass from one neuron to another. This feature is the primary risk factor that makes viral encephalitis such a serious disease and it is also the fundamental feature that makes these viruses so effective for viral tract tracing in experimental studies.

   Viral membrane proteins are essential for entry of virions into permissive cells and play prominent roles in target cell recognition, attachment, and receptor-mediated invasion (23, 24). From the standpoint of viral tracing it is important to recognize that the affinities of certain PRV membrane proteins for cellular membranes and extracellular matrix molecules restrict the spread of virions in tissues and optimize virus uptake from a restricted region. This property is adaptive from the standpoint of the life cycle of the virus and also biases the transport of virus through synaptic contacts. Glycoprotein D (gD) is the primary viral protein that binds cellular receptors and is essential for PRV virions to gain access to cells. Although there are multiple gD receptors, the nectin 1 receptor is the primary receptor for PRV virions in rodents. Nectin 1 is a member of a family of transmembrane glycoproteins that function as cell adhesion molecules (54, 55). It is present in all classes of neurons as well as endothelial cells, epithelial cells, and fibroblasts. This explains why PRV can infect any neuron irrespective of neurotransmitter phenotype and why PRV is pantropic (can infect many cell types). To date, no class of neuron that expresses nectin 1 is known to be refractory to infection by PRV. It is noteworthy that, while nectin 1 and PRV gD are absolutely required for PRV virions to infect, neither are required for subsequent passage of virus to synaptically connected neurons once infection starts (23). This property does not hold for glycoprotein B (gB), which is part of the tri-partite virion membrane fusion complex comprising gB/gH/gL that enables entry of PRV virions as well as cell-to-cell spread. Virions that do not express gB, gH, or gL cannot infect or pass from infected neurons to synaptically connected neurons. If such a null virus is grown on complementing cells (cell expressing the deleted gene), the resulting virions can infect one cell but cannot spread to other cells, despite the fact that virions are produced (56).

3. *The nonneuronal response to infection.* Infection of the nervous system elicits a defensive response in infected and local uninfected cells that biases transport of virus through synaptic connections. The demonstration that several days after infection viral antigens were detected in glia adjacent to infected neurons raised concerns that release of virus particles into the extracellular space might compromise circuit-related transport of virus through neural circuits. However, the identification and development of attenuated strains of PRV and a greater understanding of the role of the nonneuronal response to infection have largely allayed these concerns. Nevertheless, an informed understanding of this issue and how it can impact upon cell-to-cell transmission of virus remains an important issue in both the design and interpretation of viral studies. Here we summarize the basic issues as they relate to the role of glia in the specificity of viral transport through neural circuits and the potential for false negatives.

   Retrograde transport of virulent and attenuated strains of PRV through the vagus nerve has figured prominently in defining the role of the nonneuronal response to viral infection (49, 57–60). This model system (Fig. 7) has the advantage of introducing virus into the nervous system in a covert "Trojan Horse" fashion without resorting to the trauma associated with direct injections. This method allows a focal analysis of the glial responses mounted in response to viral replication and reveals a temporally organized reactive gliosis and focal recruitment of immune cells from the circulation to the site of virus replication. Collectively these nonneuronal responses isolate infected neurons prior to the appearance of any cytopathology and prevent lytic dissemination of virus through the extracellular space. Activation of astroglia and microglia adjacent to infected neurons constitutes the first phase of this response. The hypertrophy and hyperplasia characteristic of the reactive phase to infection, functions to isolate the neuron (reactive astrogliosis) and, ultimately, to cause afferents to retract from the infected neuron. Importantly, replication and transneuronal passage of progeny virus takes place before retraction of afferents from infected neurons (e.g., "synaptic stripping"; (61)). Put another way, spread of infection through a circuit is faster than the local defense response. Reactive astrogliosis is quite effective in restricting spread of virus that escapes synaptic transfer because of the high affinity of virions for this class of glia and the fact that reactive astroglia have induced a broad spectrum antiviral state that blocks replication of many viruses.

Recruitment of cells of monocytic lineage from the vasculature constitutes the second phase of the response. This directed recruitment occurs precisely in the area of the viral infection and cells recruited from the vasculature exhibit a directed migration to infected neurons. This extravasation and directed migration of immune cells contributes to a phagocytic response designed to eliminate cells advanced in viral replication. The magnitude of the nonneuronal response correlates with the virulence of the infecting strain of virus (e.g., it occurs earlier and is of greater magnitude after infection with more virulent virus).

The effectiveness of the second phase of the nonneuronal response in eliminating infected cells can lead to false negatives in viral tracing studies. A vivid example of this issue has been documented by Denes et al. in characterizing the invasiveness of a highly attenuated genetically modified strain of PRV-Bartha (Ba-DupGreen) (62). BaDupGreen invades CNS circuits with kinetics considerably slower than PRV-Bartha (63, 64). Furthermore, for this PRV recombinant, a rapid inflammatory response effectively eliminates infected neurons in a spatially and temporally organized manner. Thus, infected neurons identified at early survival times are subsequently cleared by the innate immune response and cannot be identified later after infection. These observations emphasize the importance of choice of PRV strains and temporal analysis in viral tracing studies.

4. *Virus concentration and transneuronal tracing*. The concentration of infectious virions (titer) is an essential consideration in the design and interpretation of viral transneuronal tracing studies. In general, before initiating tracing experiments, one should determine the amount of virions needed to infect the animal and spread reproducibly in a circuit of interest. The classical method is first to determine the amount of infectious virus required to infect 50% of animals (the ID50) using your injection paradigm. The ID50 depends on the animal species, the site of infection, and the virulence of the virus used. The dose/response curve is sharp with the inflection point at the ID50 concentration. At low concentrations of virus, intrinsic host defenses predominate to block infection completely, but as virus concentration increases, a point is reached where host defenses are overcome. We have found that the concentration of virions

at the site of infection dictates the number of infected neurons and subsequent transneuronal passage of progeny virus to infect synaptically linked neurons. The precise threshold may vary among neurons and tissues and is related, at least in part, to passive and active cellular defense mechanisms that impede virus entry and replication. The onset of infection of neurons also depends on the number of axon terminals available for virion uptake at the site of injection. For PRV-Bartha and its recombinants in rat, we find that the ID50 for peripheral organ injection is about 50,000 plaque-forming units (pfu; a measure of virus concentration; see Sect. 3.2) and that 10^5 pfu will infect 100% of injected animals. In contrast, infection with 10,000 pfu results in infection of about 20% of the animals (34, 65).

5. *Infection by peripheral inoculation*. The vast majority of papers on viral transneuronal tracing involve retrograde transneuronal transport of PRV from peripheral autonomic targets and skeletal muscle. Analysis of PRV invasion of the central nervous system following inoculation of peripheral autonomic targets has been particularly influential in demonstrating that transmission of virus through synapses is the primary means through which PRV moves through the nervous system. An important theme that has emerged from the studies of viral transport through autonomic circuits is the need to consider the cytoarchitecture of the injected tissue in experimental design. This is effectively illustrated in considering studies that characterized retrograde transport of PRV from the kidney and spleen (Figs. 4 and 6). The sympathetic division of the autonomic nervous system innervates each of these organs, but the density of innervation and the cytoarchitecture of the constituent tissues differ considerably. The kidney is a capsular organ with a dense parenchyma that receives a fairly dense sympathetic innervation. In contrast, the spleen is a heavily vascularized organ with an "open" circulation in which arterioles empty into sinusoids that, in turn, drain into the venous circulation. Injection of relatively small amounts (1–2 μL of a 1×10^8 pfu/mL stock) of PRV-Bartha or Bartha recombinants into the kidney produces a predictable retrograde transneuronal infection of the central preautonomic network linked to this organ (42, 66). Injection of the same amount of virus (10^5 pfu) into the spleen rarely produced an infection of central preautonomic pathways while six injections of 1 μL of the virus stock (6×10^5 pfu total) into different sites in the organ reproducibly infected the preautonomic network linked to this organ (39, 41). Failure to infect circuits linked to the spleen after injections of lower concentrations of virus may result from a "diluting" effect of the organ's open circulation and the relatively sparse neural innervation. This conclusion is supported by studies using PRV to define the organization of neural circuits

linked to the sympathetic innervation of peripheral adipose tissues. The most extensive studies in this regard involve an analysis of the neural innervation of intrascapular brown fat, which is important in rodent thermogenesis. The sympathetic neural innervation of this adipose tissue is dispersed throughout the interscapular fat pad. Several studies have demonstrated that reproducible infection of circuits linked to brown fat requires injection of large volumes of virus at multiple sites because of the sparse innervation of the tissue and the relatively dense aggregation of adipose cells that restrict the diffusion of virus from injection sites (41, 67, 68). Collectively, these observations emphasize the importance of considering the neurotropism of PRV as it relates to the cytoarchitectural features of the system under study when designing viral tracing studies.

6. *Infection by intracerebral inoculation.* The ID50 for PRV when injected directly into the brain tends to be lower than when injected into peripheral tissue. The volume of injected virus is another issue that will influence the uptake and infection after intracerebral injection. As observed in studies injecting peripheral targets, cytoarchitecture is an essential consideration for the design of studies using intracerebral injection of virus. Investigations by Vahlne et al. demonstrated that HSV has a high affinity for synaptosomes and astrocytic membrane but low affinity for membrane of neuronal perikarya (69–71). Marchand and Schwab similarly demonstrated high affinity of PRV for axon terminals of sympathetic neurons grown in vitro and further demonstrated that the binding of ^{35}S-labeled virus could be blocked by an excess of unlabeled virus or concanavalin A (71). Parametric studies of PRV-Bartha invasiveness after injection into striatum confirmed these affinities and established clear guidelines for studies using intracerebral injection of PRV for circuit analysis (43). The rich concentration of axon terminals within the striatum, combined with virion affinities for extracellular matrix molecules and glial membrane, restricted the spread of virus from the injection site and resulted in retrograde transport to neurons in the cortical mantle, midbrain tegmentum (e.g., substantia nigra pars compacta) known to project to the striatum. Furthermore, the low affinity of virions for neuronal perikarya was reflected in the very small number of neurons that were infected within the striatum, even in cases where extensive neuronal infection was observed in areas projecting to striatum. Strong support for the conclusion that onset of infection within a circuit is directly related to the concentration of injected virus and the number of axon terminals available for virus uptake was also provided by the data reported in this study. Following injection of log unit dilutions of PRV-Bartha into the same region of striatum in different animals we observed concentration-dependent changes in the progression of retrograde transneuronal infection of neural circuits.

Injection of the highest titer of virus (10^5 pfu) produced the most extensive infection while injection of lesser titers of virus resulted in a more restricted infection of the same circuitry within the same postinjection survival interval.

Several studies have now established the utility of intracerebral injection as a means of defining the functional organization of neural circuits. The power of the method for defining the reorganization of neural circuitry in animal models of injury and disease has also been demonstrated. For example, in an animal model of pediatric head trauma we defined pronounced changes of circuit organization (72). Injection of virus into the entorhinal cortex (ERC) of control rats on postnatal day 45 revealed a pattern of retrograde infection of hippocampal and neocortical circuits consistent with the well-established projections to the ERC (Fig. 5). Interestingly, equivalent injection of PRV in animals of the same age that experienced controlled cortical impact on postnatal day 17 revealed a preservation of topography in surviving regions of the neocortex and hippocampus, but a significant increase in the number of neurons incorporated into the circuit. These data and that from other studies using similar approaches clearly demonstrate that PRV can be effectively used to define the organization of neural circuits following direct injection into neural parenchyma. As in other applications, knowledge of the life cycle of the virus and the features of cytoarchitecture that influence the progression of viral replication and transport through a circuit are essential to appropriate experimental design and interpretation of data.

7. *Tracing with two isogenic recombinants* (*the dual infection paradigm*). The development of isogenic strains of PRV with comparable replication rates that express unique reporters has enabled the use of coinfection paradigms to define the organization of complex circuits. An early application of this approach identified neurons in the CNS that collateralize to synapse upon SPNs projecting to different autonomic targets (73). However, the same study also emphasized the importance of using isogenic strains of virus and conducting temporal studies to ensure that both strains of virus are moving through central circuits at equivalent rates. Kim et al. and Banfield et al. demonstrated that prior infection of a neuron with one strain of PRV can prevent infection by a second strain that reaches the cell several hours later (44, 74). Studies employing isogenic strains of PRV with matched invasive profiles have demonstrated that the dual infection paradigm can provide a powerful means of identifying neurons that collateralize to multiple output pathways (e.g., Fig. 6) (42, 75). The parameters that impact upon successful application of this experimental approach have been examined following injection of antigenically distinct isogenic recombinants into the left and right kidney (42).

8. *Conditional replication of PRV.* DeFalco et al. provided the first demonstration of conditional PRV replication to define neurons synaptically linked to phenotypically defined neurons (76). This powerful approach employs the Cre-lox site specific recombination system (77) to activate replication of a genetically modified PRV strain. DeFalco et al. constructed a derivative of PRV-Bartha that replicates only when it infects a cell expressing the Cre-recombinase (Cre) protein. This strain, PRV-2001, does not express the viral thymidine kinase (TK) gene and therefore cannot replicate in nonmitotic cells. When a cell expresses Cre, TK expression is activated and the virus replicates and spreads to all synaptically connected neurons (reviewed in (78)).

   DeFalco et al. provided the proof-of-principle for this approach for conditional replication of PRV-2001. The virus was injected into the mediobasal hypothalamus of transgenic animals that express Cre in neurons that contain the peptide NPY. Viral replication was restricted to NPY containing neurons and produced progeny virus capable of passing transneuronally to infect neurons synaptically linked to NPY-containing neurons. Subsequent studies have exploited this approach to define the circuits linked to LHRH-containing neurons in hypothalamus (79–81) and serotoninergic neurons of the dorsal raphe and raphe magnus (82).

   The conditional reporter adaptations of PRV are among the most recent, and the most powerful, advances in the use of viruses to define the functional organization of polysynaptic neural circuitry. Two groups have recently combined the transneuronal tracing capabilities of PRV with expression of fluorescent calcium indicator proteins to report on neural activity within labeled circuits (46, 83, 84). This important advance in viral tracing technology opens the door for systems based analysis of the functional activity of neurons within defined polysynaptic circuits, a powerful advance in technology that has important implications for improving understanding of the functional architecture of the CNS. Conditional reporter expression of replication competent strains of PRV have also dramatically expanded the ability to define connections to specific populations of neurons within labeled polysynaptic networks (85). This approach combines the ability to obtain targeted expression of Cre in phenotypically defined and projection-specific populations of neurons with Cre-dependent conditional expression of fluorescent reporters from the viral genome. The latter was achieved by developing a strain of PRV (PRV-263) that carries the Brainbow 1.0L cassette developed by Lichtman et al. (86) and is replication competent in all neurons (87). In the absence of Cre, infected neurons display red fluorescence from the dTomato gene of the Brainbow cassette. However, when PRV-263 infects cells expressing Cre, the dTomato gene of the Brainbow cassette is excised, enabling the expression of cyan or yellow reporters from the mCerulean or EYFP reporter genes. In proof-of-principle studies, targeted expression of Cre in brainstem catecholamine neurons combined with retrograde transneuronal infection of the renal preautonomic circuits with PRV-263 produced conditional reporter expression only in neurons synaptically connected to the Cre expressing catecholamine neurons. The availability of other technologies that restrict Cre expression to targeted populations of neurons indicates that this approach can be widely applied across functionally defined systems to characterize details of synaptology within complex circuits. These advances, and others likely to come in the future, suggest that viral tracing technology will become an increasingly powerful means of defining the means through which neurons interact to produce defined functional outcomes.

References

1. Peters A, Palay SL, Webster HD (1991) The fine structure of the nervous system, 3rd edn. Oxford University Press, New York
2. Chan-Palay V (1981) Light and electron microscopic autoradiographic techniques: radioactive amino acids, neurotransmitters, receptors, and combined methods with immunocytochemistry. In: Johnson JE (ed) Current trends in morphological techniques. CRC Press, Boca Raton, pp 53–89
3. Mesulam M-M (1981) Enzyme histochemistry of horseradish peroxidase for tracing neural connections with the light microscope. In: Johnson JE (ed) Current trends in morphological techniques. CRC Press, Boca Raton, pp 1–54
4. Schmued LC, Fallon JH (1986) Fluoro-gold: a new fluorescent retrograde axonal tracer with numerous unique properties. Brain Res 377:147–154
5. Gerfen CR, Sawchenko PE (1984) An anterograde neuroanatomical tracing method that shows the detailed morphology of neurons, their axons and terminals: immunohistochemical localization of axonally transported plant lectin, *Phaseolus vulgaris* leucoagglutinin (PHAL). Brain Res 290:219–238
6. Llewellyn-Smith IJ (1992) Retrograde tracers for light and electron microscopy. In: Bolam JP (ed) Experimental neuroanatomy: a practical approach. IRL Press, Oxford, pp 31–59
7. Smith Y (1992) Anterograde tracing with PHA-L and biocytin at the electron microscopic level. In: Bolam JP (ed) Experimental neuroanatomy: a practical approach. IRL Press, Oxford, pp 61–79
8. Horn AK, Buttner-Ennever JA (1990) The time course of retrograde transsynaptic transport of tetanus toxin fragment C in the oculomotor system of the rabbit after injection into extraocular eye muscles. Exp Brain Res 81:353–362
9. Evinger C, Erichsen JT (1986) Transsynaptic retrograde transport of fragment C of tetanus toxin demonstrated by immunohistochemical localization. Brain Res 380:383–388
10. Enquist LW, Husak PJ, Banfield BW, Smith GA (1998) Infection and spread of alpha herpesviruses in the nervous system. Adv Virus Res 51:237–347
11. Card JP (2001) Pseudorabies virus neuroinvasiveness: a window into the functional organization of the brain. Adv Virus Res 56:39–71
12. Enquist LW, Card JP (2003) Recent advances in the use of neurotropic viruses for circuit analysis. Curr Opin Neurobiol 13:603–606
13. Boldogkoi Z, Sik A, Denes A, Reichart A, Toldi J, Gerendai I, Kovacs KJ, Palkovits M (2004) Novel tracing paradigms—genetically engineered herpesviruses as tools for mapping functional circuits within the CNS: present status and future prospects. Prog Neurobiol 72:417–445

14. Kelly RM, Strick PL (2000) Rabies as a transneuronal tracer of circuits in the central nervous system. J Neurosci Methods 103: 63–71
15. Kluge JP, Mare CJ (1974) Swine pseudorabies: abortion, clinical disease, and lesions in pregnant gilts infected with pseudorabies virus (Aujeszky's disease). Am J Vet Res 35: 991–995
16. Fraser G, Ramachandran SP (1969) Studies of the virus of Aujeszky's disease. I. Pathogenicity for rats and mice. J Comp Pathol 79:435–444
17. Hagemoser WA, Kluge JP, Hill HT (1980) Studies on the pathogenesis of pseudorabies in domestic cats following oral inoculation. Can J Comp Med 44:192–197
18. Hall LB Jr, Kluge JP, Evans LE, Hill HT (1984) The effect of pseudorabies (Aujeszky's) virus infection on young mature boars and boar fertility. Can J Comp Med 48:192–197
19. McCracken RM, McFerran JB, Dow C (1973) The neural spread of pseudorabies virus in calves. J Gen Virol 20:17–28
20. Mettenleiter TC (2003) Pathogenesis of neurotropic herpesviruses: role of viral glycoproteins in neuroinvasion and transneuronal spread. Virus Res 92:197–206
21. Ramachandran SP, Fraser G (1971) Studies on the virus of Aujeszky's disease. II. Pathogenicity for chicks. J Comp Pathol 81:55–62
22. Schmidt SP, Hagemoser WA, Kluge JP, Hill HT (1987) Pathogenesis of ovine pseudorabies (Aujeszky's disease) following intratracheal inoculation. Can J Vet Res 51:326–333
23. Pomeranz LE, Reynolds AE, Hengartner CJ (2005) Molecular biology of pseudorabies virus: impact on neurovirology and veterinary medicine. Microbiol Mol Biol Rev 69:462–500
24. Mettenleiter TC, Keil GM, Fuchs W (2008) Molecular biology of animal herpesviruses. In: Mettenleiter TC, Sobrino F (eds) Animal viruses: molecular biology. Caister Academic Press, Norfolk
25. Mettenleiter TC (2000) Aujeszky's disease (pseudorabies) virus: the virus and molecular pathogenesis—state of the art. Vet Res 31:99–115
26. Card JP, Enquist LW (1995) Neurovirulence of pseudorabies virus. Crit Rev Neurobiol 9:137–162
27. Bartha A (1961) Experimental reduction of virulence of Aujeszky's disease virus. Magy Allatorv Lapja 16:42–45
28. Pickard GE, Smeraski CA, Tomlinson CC, Banfield BW, Kaufman J, Wilcox CL, Enquist LW, Sollars PJ (2002) Intravitreal injection of the attenuated pseudorabies virus PRV Bartha results in infection of the hamster suprachiasmatic nucleus only by retrograde transsynaptic transport via autonomic circuits. J Neurosci 22:2701–2710
29. Smeraski CA, Sollars PJ, Ogilvie MD, Enquist LW, Pickard GE (2004) Suprachiasmatic nucleus input to autonomic circuits identified by retrograde transsynaptic transport of pseudorabies virus from the eye. J Comp Neurol 471:298–313
30. Klupp BG, Hengartner CJ, Mettenleiter TC, Enquist LW (2004) Complete, annotated sequence of the pseudorabies virus genome. J Virol 78:424–440
31. Curanovic D, Lyman MG, Bou-Abboud C, Card JP, Enquist LW (2009) Repair of the UL21 locus in pseudorabies virus Bartha enhances the kinetics of retrograde, transneuronal infection in vitro and in vivo. J Virol 83:1173–1183
32. Dolivo M (1980) A neurobiological approach to neurotropic viruses. Trends Neurosci 3:149–152
33. Card JP, Enquist LW (1999) Transneuronal circuit analysis with pseudorabies viruses, vol 1. Wiley, San Diego
34. Sams JM, Jansen AS, Mettenleiter TC, Loewy AD (1995) Pseudorabies virus mutants as transneuronal tracers. Brain Res 687:182–190
35. McLean IW, Nakane PK (1974) Periodate-lysine-paraformaldehyde fixative. A new fixative for immunoelectron microscopy. J Histochem Cytochem 22:1077–1083
36. Enquist LW (2002) Exploiting circuit-specific spread of pseudorabies virus in the central nervous system: insights into pathogenesis and circuit tracers. J Infect Dis 186:S209–S214
37. Song CK, Enquist LW, Bartness TJ (2005) New developments in tracing neural circuits with herpesviruses. Virus Res 111:235–249
38. Loewy AD (1995) Pseudorabies virus: a transneuronal tracer for neuroanatomical studies. In: Kaplitt MG, Loewy AD (eds) Viral vectors. Gene therapy and neuroscience applications. Academic, San Diego, pp 349–366
39. Cano G, Card JP, Rinaman L, Sved AF (2000) Connections of Barrington's nucleus to the sympathetic nervous system. J Auton Nerv Syst 79:117–128
40. Cano G, Sved AF, Rinaman L, Rabin BS, Card JP (2001) Characterization of the central nervous system innervation of the rat spleen using viral transneuronal tracing. J Comp Neurol 439:1–18
41. Cano G, Passerin AM, Schiltz JC, Card JP, Morrison SF, Sved AF (2003) Anatomical

substrates for the central control of sympathetic outflow to intercapsular adipose tissue during cold exposure. J Comp Neurol 460:303–326

42. Cano G, Card JP, Sved AF (2004) Dual viral transneuronal tracing of central autonomic circuits involved in the innervation of the two kidneys in the rat. J Comp Neurol 471:462–481
43. Card JP, Enquist LW, Moore RY (1999) Neuroinvasiveness of pseudorabies virus injected intracerebrally is dependent on viral concentration and terminal field density. J Comp Neurol 407:438–452
44. Kim J-S, Moore RY, Enquist LW, Card JP (1999) Circuit-specific co-infection of neurons in the rat central nervous system with two pseudorabies virus recombinants. J Virol 75:9521–9531
45. Smith GA, Gross SP, Enquist LW (2001) Herpesviruses use bidirectional fast-axonal transport to spread in sensory neurons. Proc Natl Acad Sci USA 98:3466–3470
46. Granstedt AE, Kuhn B, Wang S-H, Enquist LW (2010) Calcium imaging of neuronal circuits in vivo using a circuit-tracing pseudorabies virus. Cold Spring Harb Protoc. doi:10.1101/pdb.prot5410
47. Li Y-W, Wesselingh S, Blevins JE (1992) Projections from rabbit caudal medulla to C1 and A5 sympathetic premotor neurons, demonstrated with phaseolus leucoagglutinin and herpes simplex virus. J Comp Neurol 317:379–395
48. O'Donnell P, Lavin A, Enquist LW, Grace AA, Card JP (1997) Interconnected parallel circuits between rat nucleus accumbens and thalamus revealed by retrograde transynaptic transport of pseudorabies virus. J Neurosci 17:2143–2167
49. Card JP, Rinaman L, Lynn RB, Lee B-H, Meade RP, Miselis RR, Enquist LW (1993) Pseudorabies virus infection of the rat central nervous system: ultrastructural characterization of viral replication, transport, and pathogenesis. J Neurosci 13:2515–2539
50. LaVail JH, Carter SR, Topp KS (1993) The retrograde tracer FluoroGold interferes with the infectivity of herpes simplex virus. Brain Res 625:57–62
51. Aston-Jones G, Chen S, Zhu Y, Ohnishi ST (2001) A neural circuit for circadian regulation of arousal. Nat Neurosci 4:732–738
52. Jasmin L, Burkey AR, Card JP, Basbaum AI (1997) Transneuronal labeling of a nociceptive pathway, the spino-(trigemino-)parabrachio-amygdaloid, in the rat. J Neurosci 17:3751–3765
53. Watson RE, Wiegand ST, Clough RW, Hoffman GE (1986) Use of cryoprotectant to maintain long-term peptide immunoreactivity and tissue morphology. Peptides 7:155–159
54. Geraghty RJ, Kummenacher C, Cohen GH, Eisenberg RJ, Spear PG (1998) Entry of alphaherpesviruses mediated by poliovirus receptor-related protein 1 and poliovirus receptor. Science 280:1618–1620
55. Milne RS, Connolly SA, Krummenacher C, Eisenberg RJ, Cohen GH (2001) Porcine HveC, a member of the highly conserved HveC/nectin 1 family, is a functional alphaherpesvirus receptor. Virology 281:315–328
56. Curanovic D, Enquist LW (2009) Directional transneuronal spread of alpha-herpesvirus infection. Future Virol 4:591–603
57. Card JP, Rinaman L, Schwaber JS, Miselis RR, Whealy ME, Robbins AK, Enquist LW (1990) Neurotropic properties of pseudorabies virus: uptake and transneuronal passage in the rat central nervous system. J Neurosci 10:1974–1994
58. Rinaman L, Card JP, Enquist LW (1993) Spatiotemporal responses of astrocytes, ramified microglia, and brain macrophages to central neuronal infection with pseudorabies virus. J Neurosci 13:685–702
59. Rinaman L, Roesch MR, Card JP (1999) Retrograde transynaptic pseudorabies virus infection of central autonomic circuits in neonatal rats. Dev Brain Res 114:207–216
60. Rassnick S, Enquist LW, Sved AF, Card JP (1998) Pseudorabies virus induced leukocyte trafficking into rat CNS. J Virol 72:9181–9191
61. Blinzinger K, Kretuzberg GW (1968) Displacement of synaptic terminals from regenerating motoneurons by microglial cells. Z Zellforsch Mikrosk Anat 85:145–157
62. Denes A, Boldogkoi Z, Hornyak A, Palkovits M, Kovacs KJ (2006) Attenuated pseudorabies virus-evoked rapid innate immune response in the rat brain. J Neuroimmunol 180:88–103
63. Boldogkoi Z, Reichart A, Toth IE, Sik A, Erdelyi F, Medveczky I, Llorens-Cortes C, Palkovits M, Lenkei Z (2002) Construction of recombinant pseudorabies viruses optimized for labeling and neurochemical characterization of neural circuitry. Mol Brain Res 109:105–118
64. Denes A, Boldogkoi Z, Uhereczky G, Hornyak A, Rusvai M, Palkovits M, Kovacs KJ (2005) Central autonomic control of the bone marrow: multisynaptic tract tracing by recombinant pseudorabies virus. Neuroscience 134: 947–963
65. Card JP, Dubin JR, Whealy ME, Enquist LW (1995) Influence of infectious dose upon productive replication and transynaptic passage of

pseudorabies virus in rat central nervous system. J Neurovirol 1:349–358

66. Schramm LP, Strack AM, Platt KB, Loewy AD (1993) Peripheral and central pathways regulating the kidney: a study using pseudorabies virus. Brain Res 616:251–262
67. Bamshad M, Song CK, Bartness TJ (1999) CNS origins of the sympathetic nervous system outflow to brown adipose tissue. Am J Physiol 45:R1569–R1578
68. Oldfield BJ, Giles ME, Watson A, Anderson C, Colvill LM, McKinley MJ (2002) The neurochemical characterisation of hypothalamic pathways projecting polysynaptically to brown adipose tissue in the rat. Neuroscience 110:515–526
69. Vahlne A, Nystrom B, Sandberg M, Hamberger A, Lycke E (1978) Attachment of herpes simplex virus to neurons and glial cells. J Gen Virol 40:359–371
70. Vahlne A, Svennerholm B, Sandberg M, Hamberger A, Lycke E (1980) Differences in attachment between herpes simplex type 1 and type 2 viruses to neurons and glial cells. Infect Immun 28:675–680
71. Marchand CF, Schwab ME (1987) Binding, uptake and retrograde axonal transport of herpes virus suis in sympathetic neurons. Brain Res 383:262–270
72. Card JP, Santone DJ, Gluhovsky MY, Adelson PD (2005) Plastic reorganization of hippocampal and neocortical circuitry in experimental traumatic brain injury in the immature rat. J Neurotrauma 22:989–1002
73. Jansen ASP, Van Nguyen X, Karpitskiy V, Mettenleiter TC, Loewy AD (1995) Central command neurons of the sympathetic nervous system: basis of the fight-or-flight response. Science 270:253–260
74. Banfield BW, Kaufman GD, Randall JA, Pickard GE (2003) Development of pseudorabies virus strains expressing red fluorescent proteins: new tools for multisynaptic labeling applications. J Virol 77:10106–10112
75. Billig I, Foris JM, Enquist LW, Card JP, Yates BJ (2000) Definition of neuronal circuitry controlling the activity of phrenic and abdominal motoneurons in the ferret using recombinant strains of pseudorabies virus. J Neurosci 20:7446–7454
76. DeFalco J, Tomishima MJ, Liu H, Zhao C, Cai X, Marth JD, Enquist LW, Friedman JM (2001) Virus-assisted mapping of neural inputs to a feeding center in the hypothalamus. Science 291:2608–2613
77. Sauer B (1987) Functional expression of the Cre-Lox site-specific recombination system in the yeast *Saccharomyces cerevisiae*. Mol Cell Biol 7:2087–2096
78. Ekstrand MI, Enquist LW, Pomerantz RJ (2008) The alpha-herpesviruses: molecular pathfinders in nervous system circuits. Trends Mol Med 14:134–140
79. Yoon H, Enquist LW, Dulac C (2005) Olfactory inputs to hypothalamic neurons controlling reproduction and fertility. Cell 123:669–682
80. Campbell RE, Herbison AE (2007) Definition of brainstem afferents to gonadotropin-releasing hormone neurons in the mouse using conditional viral tract tracing. Endocrinology 148:5884–5890
81. Campbell RE, Herbison AE (2007) Defining the gonadotrophin-releasing hormone neuronal network: transgenic approaches to understanding neurocircuitry. J Neuroendocrinol 19:561–573
82. Braz JM, Enquist LW, Basbaum AI (2009) Inputs to serotonergic neurons revealed by conditional viral transneuronal tracing. J Comp Neurol 514:145–160
83. Boldogkoi Z, Balint K, Awatramani GB, Balya D, Busskamp V, Viney TJ, Lagali PS, Duebel J, Pasti E, Tombacz D, Toth JS, Takacs IF, Scherf BG (2009) Genetically timed, activity-sensor and rainbow transsynaptic viral tools. Nat Methods 6:127–130
84. Granstedt AE, Szpara ML, Kuhn B, Wang S-H, Enquist LW (2009) Fluorescence-based monitoring of in vivo neural activity using a circuit-tracing pseudorabies virus. PLoS One 4:e6923. doi:6910.1371/journal.pone.0006923
85. Card JP, Kobiler O, McCambridge J, Ebdlahad S, Shan Z, Raizada MK, Sved AF, Enquist LW (2011) Microdissection of neural networks by conditional reporter expression from a Brainbow herpesvirus. Proc Natl Acad Sci USA 108:3377–3382
86. Livet J, Weissman TA, Kang H, Draft RW, Lu J, Bennis RA, Sanes JR, Lichtman JW (2007) Transgenic strategies for combinatorial expression of fluorescent proteins in the nervous system. Nature 450:56–63
87. Kobiler O, Lipman Y, Therkelsen K, Daubechies I, Enquist LW (2010) Herpesviruses carrying a Brainbow cassette reveal replication and expression of limited numbers of incoming genomes. Nat Commun 1:146
88. Card JP (1998) Exploring brain circuitry with neurotropic viruses: new horizons in neuroanatomy. Anat Rec 253:176–185

Chapter 12

Visualisation of Thermal Changes in Freely Moving Animals

Daniel M.L. Vianna and Pascal Carrive

Abstract

In the past decades, physiological research has progressively moved towards less invasive methods that could be used to study freely moving animals. In this chapter we describe the use of digital infrared thermography to detect changes in skin vasoconstriction, body temperature, brown adipose tissue thermogenesis, and nociception in the conscious and unrestrained rat. All these physiological phenomena can be associated with concomitant changes of the skin surface temperature, which is measured remotely through its emitted infrared radiation. With digital infrared thermography these processes can be assessed simultaneously, over any length of time, in intact animals. The relationship between each physiological process and its associated skin temperature change is reviewed, and the experimental procedures required to obtain clear and meaningful measures are explained.

Key words: Thermogenesis, Thermoregulation, Skin blood flow, Brown adipose tissue, Remote temperature measurement, Antinociception, Tail-flick test, Vasomotion, Brown fat, Stress-induced hyperthermia

1. Introduction

Physiological research has increasingly moved in the past decades from using anaesthetised preparations to studying conscious animals and their responses in a more naturalistic setting. Accordingly, there has been vigorous interest in imaging methods that can reveal physiological phenomena without the need to "cut the animal open". Many of these phenomena result in changes in skin temperature. These include skin vasomotion (1), changes in deep body temperature (2, 3), and brown adipose tissue (BAT) thermogenesis (4, 5). Because the skin is readily accessible, it presents a convenient way to study these processes noninvasively.

Emilio Badoer (ed.), *Visualization Techniques: From Immunohistochemistry to Magnetic Resonance Imaging*, Neuromethods, vol. 70, DOI 10.1007/978-1-61779-897-9_12, © Springer Science+Business Media, LLC 2012

Temperature measurement has always relied on thermometers. The main disadvantage associated with these is that the temperature is sensed in a single point, thus ignoring the variation of temperature over extended surfaces. Another limitation is that it relies on physical contact with the object to be measured. Thermometric measurements in this context would also depend on either having cables attached to the animal or the use of wireless probes.

Infrared thermography now presents an exciting alternative, where the whole visible body surface can be thermally measured without the need of physical contact, restraint, or the use of probes. It effectively captures an image of the photographed object in much the same way as visible light digital photography, the only difference being that the image is created by the heat emitted by the subject, rather than the visible light reflected by it. Hence, measurements can be performed at a distance of the subject and with as little constraints to the subject as ordinary photography.

Because infrared thermography can only measure the temperature of visible surfaces, the processes that drive changes in the skin temperature are not being measured directly, but rather inferred from their thermal impact on cutaneous temperature. Many different processes can influence skin temperature, the impact and "purity" of that influence being directly related to the proximity of the organs involved and the presence or absence of other nearby thermogenic organs. In this chapter we present the current approaches that rely on infrared thermography to study these underlying physiological processes in rats.

1.1. Applications of Infrared Thermography in Physiological Research: Measures and Significance

1.1.1. Tail and Paw Temperature and Vasomotion

Rats and other rodents have sweat glands, but these are not used in thermoregulation (6). Hence, the main adaptive mechanism for heat loss in rodents is peripheral vasomotion. The tail and plantar skin regions in rats and ears in rabbits are highly vascularised, and blood flow in these regions falls deeply in response to cold exposure and psychological stress (3, 7). Due to the scarcity of fur insulation in these areas, along with a high volume-to-surface ratio, blood flow is virtually the only physiological process influencing local temperature. Hence, vasodilation can increase tail and paw temperature to as high as 35°C, whereas vasoconstriction will lead to near-ambient temperature values (see Fig. 1a), making these organs hard to spot in a thermal image when vasoconstricted (as they will be represented in the image with the same colour coding as the background).

1.1.2. Lumbar Back Temperature and Body Temperature

Roughly the opposite occurs with the lumbar back region. It is covered with fur and overlies the biggest mass of the animal's body. It does not overlie major thermogenic organs either. Previous research suggests that the temperature of the shaved skin of the lumbar back region is a good index for body temperature, as its changes correlate well with the changes in abdominal cavity

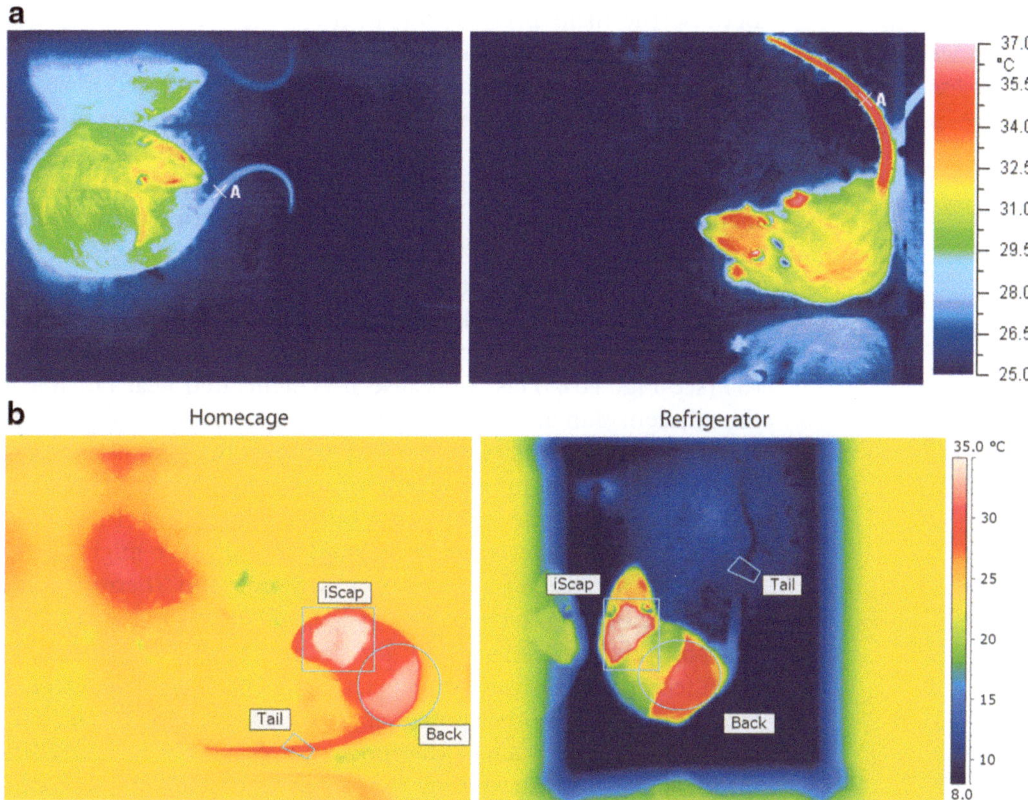

Fig. 1. (**a**) Tail vasoconstriction (*left*, marked pixel is 27.5°C) and vasodilation (*right*, marked pixel is 35.5°C) as observed by digital infrared imaging. Images were acquired with a TVS-100 camera (Avio, Japan), and converted to jpg files with the Goratec PE Professional 3.12 software. The camera was placed 70 cm above the floor of the cage, and the entire field of view (320 × 240 pixels) is shown. Notice the thermal reflections from the plexiglas walls. Please refer to the online version to see the colour reproduction of this figure. (**b**) Relative body temperature and potential brown adipose tissue (BAT) thermogenesis as observed by digital infrared imaging. *Left* and *right* images were taken at approximate ambient temperatures 25°C (homecage) and 4°C (through a hole on the top of a refrigerator), respectively, 70 cm from camera, with a ThermaCAM P45 camera (FLIR, Sweden), and analysed with the FLIR ThermaCAM Reporter 7 Pro software. The entire field of view (640 × 480 pixels) is shown. Notice that the difference between iScap and Back temperatures is greater in the refrigerator than in the homecage. Warmest pixel temperatures within area of interest (°C) are—Homecage: iScap, 34.8; Back, 33.6; Tail, 28.7; Refrigerator: iScap, 34.4; Back, 31.8; Tail, 11.1.

temperature (2) and is not directly affected by BAT thermogenesis (4). In fact changes in lumbar back skin temperature might be a more accurate indicator of changes in body temperature. When lumbar back temperature changes were compared to abdominal cavity temperature measured by a thermometric probe, we found that changes in abdominal cavity temperature lagged 5–7 min behind those of the lumbar skin, both at the onset and end of a psychological stress-induced rise in body temperature (2, 3). We do not know the reason for this lag, but we suspect that it is because of local changes in the viscera (visceral vasoconstriction?). This illustrates that there is no "pure" measure of body temperature, just local temperatures in different parts of the body, each of them with their particular bias.

1.1.3. Interscapular Temperature and Brown Adipose Tissue Thermogenesis

Because the main BAT deposit in the rat is located immediately under the interscapular skin area, BAT thermogenesis will have a major impact on skin temperature in that area. Infrared measurement of interscapular temperature (TiScap) as an index of BAT thermogenesis is not new (8). However, it was Blumberg and colleagues (4, 5), working with thermocouples, who first realised that increases in interscapular skin temperature could be offset against lumbar skin temperature (TBack) to produce a better estimate of the relative BAT contribution to body temperature. At thermoneutrality, TiScap and TBack are roughly equal, and the difference between those measures increases as ambient temperature decreases (5) (see Fig. 1b). This is because BAT-generated heat radiates to the adjacent skin and warms it up faster than the rest of the body. Blood leaving the interscapular BAT will run through the azygous vein, the heart, the lungs, where it can be assumed to cool down, and then back to the heart again before it goes back to systemic circulation (4). Alberts and colleagues (9, 10) later migrated the technique for use in conjunction with infrared thermography. These authors worked with rat pups, which cannot shiver, have no fur, and rely entirely on BAT for thermoregulation. To use the same technique in adult rats, thus, the experimenter needs to take into consideration other potential sources of thermogenesis in the interscapular area, such as thermogenesis coming from exercising interscapular muscle during locomotion, as well as shivering. A simple way of controlling for it is to have a separate group treated with a β-adrenergic antagonist (i.e. propranolol), which should block the sympathetic activation of BAT, but not somatic nerve activation of muscles (2).

1.1.4. Skin Temperature Monitoring and Tail-Flick Pain Tests

Digital infrared thermography is now fast enough that it can be used to monitor rapid changes in skin temperature in response to application of a local radiant heat stimulus. It can easily be used for example in conjunction with the tail flick test, a commonly used pain test in rodents. Thermographic recording of the tail skin reveals not only the initial temperature of the tail before the application of the noxious radiant heat stimulus but also the temporal evolution of the temperature of the skin up to the point where it reaches the nociceptive threshold and triggers the withdrawal response (11, 12). Determination of the initial temperature of the tail is important because this parameter will affect the time it takes for the skin to warm up to the nociceptive threshold. A vasoconstricted tail will take longer to reach the nociceptive threshold than a vasodilated tail, simply because it will be colder at the onset of the heat stimulus presentation. This may result in a longer withdrawal latency that could be wrongly interpreted as an increase in the nociceptive threshold. Furthermore, monitoring of the temporal evolution of the skin temperature during the heat stimulus allows extraction of the true nociceptive threshold, that is, the temperature at which the withdrawal of the tail will be triggered. This is the only way that a change in nociceptive threshold can be determined using the tail-flick test in the conscious animal. This approach was recently introduced by Le Bars and collaborators to sort out the different components of the tail-flick response and reassess the phenomenon of stress-induced analgesia (11, 12). This application of infrared thermography is likely to have important consequences for the study of pain modulation pathways.

1.2. Key Physical Principles to Bear in Mind

1.2.1. Emissivity

All objects emit infrared light. The hotter an object is, the more infrared it emits. Objects also reflect infrared coming from their surroundings, and reflectance varies according to the material the object is made of. An ideal object that would not reflect, but only emit infrared (a so-called "black body") would measure 1 in the scale of emissivity. Luckily for us, skin has very high emissivity (0.98). That means 98% of the infrared measured from skin is directly proportional to its temperature. As a result of that, it is safe to measure skin temperature in ordinary laboratory settings without having to worry about reflected infrared radiation.

1.2.2. Thermal Inertia

Physiological changes do not lead to immediate changes in skin temperature. For example, plantar arterial flow, as measured by laser Doppler flowmetry, falls to its nadir within 15 s following sympathetic stimulation, whereas plantar temperature, as measured by an infrared thermometer, only reaches its nadir after 10 min at room temperature (1, 12). Hence, for it to have a measurable thermal impact, a physiological response has to be sustained over a relatively long period of time.

1.2.3. Ambient Temperature

The temperature of the air in contact with the skin will affect its temperature as much as the temperature of the tissues under it. Hence, changes in ambient temperature will change skin surface temperature even if the temperature under the skin has not changed. For example, a sudden drop of 20°C as would occur during a cold exposure test (moving the animal from room temperature at 24°C to a fridge at 4°C) causes a sudden decrease of 3–4°C at the surface of the skin (2). The drop occurs within a few seconds and stabilises within less than 1 min because of the low thermal conductance of air relative to that of the skin. When the animal is returned to room temperature the same effect occurs, in the opposite direction. Such change in ambient temperature will also cause a slight loss of accuracy in the infrared temperature reading (the reading will skew by a few tenths of a degree), which can be corrected by adjusting the reflected and ambient temperature values in the camera settings. As this initial change in skin temperature is caused by a sudden change in ambient temperature, rather than by a physiological mechanism, it should be disregarded during data analysis.

2. Materials

2.1. Infrared Camera

Our laboratory has tested three different manufacturers and models of infrared thermographic cameras to date. Based on this experience, desirable features should include, but are not limited to:

- Automatic shutter for temperature calibration. The microbolometer (the infrared sensor) is sensitive to changes to its own temperature caused by the thermal image being projected over it. This leads to a phantom image that will be superimposed to the image of interest, leading to a skewed thermal measurement. To prevent that bias, the camera should take a picture of a known object (a blackbody with even temperature) and correct the actual measurement of this object against the expected measurement. Most cameras should have this procedure completely automated (i.e. the camera has an internal shutter that automatically covers the microbolometer, takes a picture and then calibrates the camera based on the measured bias), but we came across a cheaper model that required that process to be performed manually, which is simply impractical for our application.
- A simple system for transferring digital infrared pictures from camera to computer, such as a memory card.
- Other convenient inbuilt calibration and automation in camera hardware, such as emissivity correction, reflected and ambient temperature correction, timed picture taking, and colour contrast adjustment to minimum and maximum temperatures of interest.
- An external monitor is also desirable. While some cameras have a detachable digital screen, it is always useful to have a video or digital output available, so the experimenter can see what is being imaged and control the camera functions being displayed without having to look in the viewfinder.

2.2. Analysis Software

Most manufacturers provide proprietary thermal image file formats that can only be analysed with their proprietary software. The proprietary software could be very expensive, and the analysis, if done manually, can be time consuming. For these reasons, it is imperative to take into account the analysis software to be used when choosing the camera to be purchased. Ease of use when locating/ measuring thermal targets in sequential images, overall speed of the measuring process, and the option of exporting measurement data to a spreadsheet file (instead of using pen and paper) are desirable. Ask a salesperson to come and demonstrate their product prior to purchase.

2.3. Room Setup

- While skin emissivity is so high that reflected infrared radiation is negligible, it is advisable that potential sources of thermal energy are kept in check. The ideal situation is to have the animal placed away from thermal sources such as incandescent lighting or sun. In our setup, data acquired in the shade was comparable to data acquired directly under fluorescent lighting.
- There should be no obstruction between camera and skin. Objects that are transparent to the visible light are usually opaque in the infrared spectrum. Hence, the animal enclosure should have openings between the animal and the camera lens. Our group has successfully used boxes with 60 cm high plexiglass walls and open at the top to image rats (see Fig. 2). Other researchers (13) have placed animals on a grid floor and imaged them from below, at an angle. This setup presents the camera with the best position from which to image the tail and paws, but at the cost of leaving it dangerously close to the path of falling urine and faecal matter.
- A camera mount is highly desirable in most cases. Our lab has a rail mounted over the experimental boxes, so the camera can be slid from box to box during experiments. The rail is at a

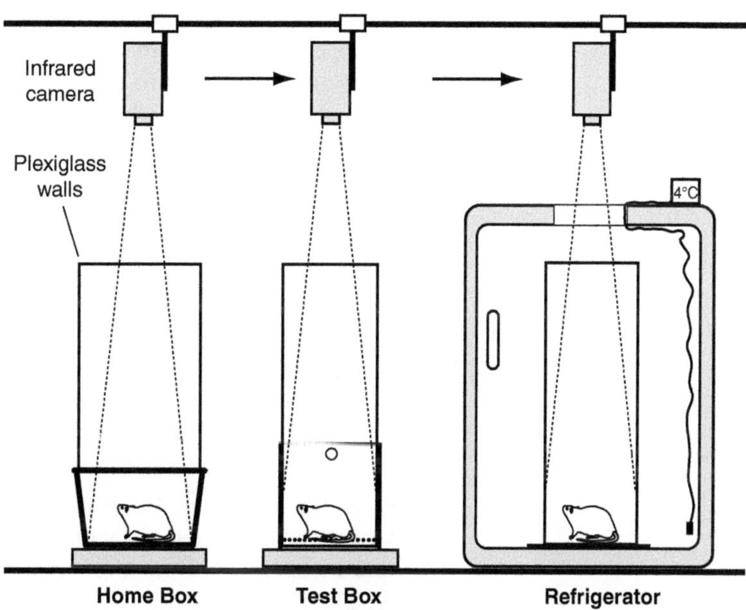

Fig. 2. Typical set up for infrared thermography in rodents. The height of the camera should be adjusted so that the field of view covers the widest area (here the home box). Lids have been replaced by plexiglass walls to allow an unobstructed view of the animal. The extended walls are high enough to prevent escape. A small window has been cut at the *top* of the refrigerator to image the animal inside the cool environment while the camera stays at room temperature.

height that allows each box to fit entirely in the image frame and stay in focus.

- Thermometer to record and monitor ambient temperature.
- Animal housing, as described in Sect. 3.1, steps 1 and 3.
- If cold exposure is intended, a small bar refrigerator is useful. Ordinary refrigerators can vary greatly their internal temperatures, so it is important to have a thermocouple fit inside it with a remote display that is placed outside. That way the experimenter can vary the temperature according to need and record the actual temperature the animals were tested on. A hole should be cut at the top of the refrigerator, so the camera can remain outside the refrigerator while imaging the animal inside it, and the animal can be placed inside the apparatus without opening the door and disturbing the internal temperature.

3. Methods

The following procedures were designed and tested with male Wistar rats, weighing 350 g or over, housed at ordinary room temperature.

3.1. Prior to Testing

1. Animals are transferred from collective housing to 38 cm (D) × 25 cm (W) × 16 cm (H) plastic single homeboxes with grid lids and free access to food and water. The homebox floor is covered with pine woodchips, and an 11 cm (ϕ) × 21 cm (D) PVC tube and a wood block (10 × 3 × 4 cm) are also present for enrichment.
2. The day prior to testing, animals are anaesthetised with halothane to have a rectangle of fur shaved off from the base of the head to roughly the caudal end of the scapulae, and from mid back down to the base of the tail.
3. If baseline measures are necessary, animals need to be housed in boxes where the infrared camera has unobstructed view of the surfaces of interest. This usually clashes with the natural preference of rats for small protected enclosures, so the more exposed the animal is, the more it will be stressed during the baseline period. We have reached a compromise by keeping the animal in its own home box, but removing everything that could block the view the day before the test. The PVC tube is removed, the food is placed on the home cage floor, and the lid is replaced by plexiglas walls extending the walls to a 60-cm height to prevent escape. A water bottle is hung from the outside of this wall, so only its tip protrudes into the home cage.

3.2. Experimental Procedure and Data Acquisition

1. The camera is fixed over the home cage, so it images the whole home cage floor. Focus should be placed on the tail, as it will be the hardest area of interest to spot when it is cold. Set the correct emissivity (0.98), ambient temperature (~25°C), intended colour palette (rainbow or iron, see Notes), and temperature scale span (~25 to ~35°C if all tests happen at room temperature, ~4 to ~35°C if cold exposure at 4°C is intended). Set the intended periodic save time (for automatic picture taking, for example, each minute) and start. Use a notebook to record animal being tested, session start time, filename of the first picture taken and number, ambient temperature, and other information such as treatment and time of injection.
2. A common experimental design includes 30-min baseline, 30-min testing, and 90-min recovery time. It is important to counterbalance the order of testing between different treatments. When administering a drug or moving the animal to a different testing cage, record again time, filename of first picture taken in that condition, etc. Make sure the injectate (intraperitoneal, subcutaneous, or intravenous) is at least at ambient temperature.
3. After finishing testing one animal, move the camera to the next. After finishing for the day, take the memory card out and move the pictures to a computer for storage and backup. It is convenient when sorting out the pictures to place files from each animal/test to a separate folder, and keeping only the ones to be analysed there, leaving out all extra images from before or after the exact time of the session.

3.3. Data Extraction

1. How much and how fast can data be extracted from the images depend on the type of software you use. The following description is designed to be used with the software ThermaCAM™ Reporter 7 Pro, which reads the FLIR Systems™ proprietary file format used in many cameras produced by this manufacturer. This software is installed as a plugin for Microsoft Word. Images taken during each rat/session are pasted into a Word document, where the experimenter can draw the outline of each area of interest (e.g. lumbar back area, mid section of the tail, interscapular skin). From these outlines the software can automatically extract the average temperature of each area, or the hottest/coldest pixel. For studies performed at or under thermoneutrality, we find that the hottest pixel is the simplest and most reliable method to extract temperature from a region of interest. Averaging is not always practicable, for it would depend on keeping the target area and angle of imaging unchanged, which is not feasible with the freely moving animal. Make sure the measured area does not include warmer pixels outside the target skin area (such as when the tail is

colder than its surrounding environment after cold exposure). The results of the entire session will be then displayed in a table that can be copy/pasted onto a spreadsheet (Microsoft Excel).

2. Data should be pasted in Excel in separate columns for each variable, i.e. TiScap, TBack, and TTail. A new column should be created with the values of TiScap-TBack.

3.4. Data Analysis

Each individual data per variable should now be exported to a statistical programme (i.e. SPSS), where repeated measures ANOVA can be performed for each intended session period (baseline, exposure, recovery).

3.5. Anticipated Results

1. Body temperature. Variations of TBack can be interpreted as variations of body temperature, as long as ambient temperature remains unchanged.
2. Tail temperature. Increases in tail temperature usually result from sustained vasodilation of tail vessels, and decreases should indicate sustained vasoconstriction. It is important to stress that brief, unsustained vasomotor events will not be long enough to cause measurable changes in tail temperature.
3. BAT thermogenesis. TiScap-TBack can be a good indicator of BAT thermogenesis once other sources of thermogenesis are ruled out. From our experience, neck muscle activity can lead to ~1°C increases of TiScap-TBack that is not affected by propranolol treatment, while BAT thermogenesis can lead to up to 3°C increases which may be reduced by propranolol (2).

4. Notes (Troubleshooting, Tips, and Tricks)

1. Missing data. As the animal moves, it is possible that some areas of interest may be concealed (e.g. paws and tail) or seen from a different angle (e.g. lumbar back when the animal rears), in which case data may be missing or erroneous. There are two ways to go about this. Either leave a gap in the data or, if the gap is small, replace by the average of the preceding and following measurements. The later is acceptable if the sampling is fast (i.e. not more than every 2 min) because thermal inertia would prevent the temperature from changing abruptly during such a small time interval.
2. Palette choice. Infrared light is not visible, so the choice of colours to represent the different temperatures in an infrared picture is totally arbitrary. Nevertheless, there are three sets of colours (palettes) that are most convenient: grey, rainbow, and iron (see Fig. 3).

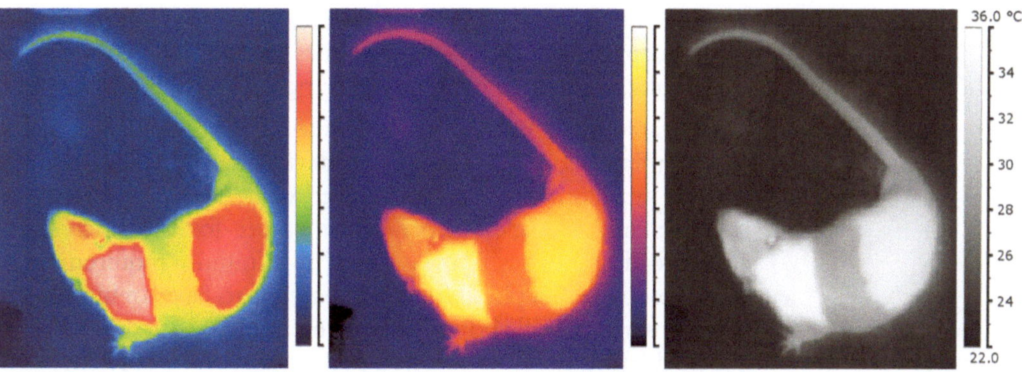

Fig. 3. Same thermal image displayed with *rainbow*, *iron*, and *grey* colour palettes. Image acquired and processed with the same equipment as in Fig. 1b, and cropped with standard image editor software. Notice that the difference between interscapular and lumbar back temperature is more evident in the *left* (*rainbow*), and less evident in the *right* (*grey*) image.

3. Grey is used mainly for print. It gives little contrast, which makes it difficult to spot small differences of temperature. Moreover, at times seeing the profile of the objects of interest is also made difficult. Usually black is used for the coldest temperature, white for the warmest, and shades of grey for everything in between.
4. Rainbow is the palette that gives most contrast, making it easiest to discern each temperature in a scale. Light red is at the top of the temperature scale, and dark blue at its bottom. It is the palette of choice for use during an experimental session, for it will allow the experimenter to easily discern temperatures with the naked eye. It is usually very good for colour prints, as long as the objects in the picture are more or less uniform in their temperature and the image is not too busy. Complex objects with too many colours being displayed at once might become too difficult to interpret. The main problem with the rainbow palette is that it becomes uninterpretable when printed in black and white.
5. Iron is the most versatile palette. Light yellow indicates the highest temperature, and dark blue the lowest. It gives more contrast than black and white, yet when printed in monochrome, all temperatures will be rendered correctly, making it an ideal option for publishing. It is also the least confusing when dealing with busy images.
6. The tail and paws are the most difficult regions to see when vasoconstricted and shot against a cold background. While it is impossible to avoid missing data in these conditions, there are steps that could be used to minimise it. Make sure the focus is on the tail in the beginning of the experiment, so it can be seen even when its temperature is close to that of the background. Rough bedding, urine, and faecal boli will often make an

uneven thermal background, and an interruption of the expected shape of those sometimes gives away a cold tail that would be invisible otherwise (see right image, Fig. 1b).

7. Infrared gun. A cheap alternative to a digital infrared camera is an infrared gun. This is a simple device that works according to the same principles as an infrared camera, except that it will give only a single temperature measure for a restricted field instead of an array of measures. The minimum requirement for it to be used in physiological research is to have an inbuilt laser pointer that will show the experimenter the exact limits of the field being measured. When purchasing, make sure the working distance of the device matches the distance to the object during the experiment.

8. Sensitivity vs. accuracy. Most digital infrared cameras will give a very acceptable sensitivity of less than 0.1°C, which means that relative changes within an experiment can be detected with a ±0.1°C error. This is sufficient resolution to study physiological responses. On the other hand, accuracy is not its strength, so a measured value of 32°C might refer to a real temperature of 30–34°C (accuracy being in the range of ±2°C). Hence, if absolute temperature measurements, rather than relative changes, are critical for an experiment, one would probably be safer to confirm the measurement made by the infrared camera with a conventional thermocouple. Regular calibration of the camera in this case is also important.

Acknowledgments

The authors wish to thank Dr. Marcio L. Vianna for his advice regarding the physical principles mentioned in the text.

References

1. Habler HJ, Stegmann JU, Timmermann L, Janig W (1998) Functional evidence for the differential control of superficial and deep blood vessels by sympathetic vasoconstrictor and primary afferent vasodilator fibres in rat hairless skin. Exp Brain Res 118:230–234
2. Marks A, Vianna DM, Carrive P (2009) Nonshivering thermogenesis without interscapular brown adipose tissue involvement during conditioned fear in the rat. Am J Physiol Regul Integr Comp Physiol 296: R1239–R1247
3. Vianna DM, Carrive P (2005) Changes in cutaneous and body temperature during and after conditioned fear to context in the rat. Eur J Neurosci 21:2505–2512
4. Blumberg MS, Sokoloff G (1998) Thermoregulatory competence and behavioral expression in the young of altricial species–revisited. Dev Psychobiol 33:107–123
5. Blumberg MS, Stolba MA (1996) Thermogenesis, myoclonic twitching, and ultrasonic vocalization in neonatal rats during moderate and extreme cold exposure. Behav Neurosci 110:305–314
6. Gordon CJ (1990) Thermal biology of the laboratory rat. Physiol Behav 47:963–991
7. Blessing WW (2003) Lower brainstem pathways regulating sympathetically mediated

changes in cutaneous blood flow. Cell Mol Neurobiol 23:527–538

8. Hayward JS, Ball EG (1966) Quantitative aspects of brown adipose tissue thermogenesis during arousal from hibernation. Biol Bull 131:94–103
9. Farrell WJ, Alberts JR (2000) Ultrasonic vocalizations by rat pups after adrenergic manipulations of brown fat metabolism. Behav Neurosci 114:805–813
10. Blumberg MS, Efimova IV, Alberts JR (1992) Thermogenesis during ultrasonic vocalization by rat pups isolated in a warm environment: a thermographic analysis. Dev Psychobiol 25:497–510
11. Benoist JM, Pincede I, Ballantyne K, Plaghki L, Le Bars D (2008) Peripheral and central determinants of a nociceptive reaction: an approach to psychophysics in the rat. PLoS One 3:e3125
12. Carrive P, Churyukanov M, Le Bars D (2011) A reassessment of stress-induced "analgesia" in the rat using an unbiased method. Pain 152:676–686
13. Cerri M, Zamboni G, Tupone D, Dentico D, Luppi M, Martelli D, Perez E, Amici R (2010) Cutaneous vasodilation elicited by disinhibition of the caudal portion of the rostral ventromedial medulla of the free-behaving rat. Neuroscience 165(3):984–995

Chapter 13

Perfusion Magnetic Resonance Imaging Quantification in the Brain

Fernando Calamante

Abstract

Cerebral perfusion, the rate of blood delivery to brain tissue, plays an important role in tissue viability and brain function. The most commonly used magnetic resonance imaging (MRI) method to assess brain perfusion and tissue haemodynamics in clinical investigations is known as dynamic susceptibility-contrast (DSC) MRI. Among the main reasons for its widespread use are its fast acquisition time, good contrast-to-noise ratio, and wealth of information available from DSC-MRI data. A description of the typical steps involved in a DSC-MRI study, the practical decisions that need to be taken, the problems that can be encountered, and the approaches that have been developed to overcome or minimise them will be the topic of this chapter. In particular, the implications of all these issues for perfusion quantification will be emphasised.

Key words: Perfusion, Cerebral blood flow, Magnetic resonance imaging, Contrast agent, Arterial input function, Deconvolution, Mean transit time, Cerebral blood volume, Dynamic susceptibility contrast

1. Introduction

Cerebral *perfusion* (also known as *cerebral blood flow*, CBF)[1] refers to the rate of blood delivery to brain tissue. It is the blood flow at the capillary level, where exchange of oxygen and nutrients, and the removal of waste products, take place. Therefore, perfusion plays an important role both in tissue viability and in brain function.

[1]In the context of Perfusion MRI, the terms perfusion, cerebral blood flow, and its acronym CBF are commonly used indistinguishably. This convention will be also used throughout this chapter.

Emilio Badoer (ed.), *Visualization Techniques: From Immunohistochemistry to Magnetic Resonance Imaging*, Neuromethods, vol. 70, DOI 10.1007/978-1-61779-897-9_13, © Springer Science+Business Media, LLC 2012

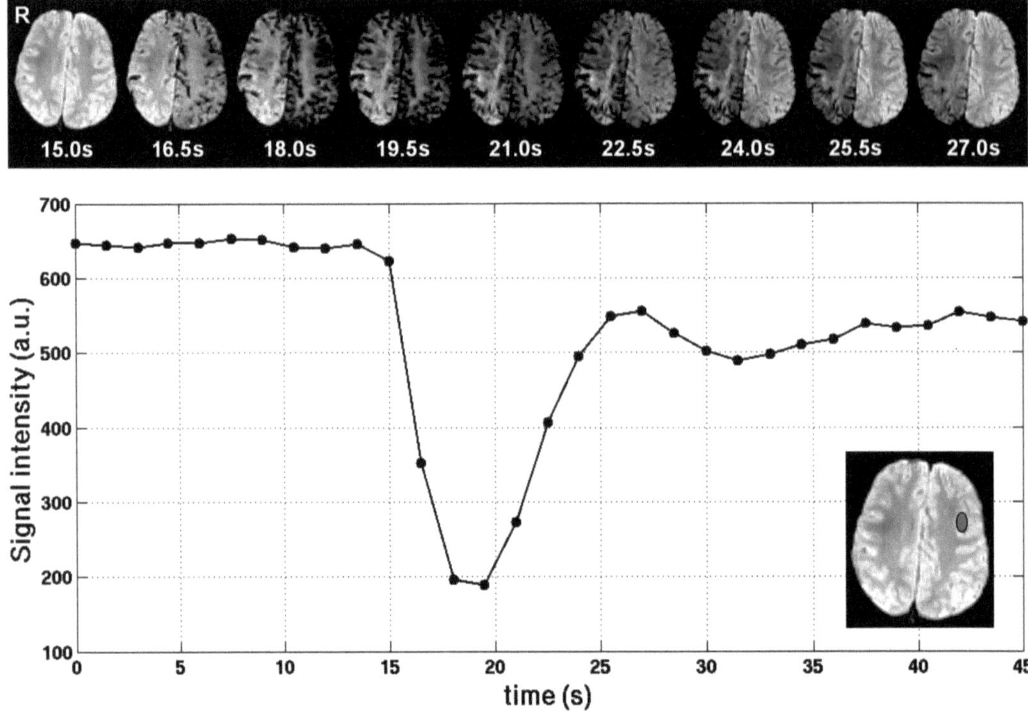

Fig. 1. Dynamic susceptibility-contrast (DSC)-magnetic resonance imaging (MRI) data on a patient with a vascular abnormality in the right anterior, posterior, and middle cerebral arteries (*left* side of the images). The patient had no infarctions at the time of the MRI examination. The nine images in the *top row* are nine time samples of the gradient-echo echo-planar images acquired during the passage of the bolus of contrast agent (TR = 1.5 s). These images show the transient decrease in signal intensity induced by the passage of the bolus. Note the asymmetric behaviour due to the abnormalities on the right major cerebral arteries. The *bottom graph* shows the signal intensity time course for a region of interest in the contralateral cortical gray matter (see inset). Three different periods can be observed: the baseline (before the arrival of the bolus to the region, approximately from 0 to 14 s), the first passage (approximately from 14 to 26 s), and the recirculation (approximately for times >26 s) (figure reproduced from Calamante (86), with permission from John Wiley & Sons, Inc).

Dynamic susceptibility-contrast (DSC) magnetic resonance imaging (MRI), also known as bolus-tracking MRI, is currently the most commonly used MRI method to assess brain perfusion and tissue hemodynamics in clinical investigations. One of the main reasons for its widespread use is that bolus-tracking MRI data can be acquired very quickly (in approximately 1 min acquisition time), while still retaining good contrast-to-noise ratio (1).

DSC MRI involves an intravenous bolus injection of a paramagnetic contrast agent (typically a gadolinium-based contrast agent), and the rapid measure of the signal changes during its passage through the brain. The bolus of contrast agent induces a transient signal-drop on T_2^*-weighted images (Fig. 1) (and see methods), which can be used to infer the time-dependent concentration of the contrast agent (Fig. 2). Quantification of perfusion using DSC-MRI involves measurement of the so-called arterial input function (AIF, which describes the contrast agent *input* to

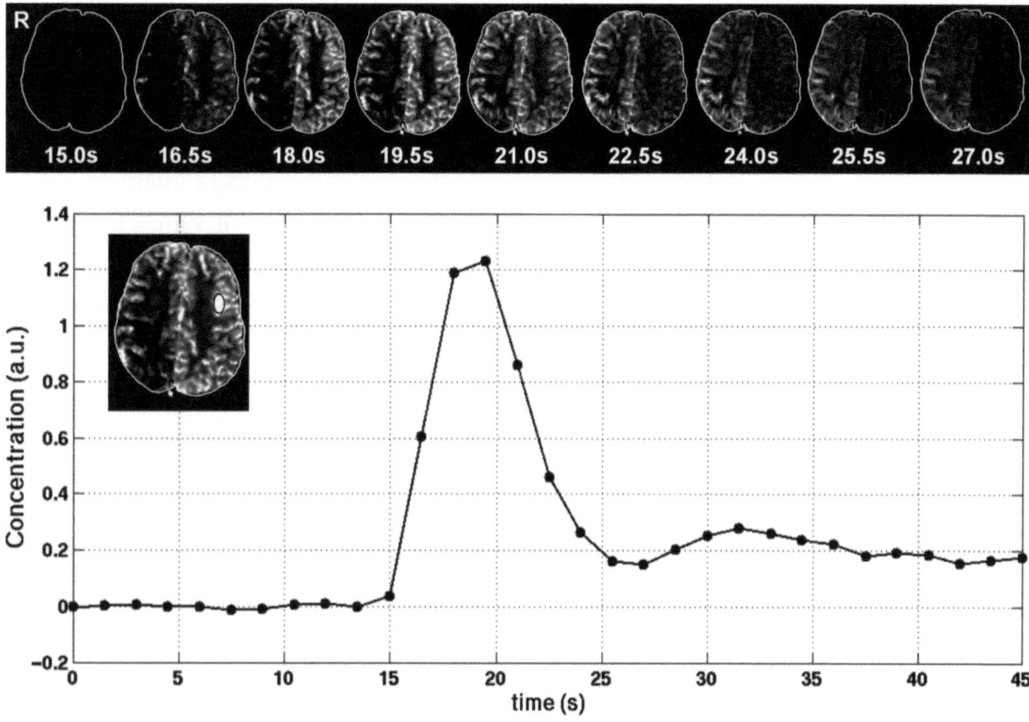

Fig. 2. Concentration time course on a patient with vascular abnormality in the right major cerebral arteries (same patient as in Fig. 1). The nine images in the *top row* are nine time samples of the concentration of the contrast agent during the passage of the bolus (TR = 1.5 s); the images show the transient increase in contrast concentration with the arrival of the bolus. Note the asymmetric behaviour due to the abnormalities on the right major cerebral arteries. The *bottom graph* shows the concentration time course for a region of interest in the contralateral cortical gray matter (see inset). As expected, the concentration is zero before the arrival of the bolus (during the baseline period) (figure reproduced from Calamante (86), with permission from John Wiley & Sons, Inc).

the tissue of interest), and a deconvolution analysis to remove, from the tissue concentration time course, the temporal spread contribution associated with the AIF (2). There are, however, many issues regarding the potential of DSC-MRI to accurately quantify perfusion (e.g. see (3, 4)), which will be discussed in this chapter.

2. Methods

A typical DSC-MRI study involves the following steps:

1. Selecting an appropriate image acquisition protocol.
2. Selecting an appropriate contrast agent injection protocol.
3. Assessing the need of motion correction as a pre-processing step.

4. Estimating the time-dependent contrast agent concentration.
5. Measuring the AIF.
6. Performing the deconvolution analysis to remove the AIF contribution.
7. Quantifying the hemodynamic parameters of interest.
8. Scaling the measurements appropriately, if values in absolute units are required.

A description of each of these steps, the practical decisions that need to be taken, the problems that can be encountered, and the approaches that have been developed to overcome or minimise them, will be discussed. In particular, the implications of all these issues for perfusion quantification will be emphasised.

2.1. Selecting an Appropriate Image Acquisition Protocol

The most commonly used data-acquisition sequence for DSC-MRI data is based on T_2^*-weighted imaging (i.e. a gradient-echo based method). In particular, since the transit time of the blood through the vasculature is only of the order of a few seconds, a fast T_2^*-weighted imaging method, such as gradient-echo echo-planar imaging (EPI (5)), is typically employed. This imaging technique allows acquisition of 15–20 slices with a time resolution of approximately 1.5 s, thus providing sufficient temporal resolution to characterise the bolus passage with good brain coverage. However, even with a fast imaging sequence such as EPI, there are a number of key parameters that must be chosen as a compromise between spatial resolution, temporal resolution, spatial coverage, and signal-to-noise ratio (SNR). The most relevant parameters are: the echo time (TE), the repetition time (TR), the flip angle (θ), the spatial resolution, and the number of slices. A set of recommendations was recently published by an international consortium of experts as part of the "Acute Stroke Imaging Research Roadmap" (6). In particular, it was recommended that TE = 35–45 ms (for studies performed at 1.5 T) or 25–30 ms (for 3 T studies), TR $\leq$ 1.5–2 s, θ = 60–90° (for 1.5 T studies) or 60° (for 3 T studies), in-plane resolution 1.8–2 mm, slice thickness 5 mm, and whole-brain coverage achieved using $\geq$12 slices. It should be noted that the choice of flip angle should not only be based on the commonly used Ernst angle criteria (i.e. $\cos\theta = \exp(-\mathrm{TR}/T_1)$), but it should be chosen to avoid introducing T_1-enhancement effects (7). In practice, the flip angle should therefore be slightly smaller than the Ernst angle (see Note 1).

Other methods based on T_2-weighted imaging (i.e. a spin-echo based acquisition), or multi-echo sequences have also been employed, although they are much less common. Similarly, alternative imaging methods to EPI have been used to achieve, for example, true full-brain coverage or higher spatial resolution. See Note 2 for a description of some of these alternative imaging methods used in DSC-MRI.

When data quality is very poor (e.g. very noisy data due to a sub-optimal choice of TE or bolus dose), the use of image denoising as a pre-processing step may be the only alternative in order to achieve usable results (8, 9). Various denoising methods have been used, from simple methods which denoise each image volume independently to more rigorous methods that consider DSC-MRI as a four-dimensional (three-dimensional space + time) data (8–10). However, any denoising method has limited power and, whenever possible, the imaging protocol should be optimised to increase data quality during the acquisition stage.

2.2. Selecting an Appropriate Contrast Agent Injection Protocol

Once the MRI protocol has been chosen, the next important step in DSC-MRI is the selection of an appropriate contrast agent injection protocol. The most relevant factors to consider in this respect are the contrast agent material, the dose, the injection site, access and rate, as well as the bolus flush. Once again, the previously mentioned international consortium has compiled very good guidelines regarding many of these factors (6). In particular, they have recommended the use of a standard gadolinium-based contrast agent, with a so-called "single" dose (for half-molar agents, corresponding to 0.1 mmoL/kg or 0.2 mL/kg body weight), administered by an intra-venous injection (preferably in the antecubital vein), using a 18–20 gauge line, at a rate of 4–6 mL/s using an MR-compatible power injector; this bolus should be followed by a 20–40 mL of saline flush. Flushing the bolus of contrast agent with saline (preferably with the same injection rate) is important to minimise the spread of the bolus of contrast agent (due to the low flow in veins).

It is important to note that these gadolinium-based contrast agents effectively behave as intravascular contrast agents, provided the blood–brain barrier (BBB, which is a physical barrier that limits the transport of substances from the blood into the central nervous system) remains intact. This assumption is essential for the validity of the common quantification kinetic model (see Sect. 2.6 below).

In recent years, there has been increasing concerns with some potential adverse reactions to contrast agents (11). Several studies have suggested a possible association between gadolinium-based contrast agents and an increased risk of nephrogenic systemic fibrosis (NSF), particularly in patients with renal insufficiency, or with glomerular filtration rate (GFR) < 30 mL/min/1.73 m^2 (11). This has led to changes in the investigation of these high-risk patient groups, for which the risk–benefit of DSC-MRI studies should be carefully considered. Furthermore, not all contrast agents were found to have the same association to increased risk, and the choice of contrast agent should be carefully considered (see Note 3 for more details).

2.3. Assessing the Need of Motion Correction as a Pre-processing Step

Although DSC-MRI has a relatively fast acquisition time (1–2 min), its data can suffer from motion-related artefacts. These artefacts do not manifest in the same way as in other conventional MRI methods (e.g. as image ghosting) because a single-shot imaging method (such as EPI) is commonly used. It is the dynamic nature of DSC-MRI that is affected by motion instead: if motion is present, different image volumes in the time-series will not match in space. It is therefore essential to assess the data for the presence of motion, particularly for the image volumes acquired in and around the passage of the bolus.

The most common source of motion is subject movement at the time of the injection. If the motion is primarily in-plane, standard motion-correction strategies have been shown to eliminate the error (12). On the other hand, if the motion has a significant through-slice component, these strategies will not work; they are not able to correct for the spin history effects due to the multi-slice two-dimensional acquisition strategies commonly employed in DSC-MRI. The use of a three-dimensional based sequence (e.g. see 3D-PRESTO MRI sequence in Note 2 below) could offer a viable solution; these 3D sequences may be worth considering if very ill or uncooperative subjects are studied.

2.4. Estimating the Time-Dependent Contrast Agent Concentration

Quantification of perfusion using DSC-MRI requires knowledge of the time-dependent concentration of the contrast agent (see step 2.6 below). However, DSC-MRI does not measure the concentration directly, and its value is estimated *indirectly* from the effect the contrast agent has on the image signal intensity. The most commonly used model is based on a linear relationship between the change in the relaxation rate ΔR_2^* (where $R_2^* = 1/T_2^*$) and the contrast agent concentration (13):

$$C(t) = k \cdot \Delta R_2^*(t) = k \cdot (R_2^*(t) - R_2^*(0)) = -\frac{k}{\mathrm{TE}} \cdot \ln\left(\frac{S(t)}{S(0)}\right) \quad (1)$$

where $C(t)$ is the time-dependent concentration (also referred to as the "peak", due to its shape, see Fig. 2), $R_2^*(0)$ indicates the baseline relaxation rate (i.e. the relaxation rate without contrast agent), $S(t)$ is the signal intensity and $S(0)$ its baseline value, and k is a proportionality constant that depends on the type of contrast agent, the magnetic field strength, the pulse sequence, and the tissue type. Since a measurement of R_2^* requires multiple images acquired with different TE*s* (and therefore, with an associated increased acquisition time), DSC-MRI data are commonly acquired with a single TE value. Therefore, it should be noted that (1) not only assumes a linear relationship between the concentration of contrast agent and the change in relaxation rate, but it also assumes that the change in relaxation rate can be estimated from the change in signal intensity measured from single-echo T_2^*-weighted images (see right-hand side of (1)). This, in turn, assumes a negligible contribution from T_1

relaxation to the changes in signal intensity. This assumption depends on several factors (7), including BBB status, sequence parameters (e.g. TR, θ), and sequence type (see also Notes 1 and 8).

The actual value of *k* cannot be easily measured in practice, and this has been one of the major criticisms for the accuracy of Perfusion MRI measurements in absolute units (compared to absolute measurements using Perfusion CT) (4). A discussion about this issue and its implication for perfusion quantification is included in Note 4.

2.5. Measuring the Arterial Input Function

One of the key steps in quantification of DSC-MRI data is the measurement of the AIF. The information about the contrast agent input to the tissue of interest is essential to be able to isolate true tissue information (e.g. cerebral perfusion) from other non-tissue related contributions, such as the specific injection protocol, cardiac output, and macro-vascular structure. For example, a wider tissue concentration time course (i.e. a wider "peak") could be due to any one, or a combination, of the following conditions: a lower tissue perfusion, and/or a slower injected bolus, and/or a larger injected bolus, and/or a slower cardiac output, and/or a longer transit route of the bolus through a collateral vascular path. The key role of the AIF is to help isolate, from all these contributions, the contribution due to tissue perfusion.

Due to practical limitations, a single AIF is commonly used for each slice (or even for the whole brain). This so-called "global AIF" is usually measured from the signal changes in and around a major artery, such as the middle cerebral artery (14). The basic assumption is that such a global AIF provides a reasonable representation of the input of the bolus to brain tissue, and contains information regarding the contribution of the injection conditions, the cardiac output, and the macro-vascular structure. However, the perceptive reader would anticipate that this assumption can lead to errors in perfusion quantification. In fact, the measurement of AIF is one of the most important possible sources of error in DSC-MRI quantification. The measured AIF can be subject to a number of possible artefacts, including: partial-volume effects (see Note 5), bolus delay and dispersion (see Note 6), non-linear effects with contrast-agent concentration (see Note 4), and truncation artefacts (see Note 7).

While the AIF used to have to be manually measured by experienced users (e.g. by manually identifying voxels with earlier and narrower "peaks"), a number of automatic methods have been developed in the last few years (15–18). These methods not only reduce demands on the need for highly trained local staff, but they have also contributed to increase the objectivity, speed, and reproducibility of the DSC-MRI measurements; they are therefore highly recommended when analysing DSC-MRI data. Figure 3 illustrates the workflow and results from one of these automatic methods.

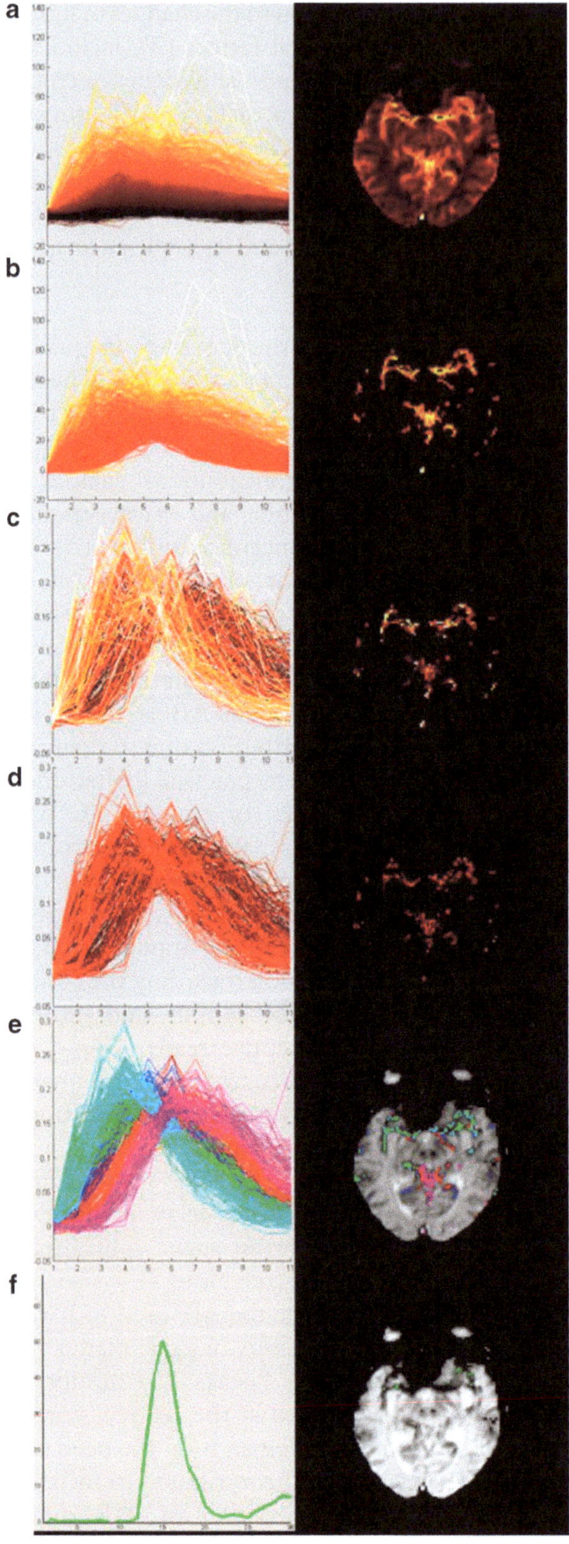
a
b
c
d
e
f

2.6. Performing the Deconvolution Analysis to Remove the AIF Contribution

Once an estimate of the AIF has been obtained, its contribution must be removed from the tissue concentration time course. This is a mathematical process known as "deconvolution", because it involves inverting a so-called convolution equation (1, 2):

$$C(t) = \alpha \cdot \mathrm{CBF} \cdot \mathrm{AIF}(t) \otimes R(t) = \alpha \cdot \mathrm{CBF} \cdot \int_0^t \mathrm{AIF}(t')R(t-t')\mathrm{d}t' \quad (2)$$

where AIF(t) is the time-dependent AIF, $R(t)$ is the tissue residue function (which determines the tissue retention of the contrast agent for the case of an ideal instantaneous bolus injection at $t=0$), α is a proportionality constant that depends on the density of brain tissue and the hematocrit levels, and $\otimes$ indicates the convolution operation (shown in its explicit integral form on the right-hand side of (2)). The convolution expression in (2) follows from indicator-dilution theory for intravascular contrast agents (19); see Note 8 for a discussion about the modifications to the model needed when the contrast agent leaks to the extravascular space (due to a leaky BBB). The convolution expression states that the tissue concentration at time $= t$ is given by the sum (or integral) of the various contributions of the contrast agent that has entered the tissue at previous times $= t'$ (i.e. $t' < t$), which is given by AIF(t'), and that still remain in the tissue at time $= t$, which is given by the factor $R(t-t')$. The *CBF* factor in (2) is included to account for the fact that the concentration in tissue is proportional to the blood delivered to the tissue (with concentration AIF(t')) per unit time (i.e. perfusion).

Many deconvolution algorithms have been proposed, including Fourier-based methods (2, 20), singular value decomposition (SVD) based methods (2, 21, 22), Tikhonov regularisation (23), iterative methods (24–26), and Bayesian methods (27, 28), to name a few. All these methods have been thoroughly tested and optimised using controlled numerical simulations (see Note 9).

A major difficulty with any deconvolution approach is that (2) is known as an "ill-posed" problem: a small amount of noise in the measured data (in our case, $C(t)$ and AIF(t)) translates into enormous errors in the calculated solution (in our case, $R(t)$ and CBF).

Fig. 3. Illustration of the workflow in an automatic arterial input function (AIF) searching algorithm based on cluster analysis. In (**a**, **b**) voxels and curves are colour coded in proportion to the area under $C(t)$. In (**a**) all concentration time curves are shown, and in (**b**) only the 10% of the curves with the largest area are included. The set of curves in (**b**) standardised to unit area are displayed in (**c**) coloured according to roughness, and in (**d**) the 25% most irregular curves have been removed. The set of curves in (**d**) are then partitioned into five clusters such that the intra-cluster variance is minimised and the inter-cluster variation is maximised. The partition is indicated in (**e**). The mean curve in the group of light-blue curves had the lowest first moment and was, therefore, picked-out by the algorithm and divided into five subclusters (not shown). The estimate of the AIF returned by the program was the mean curve with the lowest first moment, which is shown in (**f**) along with the anatomic location of the selected voxels. Please refer to the electronic version of the book for the colour version of this figure (figure kindly provided by Dr. Kim Mouridsen, and reproduced from Mouridsen et al. (18), with permission from Wiley-Liss, Inc).

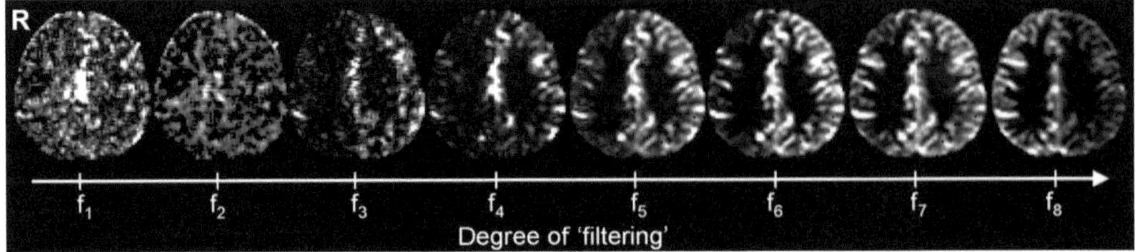

Fig. 4. Illustration of the regularisation step required in cerebral blood flow (CBF) quantification by deconvolution analysis. The eight images show the corresponding CBF maps calculated with increasing degree of regularisation (i.e. "filtering"): with increasing filtering from the *left* to *right*. The example shows the results for a patient with right internal carotid artery stenosis (*left* side of the images), and regularisation using the truncated singular value decomposition method (increasing regularisation corresponding to a higher cut-off level, and thus increasing amount of truncation). As can be seen in the figure, with little regularisation (f_1 and f_2 filter levels in the figure), the solution is completely dominated by noise. On the other hand, with too much regularisation (f_8 filter level) the estimated impulse response function is over-smoothed, leading (in this case) to the erroneous identification of the abnormal tissue: note that the normal left hemisphere is wrongly displayed as hypoperfused on the CBF map calculated with this filter level. The optimum degree of regularisation in this particular case corresponds to the f_6 filter level (figure reproduced from Calamante (86), with permission from John Wiley & Sons, Inc).

To deal with this high noise-sensitivity, deconvolution methods need some form of regularisation (e.g. see (29)): the regularisation can be seen as a "noise filter" applied during the inversion of (2). This is illustrated in Fig. 4 for the commonly used truncated SVD approach (2). For the truncated SVD method, the filter is given by the threshold used to truncate the SVD expansion; for example, Østergaard et al. (2) suggested a cut-off level corresponding to 20% of the maximum singular value of the AIF matrix obtained after discretisation of the integral expression in (2) (30). Therefore, the development of a deconvolution method for DSC-MRI can be seen, in practice, as the design of a "noise filter" and the selection of the appropriate amount of filtering. The latter can be very complex, because the optimal amount of filtering has been shown to depend on imaging parameters (e.g. TR, SNR, etc.) as well as tissue characteristics (e.g. mean transit time (MTT)) (2, 21–28, 31). Therefore, the level of filtering should ideally be adapted based on the specific data (e.g. references (23, 26, 28, 32)).

Due to their relative quicker processing speed and accuracy, SVD-based methods (particularly the variants that are delay-insensitive (21, 22)) have become very popular. However, they suffer from a number of limitations (see Note 6), and some of the other slower, but more accurate, methods may be preferable if speed is not paramount.

2.7. Quantifying the Hemodynamic Parameters of Interest

Apart from being a fast MRI method, a further characteristic that has contributed to the widespread use of DSC-MRI in clinical studies is the wealth of information that can be obtained from the acquired data. The DSC-MRI data can be used to quantify CBF, as indicated in the previous section, as well as a number of other physiological and hemodynamic-related parameters. The most commonly used parameters include:

- Cerebral blood volume (CBV) is the fraction of the tissue volume occupied by blood (e.g. normal human gray matter tissue has approximately a CBV = 4 mL/100 g) (or 4%, since the density is approximately 1 g/mL). This parameter can be calculated in various ways (33), but the most commonly used approach is based on the area under the peak:

$$\mathrm{CBV} = \alpha^{-1} \cdot \frac{\mathrm{AUC}(C)}{\mathrm{AUC}(\mathrm{AIF})} = \alpha^{-1} \cdot \frac{\int_0^{\infty} C(t)\mathrm{d}t}{\int_0^{\infty} \mathrm{AIF}(t)\mathrm{d}t} \quad (3)$$

  where AUC indicates the "area under the curve". In the case where there is leakage of the BBB, (3) needs to be modified to account for the distribution of the contrast agent to the extravascular space (see Note 8).

- MTT is the average time it takes for the blood (or in our case, for the contrast agent) to travel through the tissue vasculature, for the ideal case of an instantaneous bolus injection. It is important to note that MTT does not refer to the transit time through the voxel, but through the vasculature in the tissue (e.g. through the arterioles–capillaries–venules path). This value is approximately 4 s in normal gray matter tissue, and explains the need for fast imaging to properly characterise the passage of the bolus. The MTT can be calculated based on the central volume theorem (34): MTT = CBV/CBF. Due to the relatively low gray–white matter MTT contrast in normal tissue and the large MTT increase observed in hypoperfusion, this parameter has been very popular to identify the abnormal tissue in stroke studies.

- Time-to-maximum of the residue function (T_{max}) is a parameter that has also received increasing interest in acute stroke studies. It primarily reflects a combination of bolus delay and dispersion and, therefore, it mainly should be considered a measure of macrovascular features (35). It could play a useful role in the identification of areas subject to collateral flow. Figure 5 shows a couple of examples from patients with acute stroke: patient P1 shows an example in which T_{max} and MTT identify abnormalities of similar extent and severity; patient P2, however, shows an example of severe T_{max} lesion in the presence of mild MTT abnormality. The similar T_{max} and MTT abnormalities in P1 suggest that the severe T_{max} prolongation likely reflects a contribution also from delay or dispersion (35). However, the mild MTT abnormality in P2 (but with a severe T_{max} lesion) probably suggests a case in which sufficient collateral supply exists to maintain microvascular integrity; i.e. the

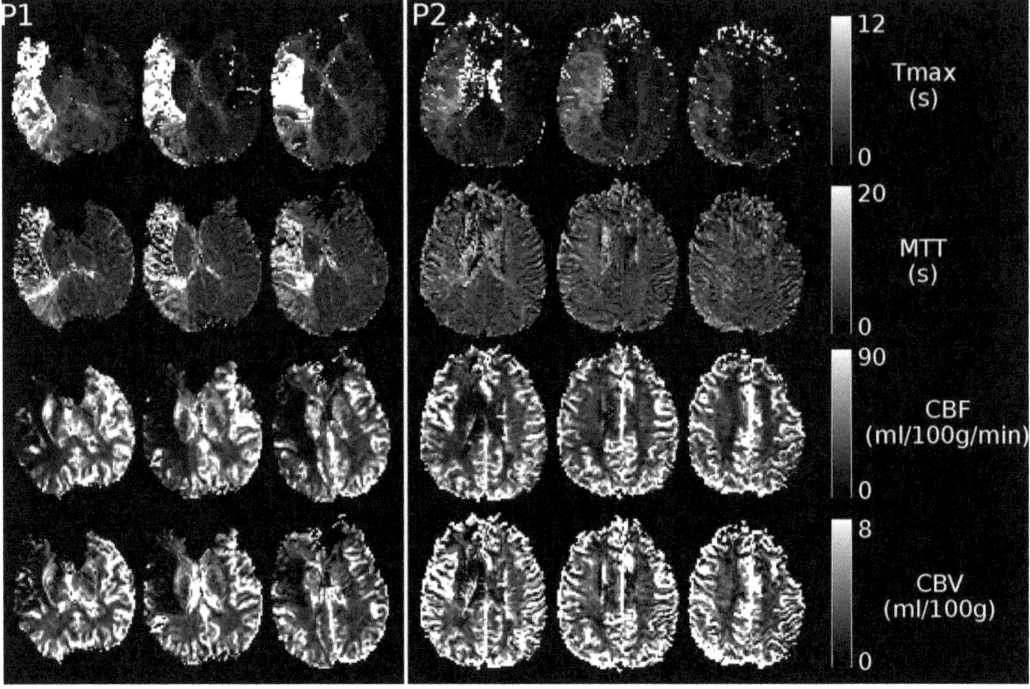

Fig. 5. In vivo DSC-MRI maps from two acute stroke patients (labelled P1 and P2); three slices are shown for each case. *Top row*, mean transit time (MTT). *Second row*, T_{max}. *Third row*, CBF. *Bottom row*, cerebral blood volume (CBV). Both patients had a right internal carotid artery occlusion (*left* side of the images), and were scanned within 6 h of symptoms onset. The AIF was automatically measured from the contralateral middle cerebral artery, and all maps were calculated using delay-insensitive oSVD (21). For display purposes, the CBF and CBV maps were scaled by setting normal white matter CBF to 22 mL/100 g/min (figure reproduced from Calamante et al. (35), with permission from American Heart Association, Inc).

T_{max} abnormality reflects almost only delay/dispersion and not MTT prolongation (benign oligemia).

- There are a number of so-called "summary parameters" (1) that can be measured directly from the DSC-MRI data without requiring a deconvolution analysis. These include: time-to-peak (TTP), bolus-arrival time (BAT), maximum-peak concentration (MPC), full-width at half-maximum (FWHM), and first moment (FM) of the tissue concentration. While these parameters can be easily calculated from the DSC-MRI data, it should be noted that they depend on the experimental conditions (injection protocol, cardiac output, etc.); this makes their interpretation less straightforward in terms of the physiological status of the tissue (36).

2.8. Scaling the Measurements Appropriately, if Values in Absolute Units Are Required

While the timing parameters (e.g. MTT, T_{max}, TTP, etc.) can be quantified in absolute units (see, however, Note 10 for possible problems), the other DSC-MRI parameters are not easily quantified in absolute units (3, 4). The main reason for this is that measurement of the contrast-agent concentration in absolute units is very

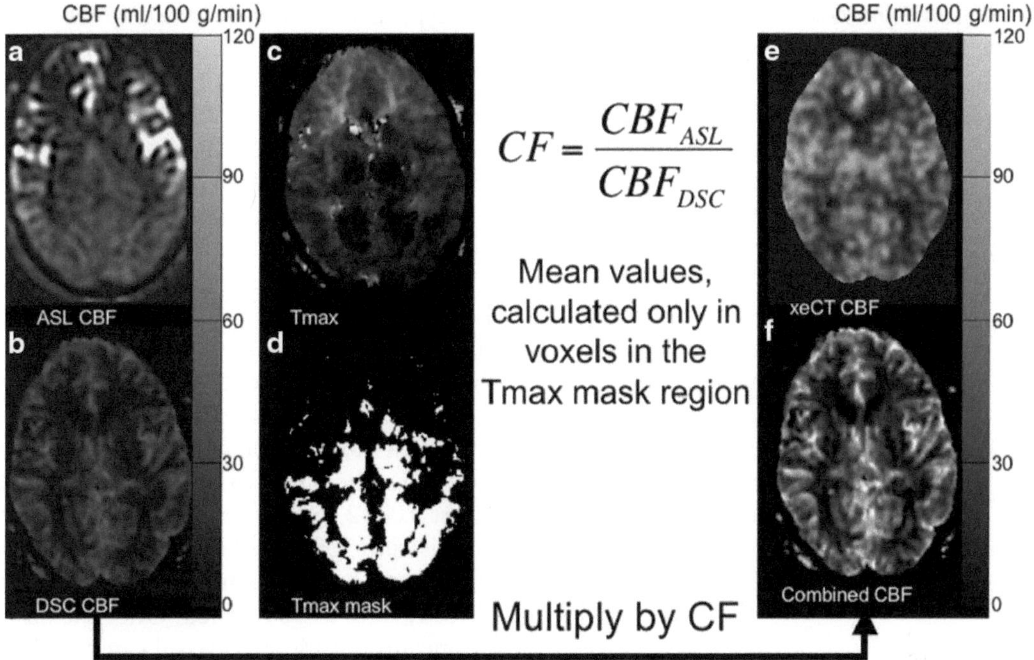

Fig. 6. Demonstration of the method of determining the correction factor (CF) for combined arterial spin labelling (ASL) and DSC CBF maps. (**a**): ASL- and (**b**) DSC-derived CBF maps in a 42-year-old man with bilateral moyamoya disease. There is marked arterial transit artifact in the bilateral anterior circulation seen on the ASL CBF maps, while the posterior circulation is unaffected. Using information from the T_{max} map (**c**) acquired during the DSC study, a thresholded T_{max} mask (**d**) can be created to include only voxels with T_{max} shorter than a pre-specified threshold (in this example, $T_{max\ thresh} = 3$ s). The CF is the ratio of the ASL-measured CBF and the DSC-measured CBF calculated only in the T_{max} masked regions. The DSC-derived CBF map is then multiplied by the CF to create the combined map (**f**), which more closely compares with the xeCT CBF map (**e**) in magnitude while more accurately depicting true CBF in the affected regions (figure kindly provided by Dr. Greg Zaharchuk, and reproduced from Zaharchuk et al. (40), with permission from Wiley-Liss, Inc).

complex (e.g. due to the exact unknown value for the proportionality constants, such as k in (1)). This has meant that DSC-MRI studies are most often quantified in relative units (e.g. relative to contralateral normal values). A number of methods have been proposed to scale the DSC-MRI values to absolute units (e.g. scale CBF values to mL/100 g/min). However, it is now well established (37, 38) that a universal scaling, as was initially done (see Note 11), is not sufficient, and that a *subject-specific* scaling is required. More recently, a number of these approaches have been proposed, such as calibrating the DSC-MRI measurements using an extra CBV map calculated from steady-state T_1 measurements (39), or using an extra CBF map obtained from arterial spin labelling (ASL) (see Note 12) (40), or a PET-based lookup table (41). If quantification in absolute units is required, the use of such a subject-specific scaling method is highly recommended; otherwise, absolute values should be interpreted with great caution. Figure 6 shows the schematic steps involved during the scaling method based on the ASL data.

2.9. DSC-MRI Analysis: Example Case

This section describes the typical steps and decisions involved *in practice* during the analysis of a DSC-MRI dataset. This section assumes that an appropriate DSC-MRI dataset has been acquired on a patient. The first step during the analysis is the assessment of the presence of image motion. A simple and efficient way to assess this is by playing the temporal image data as cine MRI. Let us assume that no significant image motion was observed. Let us also assume that there is no leakage of the BBB (i.e. the contrast agent remains intravascular); this can be easily determined by assessing the presence of signal enhancement on a T_1-weighted image acquired post-contrast injection.[2] In that case, the next step in the analysis is to convert the signal intensity data to contrast-agent concentration using (1). Therefore, we need to calculate the baseline intensity value $S(0)$ for each voxel of the image. This involves determining a time interval in the DSC-MRI data before the arrival of the bolus; at least 10–12 images are recommended to achieve a good characterisation of the baseline values (6). Furthermore, the first 2–3 image volumes sometimes need to be discarded, to allow for a steady state to be reached. Once the baseline value is known, the concentration in each voxel can be calculated using (1).

The next step in the analysis is the measurement of the AIF. As discussed in the Sect. 2.5, this is best done using an automatic searching method. Averaging the concentration time course from a number of AIF voxels identified by such a method will help increase the SNR of the AIF measurement, and thus improve quantification.

After we have measured the AIF, we are ready to perform the deconvolution analysis. Ideally, the level of regularisation should be automatically determined based on the data. However, most widely available deconvolution algorithms tend to use a fixed value for all the voxels. In that case, an appropriate value should be chosen based on the published results from numerical simulations (e.g. (2, 21, 22, 24, 31)). For example, for the commonly used truncated SVD method (2), a 20% cut-off was suggested for SNR = 20, but a 4% cut-off for a higher SNR = 4% (21). Therefore, we need to select the degree of regularisation appropriately for our own data. After deconvolution of (2), we obtain the residue function scaled by perfusion ($CBF{\cdot}R(t)$, the so-called impulse–response function). Since, in theory $R(t=0)=1$, we can calculate perfusion from the initial value of the impulse–response (or, in practice, its maximum value in order to minimise the errors due to

[2]This is one of the reasons why a post-contrast T1-weighted imaging sequence is usually included as part of a DSC-MRI protocol.

bolus dispersion (2), see Note 6). Therefore, the maximum value in each voxel can be used to create a CBF map, and the time-to-maximum to create a T_{max} map. Once CBF is calculated, we can obtain a map of MTT by calculating, on a voxel-by-voxel basis, the ratio of CBV (calculated as defined in (3)) and CBF. The CBF and CBV maps can be converted to relative maps, by dividing them to the corresponding CBF and CBV values on a normal region-of-interest (e.g. on the contralateral putamen, or the cerebellum).

As can be noticed in the description above, analysis of DSC-MRI data involves significant operator interaction. In recent years, a number of fully automated methods have been developed (42–44), which should help minimise the need for highly trained local staff. However, it is the author's opinion that a low-level operator quality control is still highly recommended, because automatic methods can never be totally flawless.

3. Applications

By far, the most common application of DSC-MRI has been in the field of stroke. In particular, it has played a major role in the MRI attempts at identifying the ischemic penumbra, using the so-called "diffusion–perfusion mismatch", i.e. an area abnormal on Perfusion MRI but normal in Diffusion MRI (45). While there is still some controversy as to the best Perfusion MRI parameter (and, even for a given parameter, the optimal threshold value to use to define the abnormal area) (6), numerous studies have assessed the role of many of these parameters in the definition of the penumbra (45). They have even been used in recent acute stroke clinical trials (e.g. (46, 47)) and are being used in the inclusion criteria in the ongoing DEFUSE-2, EXTEND, and MR RESCUE trials.

Due to the complementary information contained in the various DSC-MRI parameters, an alternative approach to using a single parameter has been to combine multiple DSC-MRI parameters, together with Diffusion MRI data and T_2-weighted images, into a predictor model of infarction (48). In these models, the relevance of each parameter to infarct prediction is determined by a training dataset and, in theory, they should provide a more optimal use of all the available information. One of the earlier methods was based on a statistical generalised linear model (GLM) algorithm (49); an illustration of the principles behind this approach is shown in Fig. 7.

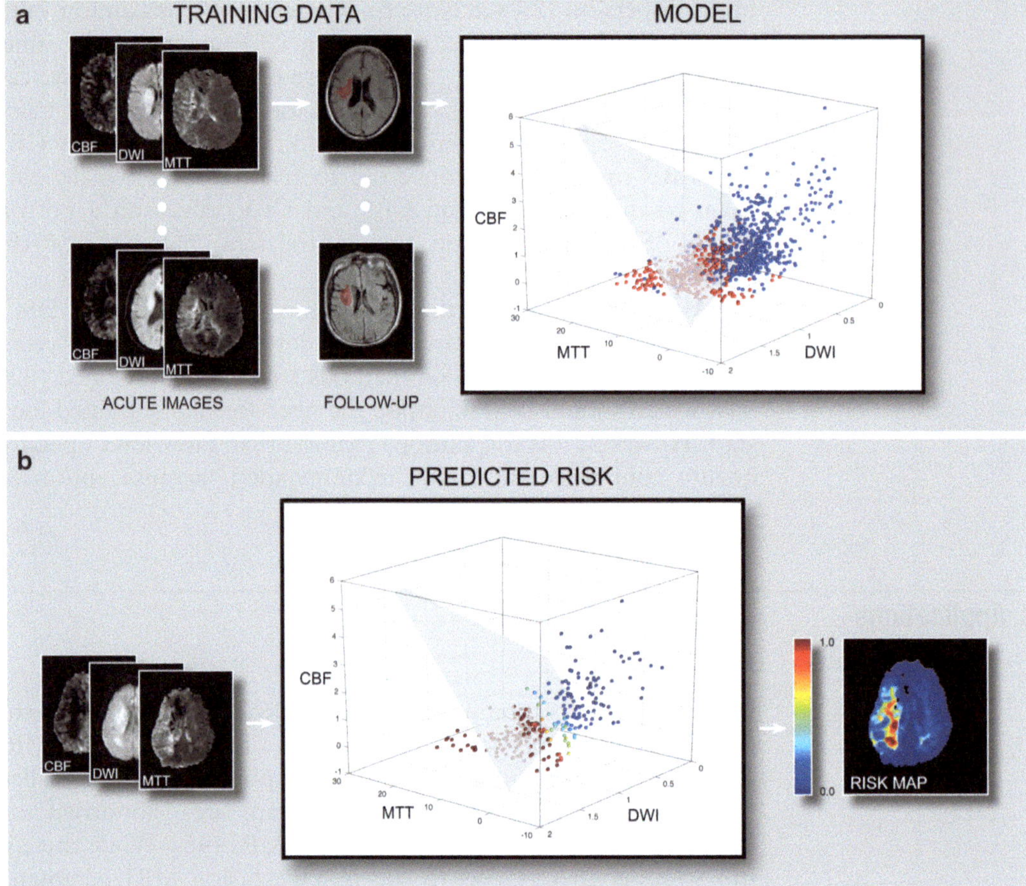

Fig. 7. Illustrative example of the use of the generalised linear model (GLM) in predictor models of infarction. (**a**) The model is constructed using a "training" dataset, i.e. MRI data from a cohort of acute stroke patients (in the example of the figure, CBF, diffusion-weighted imaging (DWI), and MTT) and the corresponding follow-up final infarction (in the example, fluid attenuated inversion recovery (FLAIR) images acquired at 3 months follow-up). In this cohort of patients, follow-up lesions are outlined by an expert (note that the follow-up images must be co-registered to the acute MRI data). The DWI, MTT, and CBF for each voxel in the training data define a position in a three-dimensional coordinate system (*top-right*), which can be colour-coded according to outcome (*red*: infarcted on follow-up, *blue*: rescued). The GLM defines the plane that best separates infarcted and rescued voxels. (**b**) The resulting GLM predictor model can now be used in a new patient: the (DWI, MTT, CBF) coordinates of each voxel in the new patient is assessed relative to the GLM plane, thus determining its infarct risk. The risk at each voxel can then be displayed as a "risk map" (*bottom right*), and use for patient management. In this way, the information of multiple maps (in this example, DWI, MTT, and CBF) is combined into a single infarct-risk map, without adding diagnostic complexity. In general, predictor models include also additional information such as T_2-weighted image intensity, CBV, T_{max}, etc. Please refer to the electronic version of the book for the colour version of this figure (figure kindly provided by Dr. Kim Mouridsen, Center of Functionally Integrative Neuroscience, Aarhus, Denmark).

4. Notes

1. Flip angle: The Ernst angle is the flip angle that gives the maximum SNR for a given time interval, when repeated radiofrequency excitations are applied. While this flip angle is commonly used in many MRI applications, its use in DSC-MRI

can lead to significant T_1-enhancement and therefore, since this effect is usually neglected (see (1) above), it can lead to quantification errors (7). Therefore, the flip angle in DSC-MRI should be optimised not only to yield high SNR, but also to limit T_1 effects. The quantitative results presented in reference (7) can be used as a guide for sequence parameter optimisation.

2. Acquisition sequence: Since the gadolinium contrast agent is a paramagnetic agent, it reduces T_2^* and T_2. Therefore, both gradient-echo and spin-echo sequences can be used for DSC-MRI. However, it has been shown that the vascular sensitivity of T_2^* and T_2 is very different: T_2 is primarily sensitised to capillary-size vessels, and T_2^* data is sensitised both to small and large vessels (13). While the capillary-specific sensitivity of spin-echo should be beneficial for measuring microvascular perfusion, the contrast-to-noise ratio of T_2-weighted DSC-MRI data is greatly reduced (due to the refocusing effect of the 180° pulse). A further problem with T_2-based DSC-MRI is that the relationship between concentration and the change in relaxation rate (see (1)) is highly non-linear (50). Due to these limitations, DSC-MRI measurements based on T_2^*-weighted imaging are still the most commonly used approach. However, the gradient-echo sequence used in DSC-MRI does not have to be limited to a single-echo sequence. Double-echo (51) and multi-echo (52) sequences allow a direct measure of ΔR_2^* without having to assume a negligible T_1 effect (see Note 1). However, except when these multi-echo methods are combined with parallel imaging to speed-up the acquisition time (52), they usually compromise the maximum number of slices (and therefore brain coverage) that can be acquired for a given TR.

   While two-dimensional EPI-based methods are by far the most commonly used in DSC-MRI, three-dimensional acquisition methods have been used to achieve true whole-brain coverage with good temporal resolution and SNR. For example, the 3D-PRESTO method can be used to acquire whole-brain data with sub-second TR (53). The drawback of these methods is that they tend to have much coarser spatial resolution.

3. NSF: A set of recommendations and guidelines for NSF prevention in the United States, Canada, and Europe were recently published (54). They pointed out that there are differences in the NSF incidence for the different gadolinium-based contrast agents. In particular, the recommendations listed as *contraindicated* for high-risk patients (e.g. GFR < 30 mL/min/1.73 m^2) the following contrast agents: Omniscan® (3–7% incidence of NSF in at-risk subjects), Magnevist® (incidence: 0.1–1%), and Optimark® (incidence: unknown). Therefore, the choice of contrast agent in this patient population should be made with caution.

4. Non-linear effects: As described in (1), the most commonly used method to calculate the contrast-agent concentration is based on a linear relationship between the concentration and the change in relaxation rate. This proportionality constant was found to be different in tissue and in the artery (50), and its actual value on a subject-by-subject basis is not known. Furthermore, several studies have now suggested that the relationship in the artery (i.e. to measure the AIF) is not even linear, but that it is better described by a quadratic relationship (55). The situation is further complicated given that this quadratic relationship is dependent on hematocrit values (55), and therefore could change from study to study:

$$\Delta R_2^*(t) = a_1 \cdot \mathrm{AIF}(t) + a_2 \frac{1.1378 \times \mathrm{Hct}}{(1-\mathrm{Hct})^2} (\mathrm{AIF}(t))^2 \quad (4)$$

where Hct is the hematocrit, $a_1 = 7.62$ mM^{-1} s^{-1} and $a_2 = 0.57$ mM^{-2} s^{-1} for 1.5 T studies (56); for 3 T studies, the equivalent coefficients are 0.49 mM^{-1} s^{-1} and 2.62 mM^{-2} s^{-1}, respectively (50). To minimise the non-linear errors, a number of studies have now used a linear relationship for the tissue concentration but a quadratic relationship for the AIF (e.g. (40, 42, 56, 57)). Even though the exact quadratic relationship may not be known for a given subject, it was shown that the quadratic model for the AIF should produce more accurate results (56). In particular, when the Hct is unknown, underestimation of Hct in (4) should result in smaller errors than overestimation.

5. Partial-volume effect: The spatial resolution of typical DSC-MRI images is $2 \times 2 \times 5$ mm^3. Therefore, partial-volume effects when measuring the AIF are almost unavoidable. Since the MRI signal is complex (i.e. it has a magnitude and a phase component), partial-volume effect between the fraction of the voxel occupied by the artery and the fraction occupied by the tissue can lead to various distortions to the measured AIF (58). Importantly, these distortions include curves that are narrower and sharper than the true AIF (59). This has two major implications: first, the errors due to PVE translate into errors in perfusion quantification using deconvolution analysis (see (2)). Secondly, the narrower curves created by the partial-volume effect are not easily avoided by either manual or automatic AIF-searching algorithms (e.g. (15–18)): they will all tend to favour such sharper curves. To avoid the first problem, an elegant correction method has been proposed that acts on the phase of MRI data (55): it uses the trajectory of the MRI data (on the complex real–imaginary plane) during the passage of the bolus to calculate the contribution from the tissue, which then needs to be subtracted out. However, this method is only

valid for arteries approximately parallel to the main magnetic field, such as is the case for the internal carotid artery. This would appear to suggest that measurement of the AIF free of partial-volume effect is not feasible in other arteries. Fortunately, this is not the case, as shown by several recent studies (60, 61). These studies have shown that the *shape* of the measured AIF can be approximately correct in certain specific areas around an artery, such as 3.5–6 mm posterior to the M1 segment of the middle cerebral artery (61). It is important to note that these measurements get the shape of the AIF correct but not its size; they can therefore only provide relative perfusion values (see also Sect. 2.8).

As mentioned before, the partial-volume effect can lead to narrower "peaked" shaped curves, which could be erroneously selected as AIF. To avoid this selection, a new criterion was recently proposed (59), which detects these shape errors based on tracer-kinetic principles for computing CBV. In particular, it was shown that the ratio of the steady-state concentration value to the area-under-the-curve of the first passage can be used to identify these narrower curves. Therefore, the incorporation of this extra criterion in the previously mentioned automatic AIF-searching algorithms may provide a robust method to measure AIF free of partial volume effects.

6. Delay and dispersion: As mentioned in Note 5, to avoid partial-volume effect, the AIF is usually measured in, or around, a major artery. Since this artery can be distant from the tissue of interest, the bolus can take longer to arrive to the true input to the tissue of interest (i.e. bolus delay) and/or the bolus can be spread in time during its transit there (i.e. bolus dispersion) (62). One of the most commonly used deconvolution algorithms, namely truncated SVD (2), has been shown to be highly sensitive to the presence of bolus delay (63), thus leading to severe errors in the presence of large bolus delay, such as it can occur in patients with vascular abnormalities (e.g. stenosis, occlusion, collateral flow). Several methods have been proposed to minimise this source of error, from shifting the curves to a common time origin before performing the deconvolution analysis (63, 64), to using alternative deconvolution algorithms that are insensitive to delays (21, 22, 26, 28). It is important to note that, while some of these methods are indeed delay-insensitive, it does not necessarily mean they produce accurate measurements of perfusion. For example, the CBF estimates from delay-insensitive SVD-variants (21, 22) depend, among other factors, on MTT and the residue function model (21). Therefore, while they do eliminate the extra confounding factor of the heterogeneous delays throughout the brain, their absolute CBF and MTT values should be interpreted with

caution. For example, Fig. 2e of reference (21) shows that the so-called oSVD method is delay-insensitive but will tend to underestimate a hypoperfusion abnormality (since the degree of CBF underestimation decreases with decreasing CBF). Nevertheless, the use of a delay-insensitive method is highly recommended when dealing with a patient population where vascular abnormalities can be important (e.g. for stroke patients).

Bolus dispersion is a more complex source of error to deal with. The major problem is that it is not easy to identify the presence of bolus dispersion in a given patient (cf. identification of the presence of bolus delay, which can be easily determined by, for example, assessing the presence of delays on a bolus arrival time map). The major problem with bolus dispersion is that this unaccounted source of bolus spread is interpreted by the quantification model (2) as occurring in the tissue (instead of in the transit of the bolus from the site of AIF measurement to the input to the tissue). This contribution is therefore incorporated into the MTT calculation, leading to an overestimation of MTT and, based on the central volume theorem (see Sect. 2.7), an underestimation of CBF (65). Errors as much as 70% CBF underestimation have been reported due to bolus dispersion in patients with vascular abnormalities (66).

From the quantification model point of view, bolus dispersion can be characterised by a vascular transport function (VTF), which describes the probability density function of vascular transit time between the site where the AIF was measured and the true input to the tissue of interest (65, 67):

$$
\begin{aligned}
C(t) = \alpha \cdot \mathrm{CBF} \cdot \mathrm{AIF}_{\mathrm{meas}}(t) \otimes \mathrm{VTF}(t) \otimes R(t) = \\
\alpha \cdot \mathrm{CBF} \cdot \mathrm{AIF}_{\mathrm{meas}}(t) \otimes R_{\mathrm{meas}}(t)
\end{aligned}
\tag{5}
$$

where $\mathrm{AIF}_{\mathrm{meas}}(t)$ is the measured AIF (in a major artery) and $R_{\mathrm{meas}}(t)$ is the calculated residue function obtained by deconvolution using the measured $\mathrm{AIF}_{\mathrm{meas}}(t)$ (65). It is important to note that, if the data are deconvolved using the measured $\mathrm{AIF}_{\mathrm{meas}}(t)$ (instead of the true AIF), the measured residue function will be distorted by the VTF, i.e. $R_{\mathrm{meas}}(t) = R(t) \otimes \mathrm{VTF}(t)$. This function has a peak shape, with $R_{\mathrm{meas}}(t{=}0){=}0$. Therefore, an estimate of CBF cannot be obtained from the *initial* value of the impulse–response function (see Sect. 2.9), but it is estimated from its *maximum* value. However, the maximum of $R_{\mathrm{meas}}(t)$ is always smaller than the maximum of $R(t)$, which explains the CBF underestimation associated with bolus dispersion. Figure 8 shows an example of dispersed residue functions in a patient with severe vascular abnormality.

In order to avoid or minimise bolus dispersion, the AIF must be measured closer to the tissue of interest. This has led to the

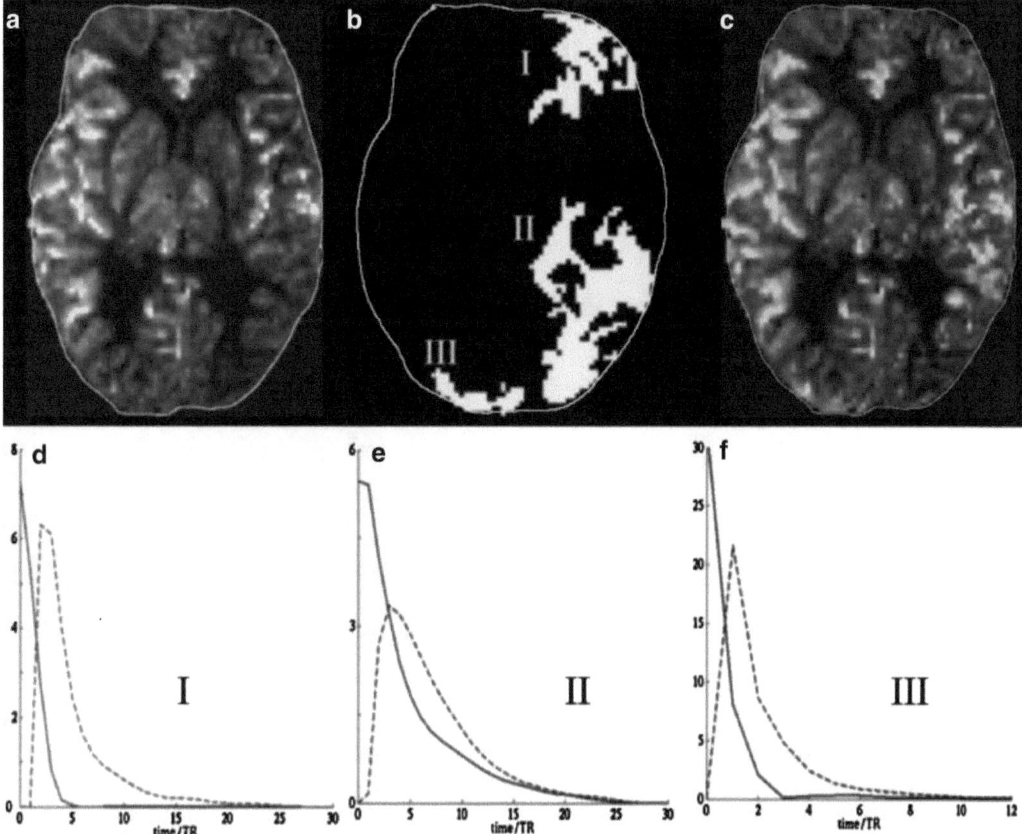

Fig. 8. DSC-MRI data from a patient with severe vascular abnormality (left internal carotid artery; *right-hand* side of the images) scanned in the chronic phase; the results were obtained using both global and regional AIFs. (**a**) CBF map calculated using a global AIF taken from a right branch of the middle cerebral artery. (**b**) Mask of regions subject to bolus delay and dispersion. (**c**) New CBF map after combining deconvolution of masked abnormal regions using the appropriate *regional* AIFs and deconvolution of the rest of the slice using a global AIF. (**d–f**) Example impulse–response functions deconvolved using the global (*dashed lines*) and regional (*solid lines*) AIFs, corresponding to the regions I, II, and III, respectively. The regions with delay/dispersion coincided with regions of abnormal CBF in (**a**); the use of the regional AIFs were found to remove the majority of the delay/dispersion within these regions. There was an average 45% increase in CBF (**c**) in region I compared with the value obtained using the global AIF (**a**). Similarly for region II: ~25% increase, and for region III: ~40% increase. After removing the delay/dispersion errors, there is still a small CBF abnormality measured in the posterior part of the left hemisphere (**c**) (figure kindly provided by Dr. Lisa Willats, and reproduced from Willats et al. (26), with permission from John Wiley & Sons, Inc).

concept of a "local AIF" (cf. the global AIF, in Sect. 2.5), in which each voxel of the brain has its own measured AIF. Several of these methods have been proposed in recent years (10, 68–70), although their role in infarct prediction in acute stroke is still controversial (71, 72). Figure 9 shows an example of a local AIF method (73) for the data of a stroke patient with internal carotid artery dissection (MRI scan performed ~9 h after symptoms onset).

7. Truncation artefact: A further potential artefact in the measurement of the AIF is the truncation of its shape. Due to the large

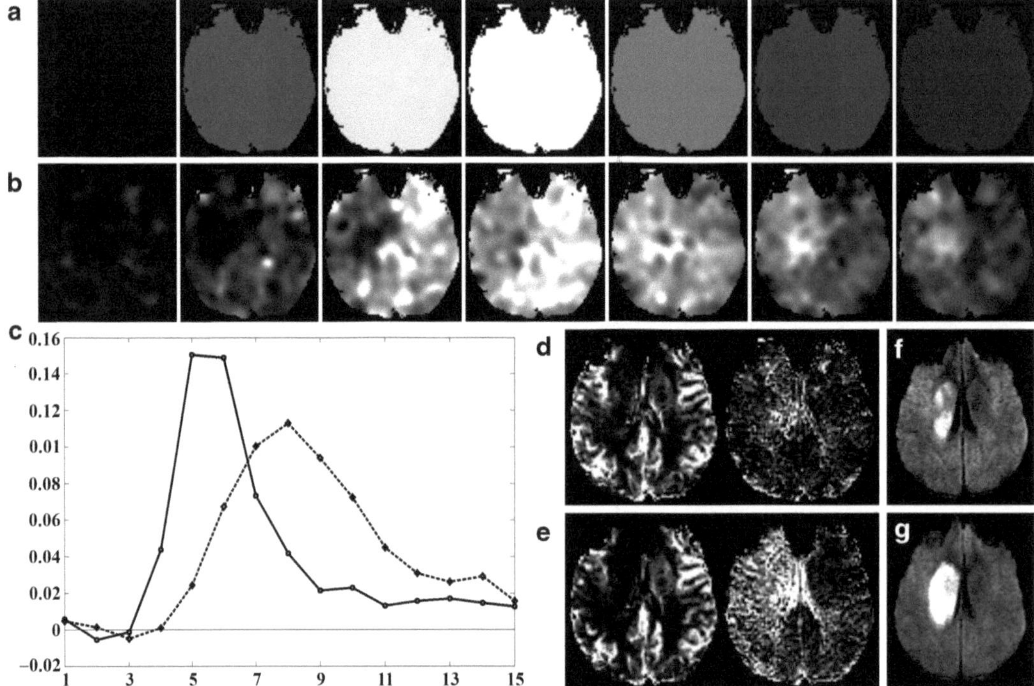

Fig. 9. Global AIF and local AIF results for the data of a patient with right internal carotid dissection (*left-hand* side of the images). The set of images are seven time points of the global AIF (**a**) and local AIF (**b**) during the passage of the bolus. The local AIF data display heterogeneity throughout the slice, with bolus delay and dispersion on the right hemisphere. This is illustrated on the local AIF curves (**c**) obtained from a region in the contralateral hemisphere (*solid line*) and from an ipsilateral region (*dashed line*). CBF and MTT maps are calculated by deconvolution using the local AIF (**d**) or global AIF data (**e**). The global AIF methodology overestimates the abnormality, with lower CBF and longer MTT (compared to the local AIF case) in regions where the local AIF was distorted. (**f**) Diffusion-weighted image acquired during the same examination; a region of abnormality (hyperintensity) can be readily appreciated. (**g**) Follow-up diffusion-weighted image showing a small lesion expansion (images kindly provided by Dr. Lisa Willats, Brain Research Institute, Melbourne, Australia).

effect of the contrast agent in arterial voxels, the signal intensity can reach the noise floor during the passage of the bolus. In that case, the AIF(t) calculated from the change in relaxation rate will have a peak shape, but with a truncated peak height. This curve can sometimes be misinterpreted as a wider bolus injection, but it is important to recognise this effect and avoid these artefactual curves when selecting the AIF voxels. One simple way to achieve this is by calculating the mean concentration time curve for the whole brain (or for a region of interest in normal gray matter), which should provide a good indication as to the expected shape of the peak (e.g. the truncated curves in the arterial voxels will appear much wider than the curves in the tissue, which is not physically possible, thus suggesting their artefactual nature). This source of error can be minimised by reducing the TE of the sequence (since this reduces the magnitude of the T_2^* effect). To avoid an associated decrease in the effect on the tissue (which would reduced the SNR of the CBF measurement), a multi-echo sequence provides the best alternative (52): the shorter echo can be used for the AIF measurement, while all the echoes can be used to measure R_2^* in the tissue.

8. BBB leakage: The theory described in the equations above assumed the contrast agent remains intravascular (i.e. the tissue has an intact BBB). On the other hand, when the BBB is disrupted, the loss of compartmentalisation of the contrast agent leads to a change in the T_2^* effect, as well as a non-negligible T_1 effect. In fact, measurement of permeability-related parameters is more commonly done using T_1-weighted imaging in the studies of patients with abnormal BBB (e.g. see reference (74)). However, if DSC-MRI studies are carried out, the effect of the contrast-agent leakage must be accounted for. The T_1-related effects can be avoided using a multi-echo sequence (51, 52), or minimised using a small pre-enhancement loading bolus (75). The latter approach was found to be efficient in avoiding a further enhancement during the second (main) bolus, provided the contrast extravasation is mild to moderate. The example shown in Fig. 10 nicely illustrates the T_1-enhancement and T_2^* effects in a patient with malignant glioneuronal tumour.

   Accounting for the T_2^*-related effects of the contrast-agent leakage is more complex. It requires a modification of the kinetic model to incorporate a term to describe the contrast extravasation. Various modifications have been proposed (e.g. (76–79)), and they all incorporate (typically, either in the expression for ΔR_2^* or for $R(t)$) a term describing the extravascular exchange. These studies have shown that the errors with the standard model can be very large for CBV quantification, but they are only moderate for CBF quantification. Therefore, if CBV or MTT measurements are needed in patients where contrast leakage is possible (e.g. in patients with tumours, multiple sclerosis, etc.), the use of these modified models is highly recommended. In particular, a good set of recommendations for measuring CBV in brain tumours was recently published by Paulson and Schmainda (77): they found that the preload-postprocessing correction and dual-echo approaches are the most robust methods for assessing CBV in brain tumours.

9. Simulations: An essential step for any deconvolution algorithm is the assessment of its accuracy and precision, as well as the characterisation of its limitations. This is commonly achieved by numerical simulations, where mathematical models for the relevant components (i.e. the AIF, $R(t)$, VTF, etc.) are assumed, and the algorithm is tested under a large range of conditions (e.g. a range of CBF, SNR, delays, TR, etc.). For example, a common model for the AIF is a gamma-variate function (2), for $R(t)$ an exponential function (2), and for VTF an exponential function or a gamma-variate function (23). A key advantage of

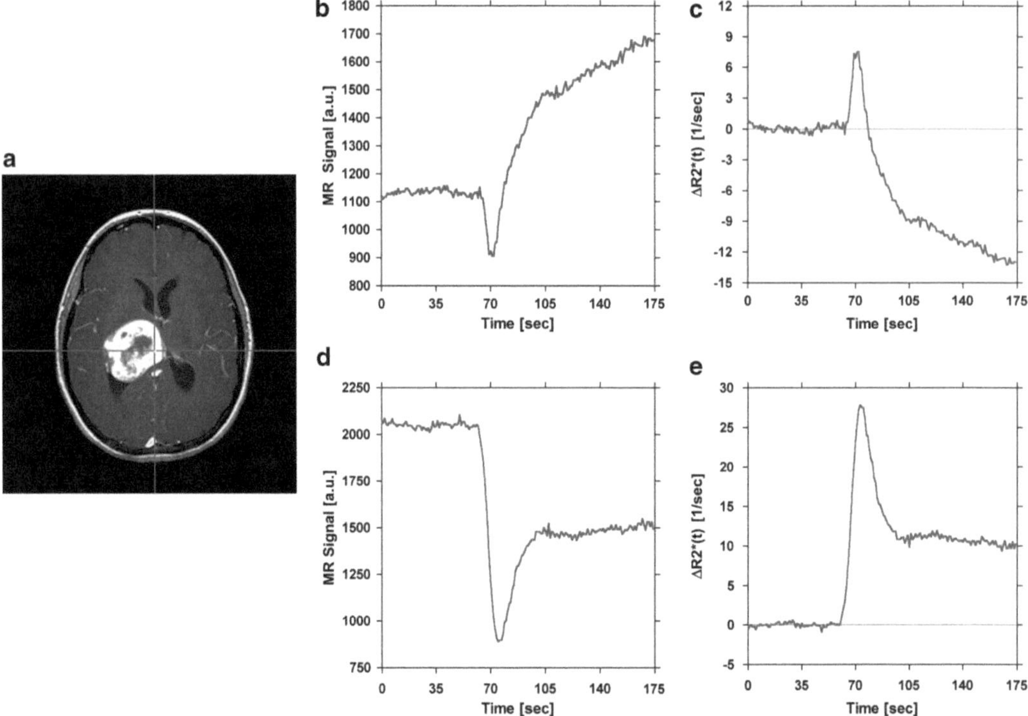

Fig. 10. (**a**) Contrast-enhanced anatomic image from a patient with malignant glioneuronal tumour. (**b–e**) MR signal- and concentration-time curves demonstrating confounding T_1 and T_2^* leakage and/or residual susceptibility effects. Data in (**b**, **c**) were collected with a gradient-echo echo-planar imaging with 90° flip angle during primary injection of standard dose of contrast agent. Data acquired during primary injection demonstrate strong T_1 leakage effect, as evidenced by the fact that post-bolus signal continues rising above its pre-bolus baseline on (**b**), which corresponds to post-bolus portion of $\Delta R_2^*(t)$ decreasing below its pre-bolus baseline level on (**c**). Data on (**d**, **e**) are same types of curves, taken from same voxel, but were acquired with a gradient-echo echo-planar imaging with 90° flip angle during secondary injection of double-dose of contrast agent after pre-load administration. Post-bolus signal remains below its pre-bolus baseline level on (**d**), which corresponds to elevated post-bolus portion of $\Delta R_2^*(t)$ on (**e**). This is consistent with a dipolar T_2 leakage effect or a residual susceptibility leakage effect. *a.u.* = arbitrary units. Please refer to the electronic version of the book for the colour version of this figure (figure kindly provided by Dr. Kathleen Schmainda, and reproduced from Paulson and Schmainda (77), with permission from Radiological Society of North America, Inc).

numerical simulation studies is that the assessment can be done in a controlled way and the true solution is known (cf. in vivo studies). These numerical studies typically involve simulation of the tissue concentration time curve using (2); these data are then converted to a hypothetically measured signal intensity time course, which after adding a suitable level of image noise, can be used to test a deconvolution algorithm by analysing the simulated signal intensity data as if they were real data.

There a number of common mistakes and limitations that can be encountered when performing these types of simulation studies which include: (1) use of the same $R(t)$ model to simulate the data and to fit the data (for model-dependent deconvolution methods); (2) exclusively testing an unrealistic range of parameters (e.g. very extreme SNR values, or CBF values, or imaging parameters); (3) test the algorithm in a very limited range of parameters (e.g. only for normal tissue parameters); (4) simulate the tissue concentration time course in (2) by performing a numerical convolution using a coarse discretisation interval (e.g. Δt=TR); (5) simulate data with very short TR values but without including the T_1 effects (i.e. using the simplified expression in (1)); and (6) assess the effect of bolus delay, but only using the particular delay values corresponding to an integer multiple of the TR.

10. Timing parameter errors: As mentioned in the Sect. 2.8, the timing parameters (e.g. MTT, T_{max}, TTP) can be quantified in absolute units (e.g. in seconds). However, these parameters can be subject to two main common sources of errors: (1) TR discretisation errors: if no model fitting of the data (e.g. using a gamma-variate function) or interpolation method is used (80), the measured timing measurement are rounded-off to a value multiple of TR (35); this leads to the measurement of discretised values, and thus rounded errors. (2) Slice-acquisition timing errors: since DSC-MRI images are commonly acquired using a multi-slice acquisition scheme, the measurement of timing parameters can be sensitive to the timing differences of slices during acquisition, which can be as large as 1 TR. It is therefore essential that these timing differences are accounted for during DSC-MRI quantification (35). Apart from these two sources of error, the timing parameters based on a deconvolution analysis (MTT and T_{max}) are obviously also susceptible to deconvolution-related errors, as mentioned above.

11. Absolute units: Until the recent methods to scale the DSC-MRI measurements on a subject-by-subject basis were proposed (39–41), quantification in absolute units had been attempted based on a more universal scaling. These approaches included: (1) use of a region of interest in normal white matter as an internal standard, by setting its value to 22 mL/100 g/min and scaling the CBF map accordingly (14). While this was initially believed to be a reasonable assumption based on early PET data, more recent studies have shown this approach to be inappropriate (37). (2) Assuming values for the various proportionality constants, to convert the DSC-MRI measurements to absolute units (e.g. (20)). However, some studies have shown the proportionality constants may vary with tissue type, subject, and pathology (3), which makes the use of a universal scaling inappropriate. (3) Use of a *common* scaling factor obtained from a previous cross-calibration study (e.g. a PET-MRI study) to scale the DSC-MRI measurements (81). However, several studies have now shown that a single conversion factor is not appropriate, and the scaling can change

between subjects and under various physiological conditions (37, 38, 82).

A further method, which is also commonly used in perfusion computed tomography (CT) (4), is based on scaling the AIF by assuming it should have, in theory, the same AUC as the venous output function (VOF, the function describing the *output* of the contrast agent from the tissue). The VOF is often measured in the sagittal sinus and, given its much larger diameter compared to the arteries used for AIF measurement, it is commonly assumed that the VOF can be measured in the absence of partial-volume effect. This method could provide, in principle, a subject-specific correction factor. However, the vessel misregistration induced by the field distortions created by the passage of the bolus itself (83) makes measurement of the VOF in the sagittal sinus very unreliable using EPI (84).

12. ASL: ASL is an alternative Perfusion MRI technique, which uses magnetically labelled blood (labelled using magnetic field gradients and radio-frequency pulses) as an endogenous contrast agent (1). This technique is therefore fully non-invasive and does not require the injection of a contrast agent. It relies on acquiring two images, one where the blood has been labelled (the so-called "label" image) and another where the blood is not labelled (the "control" image). It can be shown that the difference between these two images is proportional to perfusion (1). One of the reasons for its current limited clinical applicability is its limited contrast-to-noise ratio: the signal difference is ~1% (see a recent review (85) for a detailed discussion of the advantages and limitations of ASL).

References

1. Calamante F, Thomas DL, Pell GS et al (1999) Measuring cerebral blood flow using magnetic resonance techniques. J Cereb Blood Flow Metab 19:701–735
2. Østergaard L, Weisskoff RM, Chesler DA et al (1996) High resolution measurement of cerebral blood flow using intravascular tracer bolus passages. Part I. Mathematical approach and statistical analysis. Magn Reson Med 36:715–725
3. Calamante F (2010) Artefacts and pitfalls in perfusion MR imaging. In: Gillard J, Waldman A, Barker P (eds) Clinical MR neuroimaging, 2nd edn. Cambridge University Press, Cambridge, pp 137–155
4. Wintermark M, Sesay M, Barbier E et al (2005) Comparative overview of brain perfusion imaging techniques. Stroke 36:e83–e99
5. Stehling MK, Turner R, Masfield P (1991) Echo-planar imaging: magnetic resonance imaging in a fraction of a second. Science 254:43–50
6. Wintermark M, Albers GW, Alexandrov AV et al (2008) Acute stroke imaging research roadmap. Stroke 39:1621–1628
7. Calamante F, Vonken EPA, van Osch MJP (2007) Contrast agent concentration measurements affecting quantification of bolus-tracking perfusion MRI. Magn Reson Med 58:544–553
8. Kosior JC, Kosior RK, Frayne R (2007) Robust dynamic susceptibility contrast MR perfusion using 4D nonlinear noise filters. J Magn Reson Imaging 26:1514–1522
9. Calamante F, Connelly A (2009) Perfusion precision in bolus-tracking MRI: estimation using the wild-bootstrap method. Magn Reson Med 61:696–704

10. Calamante F, Mørup M, Hansen LK (2004) Defining a local arterial input function for perfusion MRI using independent component analysis. Magn Reson Med 52:789–797
11. Shellock FG, Spinazzi A (2008) MRI safety update 2008: part I, MRI contrast agents and nephrogenic systemic fibrosis. AJR Am J Roentgenol 191:1129–1139
12. Kosior RK, Kosior JC, Frayne R (2007) Improved dynamic susceptibility contrast (DSC)-MR perfusion estimates by motion correction. J Magn Reson Imaging 26: 1167–1172
13. Weisskoff RM, Zuo CS, Boxerman JL et al (1994) Microscopic susceptibility variation and transverse relaxation. Theory and experiment. Magn Reson Med 31:601–610
14. Østergaard L, Sorensen AG, Kwong KK et al (1996) High resolution measurement of cerebral blood flow using intravascular tracer bolus passages. Part II. Experimental comparison and preliminary results. Magn Reson Med 36:726–736
15. Murase K, Kikuchi K, Miki H et al (2001) Determination of arterial input function using fuzzy clustering for quantification of cerebral blood flow with dynamic susceptibility contrast-enhanced MR imaging. J Magn Reson Imaging 13:797–806
16. Carroll TJ, Rowley HA, Haughton VM (2003) Automatic calculation of the arterial input function for cerebral perfusion imaging with MR imaging. Radiology 227:593–600
17. Mlynash M, Eyngorn I, Bammer R et al (2005) Automated method for generating the arterial input function on perfusion-weighted MR imaging: validation in patients with stroke. AJNR Am J Neuroradiol 26:1479–1486
18. Mouridsen K, Christensen S, Gydensted L, Ostergaard L (2006) Automatic selection of arterial input function using cluster analysis. Magn Reson Med 55:524–531
19. Zierler KL (1962) Theoretical basis of indicator-dilution methods for measuring flow and volume. Circ Res 10:393–407
20. Rempp KA, Brix G, Wenz F et al (1994) Quantification of regional cerebral blood flow and volume with dynamic susceptibility contrast-enhanced MR imaging. Radiology 193:637–641
21. Wu O, Østergaard L, Weisskoff RM et al (2003) Tracer arrival timing-insensitive technique for estimating flow in MR perfusion-weighted imaging using singular value decomposition with a block-circulant deconvolution matrix. Magn Reson Med 50:164–174
22. Smith MR, Lu H, Trochet S, Frayne R (2004) Removing the effect of SVD algorithmic artifacts present in quantitative MR perfusion studies. Magn Reson Med 51:631–634
23. Calamante F, Gadian DG, Connelly A (2003) Quantification of bolus tracking MRI: improved characterization of the tissue residue function using Tikhonov regularization. Magn Reson Med 50:1237–1247
24. Vonken EP, Beekman FJ, Bakker CJ, Viergever MA (1999) Maximum likelihood estimation of cerebral blood flow in dynamic susceptibility contrast MRI. Magn Reson Med 41:343–350
25. Willats L, Connelly A, Calamante F (2006) Improved deconvolution of perfusion MRI data in the presence of bolus delay and dispersion. Magn Reson Med 56:146–156
26. Willats L, Connelly A, Calamante F (2008) Minimising the effects of bolus dispersion in bolus tracking MRI. NMR Biomed 21: 1126–1137
27. Andersen IK, Szymkowiak A, Rasmussen CE et al (2002) Perfusion quantification using Gaussian process deconvolution. Magn Reson Med 48:351–361
28. Mouridsen K, Friston K, Hjort N et al (2006) Bayesian estimation of cerebral perfusion using a physiological model of microvasculature. Neuroimage 33:570–579
29. Hansen PC (1994) Regularization tools: a MATLAB package for analysis and solution of discrete ill-posed problems. Num Algorithms 6:1–35
30. Sourbron S, Luypaert R, Morhard D et al (2007) Deconvolution of bolus-tracking data: a comparison of discretization methods. Phys Med Biol 52:6761–6778
31. Murase K, Shinohara M, Yamazaki Y (2001) Accuracy of deconvolution analysis based on singular value decomposition for quantification of cerebral blood flow using dynamic susceptibility contrast-enhanced magnetic resonance imaging. Phys Med Biol 46:3147–3159
32. Liu HL, Pu Y, Liu Y et al (1999) Cerebral blood flow measurement by dynamic contrast MRI using singular value decomposition with an adaptive threshold. Magn Reson Med 42:167–172
33. Perkiö J, Aronen HJ, Kangasmäki A et al (2002) Evaluation of four postprocessing methods for determination of cerebral blood volume and mean transit time by dynamic susceptibility contrast imaging. Magn Reson Med 47:973–981
34. Stewart GN (1894) Researches on the circulation time in organs and on the influences which affect it. Part I-III. J Physiol 15:1–89
35. Calamante F, Christensen S, Desmond PM et al (2010) The physiological significance of the time-to-maximum (Tmax) parameter in perfusion MRI. Stroke 41:1169–1174

36. Perthen JE, Calamante F, Gadian DG, Connelly A (2002) Is quantification of bolus tracking MRI reliable without deconvolution? Magn Reson Med 47:61–67
37. Mukherjee P, Kang HC, Videen TO et al (2003) Measurement of cerebral blood flow in chronic carotid occlusive disease: comparison of dynamic susceptibility contrast perfusion MR imaging with positron emission tomography. AJNR Am J Neuroradiol 24:862–871
38. Grandin C, Bol A, Smith A et al (2005) Absolute CBF and CBV measurements by MRI bolus tracking before and after acetazolamide challenge: repeatability and comparison with PET in humans. Neuroimage 26:525–535
39. Shin W, Horowitz S, Ragin A et al (2007) Quantitative cerebral perfusion using dynamic susceptibility contrast MRI: evaluation of reproducibility and age- and gender-dependence with fully automatic image postprocessing algorithm. Magn Reson Med 58:1232–1241
40. Zaharchuk G, Straka M, Marks MP et al (2010) Combined arterial spin label and dynamic susceptibility contrast measurement of cerebral blood flow. Magn Reson Med 63:1548–1556
41. Zaro-Weber O, Moeller-Hartmann W, Heiss WD, Sobesky J (2010) A simple positron emission tomography-based calibration for perfusion-weighted magnetic resonance maps to optimize penumbral flow detection in acute stroke. Stroke 41:1939–1945
42. Straka M, Albers GW, Bammer R (2010) Real-time diffusion-perfusion mismatch analysis in acute stroke. J Magn Reson Imaging 32:1024–1037
43. Bjørnerud A, Emblem KE (2010) A fully automated method for quantitative cerebral hemodynamic analysis using DSC-MRI. J Cereb Blood Flow Metab 30:1066–1078
44. Kim J, Leira EC, Callison RC et al (2010) Toward fully automated processing of dynamic susceptibility contrast perfusion MRI for acute ischemic cerebral stroke. Comput Methods Programs Biomed 98:204–213
45. Kane I, Sandercock P, Wardlaw J (2007) Magnetic resonance perfusion diffusion mismatch and thrombolysis in acute ischaemic stroke: a systematic review of the evidence to date. J Neurol Neurosurg Psychiatry 78:485–491
46. Albers GW, Thijs VN, Wechsler L et al (2006) MRI profiles predict clinical response to early reperfusion: the diffusion-perfusion imaging evaluation for understanding stroke evolution (DEFUSE) study. Ann Neurol 60:508–517
47. Davis SM, Donnan GA, Parsons MW et al (2008) Effects of alteplase beyond 3h after stroke in the echoplanar imaging thrombolytic evaluation trial (Epithet): a placebo controlled randomised trial. Lancet Neurol 7:299–309
48. Ostergaard L, Jónsdóttir KY, Mouridsen K (2009) Predicting tissue outcome in stroke: new approaches. Curr Opin Neurol 22:54–59
49. Wu O, Koroshetz WJ, Ostergaard L et al (2000) Predicting tissue outcome in acute human cerebral ischemia using combined diffusion- and perfusion-weighted MR imaging. Stroke 32:933–942
50. Kjolby BF, Østergaard L, Kiselev VG (2006) Theoretical model of intravascular paramagnetic tracers effect on tissue relaxation. Magn Reson Med 56:187–197
51. Vonken EPA, van Osch MJP, Baker CJG et al (1999) Measurement of cerebral perfusion with dual-echo multi-slice quantitative dynamic susceptibility contrast MRI. J Magn Reson Imaging 10:109–117
52. Newbould RD, Skare ST, Jochimsen TH et al (2007) Perfusion mapping with multiecho multishot parallel imaging EPI. Magn Reson Med 58:70–81
53. Pedersen M, Klarhofer M, Christensen S et al (2004) Quantitative cerebral perfusion using the PRESTO acquisition scheme. J Magn Reson Imaging 20:930–940
54. Leiner T, Kucharczyk W (2009) NSF prevention in clinical practice: summary of recommendations and guidelines in the United States, Canada, and Europe. J Magn Reson Imaging 30:1357–1363
55. van Osch MJ, Vonken EJ, Viergever MA et al (2003) Measuring the arterial input function with gradient echo sequences. Magn Reson Med 49:1067–1076
56. Calamante F, Connelly A, van Osch MJP (2009) Non-linear $\Delta R2^*$ effects in perfusion quantification using bolus-tracking MRI. Magn Reson Med 61:486–492
57. Zaharchuk G, Bammer R, Straka M et al (2009) Improving dynamic susceptibility contrast MRI measurement of quantitative cerebral blood flow using corrections for partial volume and nonlinear contrast relaxivity: a xenon computed tomographic comparative study. J Magn Reson Imaging 30:743–752
58. van Osch MJP, van der Grond J, Bakker CJG (2005) Partial volume effects on arterial input functions: shape and amplitude distortions and their correction. J Magn Reson Imaging 22:704–709
59. Bleeker EJW, van Osch MJP, Connelly A et al (2011) New criterion to aid manual and automatic selection of the arterial input function in

dynamic susceptibility contrast MRI. Magn Reson Med 65:448–456

60. Duhamel G, Schlaug G, Alsop DC (2006) Measurement of arterial input functions for dynamic susceptibility contrast magnetic resonance imaging using echoplanar images: comparison of physical simulations with in vivo results. Magn Reson Med 55:514–523
61. Bleeker EJ, van Buchem MA, van Osch MJ (2009) Optimal location for arterial input function measurements near the middle cerebral artery in first-pass perfusion MRI. J Cereb Blood Flow Metab 29:840–852
62. Calamante F, Gadian DG, Connelly A (2002) Quantification of perfusion using bolus tracking MRI in stroke. Assumptions, limitations, and potential implications for clinical use. Stroke 33:1146–1151
63. Calamante F, Gadian DG, Connelly A (2000) Delay and dispersion effects in dynamic susceptibility contrast MRI: simulations using singular value decomposition. Magn Reson Med 44:466–473
64. Ibaraki M, Shimosegawa E, Toyoshima H et al (2005) Tracer delay correction of cerebral blood flow with dynamic susceptibility contrast-enhanced MRI. J Cereb Blood Flow Metab 25:378–390
65. Calamante F (2005) Bolus dispersion issues related to the quantification of perfusion MRI data. J Magn Reson Imaging 22:718–722
66. Calamante F, Willats L, Gadian DG, Connelly A (2006) Bolus delay and dispersion in perfusion MRI: implications for tissue predictor models in stroke. Magn Reson Med 55:1180–1185
67. Calamante F, Yim PJ, Cebral JR (2003) Estimation of bolus dispersion effects in perfusion MRI using image-based computational fluid dynamics. Neuroimage 19:341–353
68. Lorenz C, Benner T, Lopez CJ et al (2006) Effect of using local arterial input functions on cerebral blood flow estimation. J Magn Reson Imaging 24:57–65
69. Grüner R, Bjørnarå B, Moen G et al (2006) Magnetic resonance brain perfusion imaging with voxel-specific arterial input functions. Magn Reson Med 23:273–284
70. Lee JJ, Bretthorst GL, Derdeyn CP et al (2010) Dynamic susceptibility contrast MRI with localized arterial input functions. Magn Reson Med 63:1305–1314
71. Lorenz C, Benner T, Chen PJ et al (2006) Automated perfusion-weighted MRI using localized arterial input functions. J Magn Reson Imaging 24:1133–1139
72. Christensen S, Mouridsen K, Wu O et al (2009) Comparison of 10 perfusion MRI parameters in 97 sub-6-hour stroke patients using voxel-based receiver operating characteristics analysis. Stroke 40:2055–2061
73. Willats L, Christensen S, Ma H et al (2011) Validating a local Arterial Input Function method for improved perfusion quantification in stroke. J Cereb Blood Flow Metab 31:2189–2198
74. Jackson A, Buckley DL, Parker GJM (eds) (2005) Dynamic contrast-enhanced magnetic resonance imaging in oncology. Springer, Heidelberg
75. Sorensen AG, Reimer P (2000) Cerebral MR perfusion imaging. Principles and current applications. Thieme, Stuttgart
76. Boxerman JL, Schmainda KM, Weisskoff RM (2006) Relative cerebral blood volume maps corrected for contrast agent extravasation significantly correlate with glioma tumor grade, whereas uncorrected maps do not. AJNR Am J Neuroradiol 27:859–867
77. Paulson ES, Schmainda KM (2008) Comparison of dynamic susceptibility-weighted contrast-enhanced MR methods: recommendations for measuring relative cerebral blood volume in brain tumors. Radiology 249:601–613
78. Vonken EPA, van Osch MJP, Baker CJG et al (2000) Simultaneous qualitative cerebral perfusion and Gd-DTPA extravasation measurements with dual-echo dynamic susceptibility contrast MRI. Magn Reson Med 43: 820–827
79. Quarles CC, Ward BD, Schmainda KM (2005) Improving the reliability of obtaining tumor hemodynamic parameters in the presence of contrast agent extravasation. Magn Reson Med 53:1307–1316
80. Salluzzi M, Frayne R, Smith MR (2006) Is correction necessary when clinically determining quantitative cerebral perfusion parameters from multi-slice dynamic susceptibility contrast MR studies? Phys Med Biol 51:407–424
81. Østergaard L, Johannsen P, Poulsen PH et al (1998) Cerebral blood flow measurements by magnetic resonance imaging bolus tracking: comparison with (O-15) H_2O positron emission tomography in humans. J Cereb Blood Flow Metab 18:935–940
82. Lin W, Celik A, Derdeyn C et al (2001) Quantitative measurements of cerebral blood flow in patients with unilateral carotid artery occlusion: a PET and MR study. J Magn Reson Imaging 14:659–667

83. Hou L, Yang Y, Mattay VS et al (1999) Optimization of fast acquisition methods for whole-brain relative cerebral blood volume (rCBV) mapping with susceptibility contrast agents. J Magn Reson Imaging 9:233–239
84. Knutsson L, Ståhlberg F, Wirestam R (2010) Absolute quantification of perfusion using dynamic susceptibility contrast MRI: pitfalls and possibilities. MAGMA 23:1–21
85. Petersen ET, Zimine I, Ho YC, Golay X (2006) Non-invasive measurement of perfusion: a critical review of arterial spin labelling techniques. Br J Radiol 79:688–701
86. Calamante F (2008) Measuring cerebral perfusion using magnetic resonance imaging. In: Yim PJ (ed) Vascular hemodynamics: bioengineering and clinical perspectives. Wiley, New Jersey

INDEX

Emilio Badoer (ed.), *Visualization Techniques: From Immunohistochemistry to Magnetic Resonance Imaging*, Neuromethods, vol. 70, DOI 10.1007/978-1-61779-897-9, © Springer Science+Business Media, LLC 2012

F

G

H

I

J

K

L

M

N

P

R

S

T

U

V

W

Z

MIX
Papier aus verantwortungsvollen Quellen
Paper from responsible sources
FSC® C105338

If you have any concerns about our products,
you can contact us on
ProductSafety@springernature.com

In case Publisher is established outside the EU,
the EU authorized representative is:
Springer Nature Customer Service Center GmbH
Europaplatz 3, 69115 Heidelberg, Germany

Printed by Libri Plureos GmbH
in Hamburg, Germany